D1043321

FIXING THE SKY

COLUMBIA STUDIES
IN INTERNATIONAL
AND GLOBAL HISTORY

COLUMBIA STUDIES IN INTERNATIONAL AND GLOBAL HISTORY

The idea of "globalization" has become a commonplace, but we lack good histories that can explain the transnational and global processes that have shaped the contemporary world. Columbia Studies in International and Global History will encourage serious scholarship on international and global history with an eye to explaining the origins of the contemporary era. Grounded in empirical research, the titles in the series will also transcend the usual area boundaries and will address questions of how history can help us understand contemporary problems, including poverty, inequality, power, political violence, and accountability beyond the nation-state.

Cemil Aydin, *The Politics of Anti-Westernism in Asia: Visions of World Order in Pan-Islamic and Pan-Asian Thought*

Adam M. McKeown, *Melancholy Order: Asian Migration and the Globalization of Borders*

Patrick Manning, *The African Diaspora: A History Through Culture*

FIXING THE SKY

THE CHECKERED HISTORY OF WEATHER AND CLIMATE CONTROL

JAMES RODGER FLEMING

COLUMBIA UNIVERSITY PRESS / NEW YORK

COLUMBIA UNIVERSITY PRESS
Publishers Since 1893
New York Chichester, West Sussex

Library of Congress Cataloging-in-Publication Data
Fleming, James Rodger.
Fixing the sky : the checkered history of weather and climate control /
James Rodger Fleming.
p. cm. — (Columbia studies in international and global history)
Includes bibliographical references and index.
ISBN 978-0-231-14412-4 (cloth : alk. paper) ISBN 978-0-231-51306-7 (e-book)
1. Weather control—History. 2. Nature—Effect of human beings on. I. Title. II. Series.

QC928.F54 2010
551.68—dc22 2010015482

Columbia University Press books are printed on permanent and durable acid-free paper.
This book is printed on paper with recycled content.
Printed in the United States of America

c 10 9 8 7 6 5 4 3 2 1

Designed by Lisa Hamm

"Give me a lever long enough and a place to stand, and I will move the world" (Archimedes), but where will it roll? (*MECHANICS MAGAZINE*, 1824, AND ORIGINAL ART)

The middle course is safest and best.

—HELIOS

CONTENTS

PREFACE

THE possibility but not the desirability of weather and climate control entered my consciousness in the early 1970s. I was a graduate student in atmospheric science at Colorado State University and one of my air force colleagues proposed to zap clouds with laser beams to make them more energetic. I think he was wondering if a focused beam of radiation could destroy hailstones, perhaps like microwave medical treatment for kidney stones, although I suspect his patrons had other ideas. I also noted a widespread ambivalence among most of my professors toward cloud seeding, which was not part of the main curriculum. My own project involved putting cirrus clouds into a computerized tropical atmosphere and modeling their radiative effects. This was an early exercise in climate modeling. The way clouds interact with sunlight and heat radiation was very speculative then and is still an open question, even as some climate engineers propose to "manage" solar radiation.

Colorado State University also supported my training as a high-altitude observer, which qualified me to work with the National Center for Atmospheric Research in Boulder in its instrumented glider project. This was an ideal job for someone who loved clouds, since we were able to slip quietly and unobtrusively into them, spend up to several hours spiraling in their growing updrafts and collecting data, and then exit the tops of budding thunderstorms for absolutely spectacular views. This project was conducted in 1973 over the Continental Divide near Leadville, Colorado. Its poignant but unintended link to weather control came one evening, unexpectedly, when the Colorado state police visited our lodgings and informed us that someone had thrown a Molotov cocktail into the hangar and had burned up one wing of the glider. Apparently, some of the locals mistakenly thought that we were engaged in cloud seeding, or, colloquially, "stealing their sky water."

In my next project, at the University of Washington, I flew in a World War II–era surplus bomber equipped with cloud physics equipment to investigate the claims of the weather modification community. The problem was, we usually flew only when the weather was bad, so we were buffeted quite frequently by Pacific storms. Early one morning, after a particularly harrowing night of flying, the airplane clipped off the top of a pine tree during our landing approach. I recall pulling a 2-inch-diameter branch out of the equipment slung below the plane and deciding, pretty much there and then, that I would be seeking other, safer modes of engagement with the atmosphere.

The history of science and technology, including its relevance to public policy, became for me that mode of engagement. I received my doctorate in history at Princeton University with a dissertation on the history of meteorology in America. Since then, I have had more than a passing interest in the history of weather and climate modification and have written several essays on the topic. I also remain deeply involved in issues involving climate change history and public policy. Today's climate engineers are championing an approach to the problem of what to do about climate change that arouses my deepest suspicions regarding technological fixes. It is a seriously flawed and speculative undertaking that typically involves impractical or even dangerous schemes to "fix the sky." Proposals include "solar radiation management" and other forms of planetary leverage, including thermodynamically impractical schemes for large-scale carbon capture and sequestration. The proposals are typically supported by scribbling back-of-the-envelope calculations or tinkering with simple computer models that are just not good enough. It is an approach that is oblivious to the checkered history of weather and climate control.

ACKNOWLEDGMENTS

MOST of the research for this book was conducted during a research leave supported by Colby College, the Smithsonian National Air and Space Museum, the American Association for the Advancement of Science, the National Academy of Sciences, and the Woodrow Wilson International Center for Scholars. Early versions of this work were presented at a number of venues. I extend my profound gratitude to my hosts, sponsors, listeners, and questioners.

Special thanks are due to Ralph Cicerone and Lee Hamilton and their staffs; British Broadcasting Corporation filmmakers Fiona Scott and Weini Tesfu; 4th Row Films producers and directors Robert Greene and Greta Wink; Wilson Center Environmental Change and Security director Geof Dabelko, *Wilson Quarterly* editor Steve Lagerfeld and assistant editor Rebecca Rosen, Wilson Center press director Joe Brinley, and Wilson Center librarian Janet Spikes; National Academy of Science archivists Dan Barbiero and Janice Goldblum; Colby College librarians Susan Cole, Margaret Menchen, Darylyne Provost, and Alisia Wygant Rizett; Len Bruno and the staffs of the Library of Congress, the

Smithsonian Institution Libraries, and the NOAA Central Library; and a host of other archivists and librarians.

The following individuals provided extremely helpful perspectives and welcome encouragement: Stephen Cole, NASA Goddard; Lindsay Collins, Brooke Larson, and Theda Perdue, Woodrow Wilson Center; Matthew Connelly, Columbia University; Bill Frank and Charles Hosler, Pennsylvania State University; Justin Grubich and Alexey Voinov, AAAS Policy Fellows; Vladimir Jankovic, Manchester University; D. Whitney King, Ursula Reidel-Schrewe, and Thomas Shattuck, Colby College; Roger Launius, David DeVorkin, Martin Collins, and Michael Neufeld, Smithsonian National Air and Space Museum; Milton Leitenberg, University of Maryland; Michael MacCracken and John Topping, Climate Institute; Alan Robock, Rutgers University; Richard Somerville, Scripps Institution of Oceanography; and Roger Turner, University of Pennsylvania.

Gregory Cushman and students in his graduate seminar at the University of Kansas read and commented on an early draft of the manuscript; so, too, did Roger Launus, Alan Robock, and several unnamed reviewers. Research assistants Noah B. Bonnheim, Alice W. Evans, Amie R. Fleming, Ashley J. Oliver, Mette Fog Olwig, Mandy Reynolds, and James S. Westhafer made essential contributions to the process. My editor, Anne Routon, greatly facilitated production of the book.

FIXING THE SKY

INTRODUCTION

In facing unprecedented challenges, it is good to consider historical precedents.

AS alarm over global warming spreads, some climate engineers are engaging in wild speculation and are advancing increasingly urgent proposals about how to "control" the Earth's climate. They are stalking the hallways of power, hyping their proposals, and seeking support for their ideas about fixing the sky. The figures they scribble on the backs of envelopes and the results of their simple (yet somehow portrayed as complex) climate models have convinced them, but very few others, that they are planetary saviors, lifeboat builders on a sinking *Titanic*, visionaries who are taking action in the face of a looming crisis. They present themselves as insurance salesmen for the planet, with policies that may or may not pay benefits. In response to the question of what to do about climate change, they are prepared to take ultimate actions to intervene, even to do too much if others, in their estimation, are doing too little.

These climate engineers share a growing concern that something is terribly wrong with the sky. They are convinced that the climate system is headed into uncharted territory, carbon mitigation will fail or at least move too slowly to avert an environmental disaster, and adaptation will be too little, too late. Some simply place more faith in engineering solutions than in human agreements.

They have come to the conclusion that the twenty-first century will be "geotechnic"—that the atmosphere is humanity's aerial sewer, sorely in need of treatment, and the Earth needs a thermostat or perhaps global air-conditioning. They seek a technological fix through geoengineering, which they loosely define as the intentional large-scale manipulation of the global environment. Some have called it the "ultimate technological fix"; critics say it has unlimited potential for planetary mischief. Shade the planet by launching a solar shield into orbit. Shoot sulfates or reflective nanoparticles into the upper atmosphere, turning the blue sky milky white. Make the clouds thicker and brighter. Fertilize the oceans to stimulate massive algae blooms that turn the blue seas soupy green. Suck carbon dioxide (CO_2) out of the air with hundreds of thousands of giant artificial trees. Flood the Sahara and the Australian outback to plant mega-forests of eucalyptus trees. Surround the Arctic sea ice with a white plastic flotilla. While all this may sound like science fiction, it is actually just the latest set of installments in the perennial story of weather and climate control. For more than a century, scientists, soldiers, and charlatans have hatched schemes to manipulate the weather and climate. Like them, today's aspiring climate engineers wildly exaggerate what is possible, while scarcely considering the political, military, and ethical implications of attempting to manage the world's climate. This is not, in essence, a heroic saga about new scientific discoveries that can save the planet, as many of the participants claim, but a tragicomedy of overreaching, hubris, and self-delusion. At a National Academy of Sciences meeting in June 2009 on geoengineering, planetary scientist Brian Toon told the audience that we do not have the technology to engineer the planet. We do not have the wisdom either.[1] Global climate engineering is untested and untestable, and dangerous beyond belief.

The latest resurgence of interest in geoengineering dates to an editorial written in 2006 by Nobel laureate Paul Crutzen, "Albedo Enhancement by Stratospheric Sulfur Injections: A Contribution to Resolve a Policy Dilemma?"[2] The question mark is well placed, for far from solving a policy dilemma, he actually opened a can of worms, albeit from a historical pantry filled with such cans. Crutzen's basic message, that "research on the feasibility and environmental consequences of climate engineering . . . should not be tabooed," was but the latest round in an ancient quest for ultimate control of the atmosphere—a quest with very deep roots in traditional cultures, practices, myth, fiction, and history.

Ever since Archimedes, engineers have been excited about technological leverage, but they have never had the "standing" or the ability to predict all or even most of the consequences of their actions. This is a perennial issue. Yet today's geoengineers exude a false confidence when they proclaim that their tools and techniques have now matured to the extent that fixing the sky—cooling the planet, saving humanity, and minimizing unwanted side effects, whether physical or moral—is now both possible and desirable. How did we arrive at this situation?[3]

This book examines historical and current ways of thinking about weather and climate control. It includes stories from a long and checkered history and a dizzying array of contemporary ideas—most of them wildly impractical. It examines the proposals and actual practices of a large number of dreamers, militarists, and outright charlatans, of rain kings and queens, of weather warriors and climate engineers, both ancient and modern. It provides scholars and the general public with new perspectives that are missing from the technically oriented or policy-oriented conversations about control. This book is based on research in original manuscript and document collections; it also contains fresh interpretations of existing work. It is an extended essay arguing for the relevance of history, the foolishness of quick fixes, and the need to follow a "middle course" of expedited moderation in aerial matters, seeking neither to control the sky nor to diminish the importance of environmental problems we face.

This history is located within a long tradition of imaginative and speculative literature involving the "control" of nature. Early efforts to exercise some form of control over the environment included seeking shelter from the elements, using fire for warmth, herding animals, cultivating plants, and moving and storing fresh water. Yet control of the heavens remains far beyond the ability of mortals. Our ancestors either bowed or cowered before the ancient sky gods, while the mythological figures of classical antiquity met with tragedy when they sought to exceed mortal limits. Many societies, seeking a measure of influence over the vagaries of the sky, invested their rulers or shamans with the title "rain king" or "rain queen" and charged them with ceremonial duties of vast significance not only for upholding the physical well-being and prosperity of the tribe but also for maintaining the proper relationships between heaven and Earth.

Since the seventeenth century, the Baconian expectation that increasing knowledge would lead to new technologies "for the common good" has been widely applied to all scientific fields, including, notably, meteorology and climatology. For several centuries now, planners, politicians, scientists, and soldiers have proposed schemes for the purposeful manipulation of weather and climate, usually for commercial or military purposes. Their stories have tragic, comedic, and heroic aspects. Control of weather and climate is a perennial issue rooted in hubris and tragedy; it is a pathological issue, illustrating what can go wrong in science; and it is a pressing public policy issue with widespread social implications.

Enlightenment philosophers supposed that the climate of Europe had moderated since Roman times in response to human activity. Thomas Jefferson thought that clearing the forests, draining the marshes, and cultivating the land would improve the American climate.[4] In the 1840s, James Espy, the first meteorologist in U.S. government service, proposed rainmaking by lighting huge fires to stimulate convective updrafts.[5] The following era in rainmaking was dominated by artillerists and "rain fakers," the so-called pluviculturalists.[6]

Nineteenth-century climatologists could find no trends in the weather records beyond variability and temporarily quashed the notion that humans can influence climate. Yet by mid-century, geologists had discovered great changes, ice ages and interglacial epochs, in the record of the rocks. The two timescales (the human historical and the geological) and the two agencies (anthropogenic forces and natural forces) were reunited in a new form at the dawn of the twentieth century by the Swedish meteorologist Nils Gustaf Ekholm (1848–1923), who wrote about "the climate of the geological and historical past." Ekholm regarded variations in carbon dioxide concentration as the principal cause of climatic variations, citing the "elaborate inquiry on this complicated phenomenon" made by his colleague Svante Arrhenius. He explained that carbon dioxide is a key player in the greenhouse effect and that this conclusion is based on the earlier work of Joseph Fourier, John Tyndall, and others. By his estimates, an increase in carbon dioxide would heat high latitudes more than the tropics and would create a warmer, more uniform climate over the entire Earth; a tripling of atmospheric carbon dioxide levels would raise global temperatures by 7 to 9°C (12 to 16 °F).

According to Ekholm, the secular cooling of the originally hot Earth was the principal cause of variation in the quantity of carbon dioxide in the atmosphere. As the Earth cooled, the oceans sequestered great amounts of carbon into limestone and other calcium carbonate deposits, reducing the amount of carbon dioxide in the air. This caused temperatures to fall, triggering a chain reaction of feedback mechanisms that lowered carbon dioxide levels even further. Other processes added carbon dioxide to the air. Volcanic emissions, mountain uplift, and changes in sea level and plant cover produced the periodical variations evident in the geological record.

Ekholm pointed out that humanity was now playing a role in these geological processes. He held that over the course of a millennium the accumulation in the atmosphere of CO_2 from the burning of pit coal would "undoubtedly cause a very obvious rise of the mean temperature of the Earth." He also thought this effect could be accelerated by burning coal exposed in shallow seams or perhaps decreased "by protecting the weathering layers of silicates from the influence of the air and by ruling the growth of plants." Ekholm pointed to the grand possibility that by such means it might someday be possible "to regulate the future climate of the Earth and consequently prevent the arrival of a new Ice

Age." In this scenario, climate warming by enhanced coal burning would be pitted against the natural changes in the Earth's orbital elements, recently identified by James Croll, or the secular cooling of the Sun, as pointed out by Lord Kelvin (William Thomson). Ekholm concluded, "It is too early to judge of how far Man might be capable of thus regulating the future climate. But already the view of such a possibility seems to me so grand that I cannot help thinking that it will afford Mankind hitherto unforeseen means of evolution."[7]

Arrhenius popularized Ekholm's observations in his book *Worlds in the Making,* noting that "the slight percentage of carbonic acid in the atmosphere may by the advances of industry be changed to a noticeable degree in the course of a few centuries." Arrhenius considered it likely that in future geological ages, the Earth would be "visited by a new ice period that will drive us from our temperate countries into the hotter climates of Africa." On the timescale of hundreds to thousands of years, however, Arrhenius agreed with Ekholm that a "virtuous circle" could be defined in which the burning of fossil fuels could help prevent a rapid return to the conditions of an ice age and could perhaps inaugurate a new carboniferous age of enormous plant growth.

Yet in the early decades of the twentieth century, the carbon dioxide theory of climate change, along with the human influence theory, fell out of favor with most scientists. Most scientists believed that at current atmospheric concentrations, carbon dioxide already absorbed all the available long-wave radiation; thus any increases would not change the radiative heat balance of the planet. The person responsible for reviving the ideas of Arrhenius and Ekholm and placing them on a revised scientific basis was Guy Stewart Callendar (1898–1964), a British steam and defense engineer. In 1938 Callendar reformulated the carbon dioxide theory by arguing that rising global temperatures and increased fossil fuel burning were closely linked. He compiled weather data from stations around the world that clearly indicated a climate warming trend of 0.5°C (0.9°F) in the early decades of the twentieth century. His estimate of 290 parts per million for the nineteenth-century background concentration of CO_2 is still a valid estimate, and he documented an increase of 10 percent between 1900 and 1935, which closely matched the amount of fuel burned. On the basis of new understanding of the infrared spectrum and calculations of the absorption and emission of radiation by trace gases in the atmosphere, Callendar established the carbon dioxide theory of climate change in its recognizably modern form, reviving it from its earlier, physically unrealistic and moribund status. Today the theory that global climate change can be attributed to an enhanced greenhouse effect resulting from elevated levels of carbon dioxide in the atmosphere from anthropogenic sources and activities is called the Callendar effect.[8]

The dawn of aviation brought new needs and challenges, with fog dispersal taking center stage. A number of ineffective efforts using chemical and electrical means preceded the massive World War II fog-clearing project FIDO (Fog Investigation and Dispersal Operation), which allowed British Royal Air Force and Allied planes to take off and land when the Germans were grounded. With national survival at stake, it did not matter that it required burning 6,000 gallons of gasoline to land one airplane in the fog.

After World War II, promising discoveries in "cloud seeding" at the General Electric Corporation rapidly devolved into questionable practices by military and commercial rainmakers seeking to control the weather. At the same time, hopeful developments in digital computing led to speculation that a perfect machine forecast of weather and climate could lead to perfect understanding and control. During the cold war, speculation about geoengineering by the Soviets promoted a chilling vision (to Westerners) of global climate control. Geoscientific speculators in the West returned the favor.

By 1962 the results of early computer simulations of the general circulation of the atmosphere and the first satellite estimates of the Earth's heat budget led Harry Wexler, head of research at the U.S. Weather Bureau, to warn a United Nations symposium on the environment of the "inherent risk" in attempted climate control "of irremediable harm to our planet or side effects counterbalancing the possible short-term benefits."[9] Yet only three years later, the President's Science Advisory Committee reported that scientists might soon need to increase the Earth's albedo, or planetary brightness, deliberately in response to increased warming from carbon dioxide emissions.[10]

During the hot summer of 1988, with Yellowstone National Park in flames and global warming in the headlines, an international scientific conference sponsored by the UN and the World Meteorological Organization recommended reductions of carbon dioxide emissions to 20 percent below 1988 levels, to be achieved by 2005. Today, we are nowhere near reaching that goal. Experts advise that reductions of greenhouse gas emissions of at least 80 percent will be necessary, while popular cries of "Stop global warming" and "Control climate change" are becoming more and more widespread. Invoking the unlikelihood that such reductions will be accomplished voluntarily and the fear of passing a climate "tipping point," some modern-day climate engineers are suggesting that they can provide cheap, reliable technological "fixes" for the climate system through macroengineering options that include "solar radiation management," carbon capture and sequestration, and other invasive techniques of "planetary surgery."

Weather and climate are intimately related: weather is the state of the atmosphere at a given place and time, while climate is the aggregate of weather conditions over time. A vast body of scientific literature addresses these interactions. In addition, historians are revisiting the ancient but elusive term *Klima*, seeking to recover its multiple social connotations.[11] Weather, climate, and the climate of opinion matter in complex ways that invite—some might say require or demand—the attention of both scientists and historians.

Yet some may wonder how weather and climate are interrelated rather than distinct. Both, for example, are at the center of the debate over greenhouse warming and hurricane intensity.[12] A few may claim that rainmaking, for example, has nothing to do with climate engineering, but *any* intervention in the Earth's radiation or heat budget (such as managing solar radiation) would affect the general circulation and thus the location of upper-level patterns, including the jet stream and storm tracks. Thus the weather itself would be changed by such manipulation. Conversely, intervening in severe storms by changing their intensity or their tracks or modifying weather on a scale as large as a region, a continent, or the Pacific basin would obviously affect cloudiness, temperature, and precipitation patterns, with major consequences for monsoonal flows and ultimately the general circulation. If repeated systematically, such interventions would influence the overall heat budget and the climate.

In the 1950s, Irving Langmuir sought to cause changes in the seasons and the climate of large regions such as the North American continent and the Pacific Ocean by massive seeding of weather systems. Three decades earlier, L. Francis Warren tried to develop a system of universal weather control using electrified sand. In the 1840s, James Espy's proposed large fires were intended to act as artificial volcanoes, triggering regular rains along the entire eastern seaboard to change the climate and improve the health of the region, while Thomas Jefferson speculated on climate engineering at the dawn of the nineteenth century and thought that the sum total of American agricultural practices would surely change local weather and warm the entire continent. Thus, both by definition and in historical practice, weather and climate occupy a continuous spectrum ranging from local to global scales and from short- to long-term temporal changes. As Harry Wexler liked to point out, if you change the weather repeatedly on a large spatial scale, you are changing the climate, and vice versa.

I have set down in writing my ideas about fixing the sky—primarily historical ideas about mending, repairing, or somehow improving perceived defects in the weather or in climate systems—but fixing the sky has many, many other possible meanings. In the *Oxford English Dictionary*, the "sky" is the apparent arch or vault of heaven, whether covered with clouds or clear and blue; it may be the climate or clime of a particular region, nowadays usually designated more globally than locally. The appearance of the sky is variously sunny, starry, hazy, overcast, azure, copper, even milky white. Fog or cloud modification involves fixing the sky. Sky gods and goddesses, sky-shades, and sky-fliers (the overly ambitious) have all played their roles in this seemingly limitless and often extravagantly fanciful history.

A "fix" is a predicament, difficulty, dilemma, or a "tight place." It refers to a heroic intervention to help the hopeless and make things right again. It can also be a certified position at sea, in the air, or on the trading floor; a dose of narcotics for an addict; or an illegal bribe or illicit arrangement. A fix is a measure undertaken to resolve a problem, an easy remedy, sometimes known as a "quick fix," which connotes an expedient but temporary solution that fails to address underlying problems. It can be a "tech fix" that emphasizes the engineering aspect rather than the social dimensions of an issue. Something "fixed" is not changing or vacillating; it possesses stability and consistency, even if it is a steady, concentrated, unwavering, or mesmerizing fixed gaze. When the chemist Joseph Black discovered what we now call carbon dioxide, he called it "fixed air" because of its stable properties—ironic now that this compound is the volatile core of all environmental discussion. Plants are good at fixing carbon into their tissues through photosynthesis, but we have yet to learn how to capture, fix, and sequester carbon dioxide underground or in ocean trenches. Sporting events and elections can be "fixed" by illegal means, bulls by legal means; the unattached can be "fixed up" with likely partners.

In 1966 physicist Alvin Weinberg coined the term "technological fix." Since then, it has come to connote simplistic or stopgap remedies to complex problems, partial solutions that may generate more problems than they solve. Placing more faith in technology than in human nature, Weinberg offered engineering as an alternative to conservation or restraint. We face this dilemma with technological fixes for global warming, although those who propose such ideas are quick to say that they are only buying time until more reasonable forms of mitigation and adaptation can take effect.[13]

In a practical way, humans have long practiced a form of climate control in their technologies of clothing and shelter. By controlling the heat and moisture budgets within a centimeter of the skin surface, humans can function in even the harshest weather conditions. Mountain climbers, polar explorers, even the French Foreign Legion represent extreme examples of what we all do—clothe ourselves according to expected environmental conditions.

Controlling the heat budget (and to some extent the moisture budget) within small, enclosed spaces allows humans to live, work, and play in relative comfort and safety in most weather conditions and climate zones. As Ralph Waldo Emerson said, "Coal is a portable climate. . . . Watt and Stephenson whispered in the ear of mankind their secret, that a half-ounce of coal will draw two tons a mile, and coal carries coal, by rail and by boat, to make Canada as warm as Calcutta, and with its comfort brings its industrial power."[14] Just one century ago, industrial power was applied to cooling, drying, and purifying the air when Willis H. Carrier invented an industrial air-conditioning system. Carrier's invention has now infiltrated all aspects of modern life. It is doubtful whether the American Sun Belt would be growing as it is today without the widespread use of home, auto, and industrial air-conditioning. As these brief examples indicate, controlling the weather and climate is something we all do (on a small scale), while some fantasize about it on a large scale. Clark Spence, in his entertaining book *The Rainmakers*, surveyed the sometimes fantastic and always quixotic history of scientific weather modification before World War II. Here those stories are expanded and continued after 1945.

While many works in the history of science and technology have been crafted in a heroic mode—great men with great ideas "standing on the shoulders of giants"—and environmental histories are often written as tragedies, the history of weather and climate control is best told by invoking a broader range of approaches, including a mixture of the tragic and comedic genres. Most of the rainmakers and climate engineers portray their activities as heroic and dramatic attempts to rescue humanity from a recalcitrant sky by exercising control over it; however, their efforts often have commercial or military dimensions and almost always fall far short of the stated goals. Here is where tragicomedy—or perhaps just comedy—best captures the flawed anti-heroics of those who would seek to fix the sky or control the weather and climate. In this book, I present a comedy of ideas extending from the mythological past to the present, with the common denominator being farce, and sometimes satire, especially when the hype becomes too great. Most of the stories emphasize the perennial nature of the claims, the hubris and ineptitude of the protagonists, the largely pathological science on which they are based, the opportunistic appeals to new technologies, the false sense that macro-engineering will solve more problems than it creates, and the ineptitude of the protagonists.

The trinity of understanding, prediction, and control undergirds the dominant fantasies of both science and science fiction. Understanding often involves reducing a complex phenomenon to a set of basic laws or mechanisms. This may even involve extreme "molecular reductionism"—for example, in the treatment of silver iodide (AgI) as a "trigger" mechanism for widespread weather modification or of carbon dioxide, today's environmental molecule of choice, as an international symbol of human intervention in the climate system, signaling and codifying both affluence and apprehension.

Prediction introduces the time dimension in which the future state of a natural phenomenon is specified. If you understand a phenomenon, scientists say, you should be able to predict its behavior. But while rather precise prediction of the appearances of the sky was practiced in antiquity, weather prediction and basic climate modeling were not possible until the mid-1950s when digital computing provided our first glimpses of the possibility of handling the extreme complexity of this nonlinear system.

Control is the third member of the trinity, but understanding does not imply either predictability or control. If you know from observation that horses need pasture and fresh water, you may predict that a wild herd will gather in the grassy fields near the river. Capturing them, taming them, and bending them to your will, however, is a far more difficult undertaking. For some, in the age of digital computing, Earth observations from space, and extremely precise measurement of atmospheric chemical species, controlling the weather and climate is more desirable than merely observing or predicting it. Some think that this is now possible and that science and technology have given us an Archimedean set of levers with which to move the Earth. This book examines these ancient, perennial, and contemporary quests and questions by placing recent developments in the context of the deeper past.

Chapter 1, stories of control, highlights imaginative and speculative literature on the control of nature. It draws from the classical tradition, including Phaethon's blunder, Milton's *Paradise Lost*, and Dante's *Divine Comedy*, among others. The examples indicate that myth, magic, religion, and legend are not relics of the past but constitute deep roots and living sparks of contemporary practices. An excursion into early geoscientific fiction follows, demonstrating the affinities between the genre of science fiction and the fantasies of the cloud and climate controllers. The works of famous authors such as Jules Verne, Mark Twain, and Kurt Vonnegut serve to anchor the analysis of a host of lesser-known but still important, enlightening, and entertaining early fiction. Tales of the rainmakers, including the well-known play *The Rainmaker*, by N. Richard Nash, appear alongside popularizations from the television series *Sky King* and comics from Warner Brothers and Walt Disney. Here, as in Twain, the comedic genre clearly trumps the heroic and the tragic. It is also clear that fiction writing has a moral core that is missing from the speculative proposals of scientists and engineers. Moreover, the writers tend to employ female voices to remind their predominantly male protagonists of their ethical excesses.

Scientific rainmakers take the historical stage in chapter 2. The story of control begins with the aspirations of Sir Francis Bacon and continues as a legacy of the historiographically contentious scientific revolution. Enlightened dreamers, enamored by the notion of progress, enthusiastically sought to understand, predict, and ultimately control the weather and climate. But did they reveal nature's deepest secrets or abuse our deepest sensibilities? The distinguished American meteorologist James Espy wanted to control rainfall with great fires, a problematic goal that, if ever accomplished, would have raised immense ethical dilemmas. Another group wanted to cannonade the clouds to wring out their moisture, but succeeded mainly in entertaining onlookers with pyrotechnic displays. The notion of progress was such a heady surety that it seemed that anything was possible; not even the sky was the limit. Surely things are much different now. Or are they?

Chapter 3 examines the rain fakers, the charlatans or confidence men who lived by their wits and accepted payment from desperate and gullible farmers for their questionable services. Hail shooting falls into this category, as do the Kansas and Nebraska proprietary rainmakers of the 1880s and 1890s. Charles Hatfield, the "moisture accelerator," was a charlatan's charlatan who mixed his proprietary chemicals and dispensed them from high towers at considerable profit in the first three decades of the twentieth century. George Ambrosius Immanuel Morrison Sykes, who scammed the Belmont Park racetrack in the 1930s; Wilhelm Reich, who scammed his followers in the 1950s; Irving Krick, who practiced commercial cloud seeding over most of the American West; and the Provaqua project, which just recently tried to scam the citizens of Laredo, Texas, serve to illustrate the perennial nature of these questionable but humorous (at least from a distance) practices. Ironically, the rainmakers and the rain fakers employed surprisingly similar techniques, although the former actually believed in what they were doing, while the latter clearly did not.

Chapter 4 focuses on fog removal in the era of early aviation. As the airplane provided a new platform for aerial experimentation, it also raised the stakes for aviation safety and military efficiency. Teams of experimenters, some working largely on their own and some with the full support of governments, tried electrical, chemical, and physical methods of fog removal. These included attacking clouds with electrified sand, spraying calcium chloride on airports, and burning hundreds of thousands of gallons of gasoline in a brute-force effort to keep the Royal Air Force aloft and return its pilots safely. The chapter ends with a look at the "airs of the future," both indoors and out. The rising popularity of air-conditioning in the 1930s was an approach to weather and climate control that has since reached the level of domed stadiums and indoor shopping malls, falling just shy of totally air-conditioned cities. Also in the 1930s and early 1940s, meteorologists shared their visions of technological breakthroughs in the coming decades leading to perfect forecasts and the holy grail of weather control.

Chapter 5 examines the defining characteristics of "pathological science" established by Nobel laureate Irving Langmuir and then proceeds to indict him on his own criteria. Langmuir, and to some extent his associates at General Electric, was an overenthusiastic supporter of weather control by using dry ice and silver iodide as cloud-seeding agents. When the military took over the project, the stage was set for heavy-handed intervention in hurricanes, large-scale tests with few controls, and sweeping but unsupportable claims. As the technique spread around the world, a host of commercial cloud seeders, personified by Irving Krick, made their living at the expense of those in need of rainfall. The chapter concludes with stories of meteorological disasters in England and the former Soviet Union attributed to but not proved to have been caused by cloud seeding.

The mood darkens considerably in chapter 6 as military themes take center stage. What are the historical dimensions of military interest and involvement in the weather; how were the clouds weaponized, especially in the cold war era; and how did a race for weather control domination emerge between the United States and the Soviet Union? The sordid episode of rainmaking in Vietnam over the Ho Chi Minh Trail and the ban by the United Nations on environmental warfare quashed much of this enthusiasm in the 1970s. Yet the weather and climate warriors are with us still, preparing to "own" and manipulate the weather over the battlefields of the future and seeking to control the evolving nature of climate change in the interest of national security.

Chapter 7 examines climate fears, climate fantasies, and the possibility of global climate control between 1945 and 1962. It illuminates technical, scientific, social, and popular issues and moves us beyond the timeworn origin stories of numerical weather prediction into a new field of numerical climate control—a marketplace of wild ideas, a twentieth-century Hall of Fantasy, or even Twilight Zone, whose boundaries are those of imagination. It does so by examining some of the chemists, physicists, mathematicians, and, yes, meteorologists who tried to "interfere" with natural processes. They intervened not with dry ice or silver iodide, but with the new Promethean possibilities of climate tinkering using digital computing, satellite remote sensing, and nuclear power. Key players include Vladimir Zworykin, the inventor of television, and the noted mathematician John von Neumann, both of whom were seeking a perfect forecasting machine, and Harry Wexler, who imagined cutting a hole in the stratospheric ozone layer and issued a clear warning about the coming dangers of climate engineering.

Finally, chapter 8 examines recent and current ideas and proposals regarding geoengineering. Driven by the fear of global warming and their underlying certainty that mitigation and adaptation will not be sufficient to prevent a climate catastrophe, the climate engineers are pushing for the authority and the wherewithal to go beyond paper studies and computer models to field trials and full-scale implementation as a technological fix for global warming. But in their quest to create a "planetary thermostat," they lack a widespread following and appear to most mainstream scientists, environmentalists, and policymakers as overly aggressive in their vocal advocacy for untested, and perhaps untestable, practices. It is likely that humanity as a whole has done too little in response to the problem, but the climate engineers are seeking to do too much.

Attempted control of the environment may not be a good thing, especially when it is based on simplistic assumptions (for example, that hurricanes may be readily redirected or that basic radiation physics controls the Earth's climate) or when it exceeds the knowledge base or verges on science fantasy. Like the pseudo-scientific rainmakers of yore, today's aspiring climate engineers wildly exaggerate what is possible and scarcely consider the political or ethical implications of attempting to manage the world's climate—with potential consequences far greater than any of their predecessors were ever likely to face.

Who has the moral right to modify the weather or the global climate? Where will a global thermostat be located, and who gets to control it? Will climate engineering reduce incentives to mitigate carbon emissions? What about unknown side effects? Should it be commercialized? What if nations or companies do it unilaterally? Does it violate existing treaties? Why is the military so interested? Once it begins, can we ever stop it? How will weather and climate engineering alter fundamental human relationships to nature?

This book is grounded in the practices of the past and provides perspectives on the largely fantastic claims of the current batch of geoscientific speculators, collectively known as the climate engineers, who are proposing to cool the planet in response to fears of global warming. In facing the unprecedented challenges posed by humanity's current confrontation with the elements—a situation exacerbated by world population, a host of aerial effluents, and generally rising affluence—it is good to seek historical precedents.

The current generation of climate engineers is not the first to consider planetary-scale environmental manipulation. Indeed, they are heirs to a long and checkered history of weather and climate control populated by a colorful cast of dreamers and losers. If this history provides new insights, raises new issues, provokes new controversies, or serves to inform social, ethical, and public policy considerations, I will deem the effort worthwhile. The goal is the articulation of perspectives fully informed by history and the initiation of a dialogue that uncovers otherwise hidden values, ethical implications, social tensions, and public apprehensions surrounding our past and current environmental situations.

1

STORIES OF CONTROL

The General Electric Company was *science fiction.*

—KURT VONNEGUT

THROUGHOUT history and across cultures, civilizations have told stories about gods and heroes who have attempted to control that which may be largely uncontrollable, including phenomena both above and below the horizon. There are many sources for such stories. Myth, religion, and traditional practices form the foundations of culture and are often invoked when people seek group solidarity—for example, when the expected rains fail to arrive or a violent storm rages. Stories drawn from Greek mythology, the Western canon, Native American rainmaking, and recent fiction are presented here, followed by examples of geo-science fiction before about 1960—drawn from the pages of pulp fiction, the stage and silver screen, and the boob tube—that serve to illuminate popular culture. But much more than edification is at stake. Storytelling skirts the borderlands between fact and fantasy and acknowledges their reciprocal relationship. Here the comic and tragicomic genres provide fresh insights into the speculative practices of the meteorological Don Quixotes and Rube Goldbergs of the past. Such storytelling clearly trumps the heroic and the tragic genres so typical in the literature of science studies. It is an excursion that historians of science, technology, and environment have only recently begun to take.

Phaethon's Blunder

In uncovering the deeper cultural roots of weather and climate engineering, it is instructive to consider the wisdom invested in mythological stories, since whether we realize it or not, much of Western civilization rests on these foundations. In Greek mythology, the youth Phaethon lost control of the Sun chariot, and his recklessness caused extensive damage to the Earth before Zeus shot him out of the sky. The story began when Phaethon, mocked by a schoolmate for claiming to be the son of Helios, asked his mother, Clymene, for proof of his heavenly birth. She sent him east toward the sunrise to the awe-inspiring palace of the Sun god in India. Helios received the youth warmly and granted him a wish. Phaethon immediately asked his father to be permitted for one day to drive the chariot of the Sun, causing Helios to repent of his promise, since the path of the zodiac was steep and treacherous and the horses were difficult, if not impossible, to control. Helios replied prophetically, "Beware, my son, lest I be the donor of a fatal gift; recall your request while yet you may. . . . It is not honor, but destruction you seek. . . . I beg you to choose more wisely."[1] But the youth held to his demand, and Helios honored his promise. At the break of dawn, the horses were harnessed to the resplendent Sun chariot, and Helios, with a foreboding sigh, urged his son to spare the whip, hold tight the reins, and "keep within the limit of the middle zone," neither too far south or north, nor too high or low: "The middle course is safest and best" (63).

Now Phaethon spared the whip; that was not the problem. A bigger problem was that the youth was a lightweight (literally) and the horses sensed this, "and as a ship without ballast is tossed hither and thither on the sea, so the chariot, without its accustomed weight, was dashed about as if empty" (64). Also Phaethon was a completely inexperienced driver without a clue as to the proper route to take: "He is alarmed, and knows not how to guide them; nor, if he knew, has he the power" (64). The chariot veered out of the zodiac, with hapless Phaethon looking down on the vast expanse of the Earth, growing pale and shaking with terror. He repented of his request, but it was too late. The chariot was borne along "like a vessel that flies before a tempest," and Phaethon, losing control completely, dropped the reins. So much for the middle course:

> The horses, when they felt them loose on their backs, dashed headlong, and unrestrained went off into unknown regions of the sky, in among the stars, hurling the chariot over pathless places, now up in the high heaven, now down almost to earth. . . . The clouds begin to smoke, and the mountain tops take fire; the fields are parched with heat, the plants wither, the trees with their leafy branches burn, the

1.1 Phaethon, from the series *The Four Disgracers* (1588) by Hendrick Goltzius (Netherlandish, 1558–1617). Icarus, Ixion, and Tantalus are also called "disgracers" for their overweening ambition.

> harvest is ablaze! But these are small things. Great cities perished, with their walls and towers; whole nations with their people were consumed to ashes! The forest-clad mountains burned. . . . Then Phaeton beheld the world on fire, and felt the heat intolerable. The air he breathed was like the air of a furnace and full of burning ashes, and the smoke was of a pitchy darkness. (65–66)

With the Earth on fire, the oceans at risk, and the poles smoking, Atlas did more than shrug—he fainted. The Earth, overcome with heat and thirst, implored Zeus to intervene, "lest sea, Earth, and heaven perish, [and] we fall into ancient Chaos. Save what yet remains to us from the devouring flame. O, take thought for our deliverance in this awful moment" (67). Zeus responded by shooting the devious charioteer out of the sky with a fatal lightning bolt as Helios looked on in shock and dismay (figure 1.1). A utilitarian ethic applies in the myth. After much of the Earth was incinerated, Phaethon was killed by a higher authority to avoid further damage. And rightly so.

The story of Phaethon was invoked in 2007 by the noted meteorologist Kerry Emanuel to frame a short discussion of contemporary climate change science and politics. Emanuel, widely known and respected for his hurricane studies, called attention to a growing scientific consensus on climate change prominently and authoritatively spearheaded by today's Intergovernmental Panel on Climate Change (IPCC) reports, yet he admitted pointedly that "we are . . . conscious of our own collective ignorance of how the climate system works."[2] Abruptly returning to myth, Emanuel ended his essay, "Like it or not, we have been handed Phaeton's reins, and *we will have to learn how to control climate* if we are to avoid his fate" (69; emphasis added). Emanuel thus advocated repeating Phaethon's blunder. Think underage driver of gasoline tanker, taken with father's permission, veers out of control in reckless, high-speed chase before being subdued by the authorities. Or more globally, geoengineering project given the green light last year results in the collapse of the Indian monsoon, leaving millions starving.

What about Emanuel's final thought—that we "will have to learn how to control climate"? That is the subject of the final chapter of this book. Cambridge scientist Ross Hoffman has proposed a speculative "star wars" system to redirect hurricanes by beaming lasers at them from satellites—assuming one knew where the storm was originally headed and that there would be no liabilities along its new path. Is this an example of Phaethon's reins? Since the Sun god Helios was directly involved, what about other means of "solar radiation management," such as Nobel laureate Paul Crutzen's suggestion (made in 2006, but actually a much older idea) to cool the Earth by injecting sulfates or other reflective aerosols into the tropical stratosphere? There are many, many more such dangerous and expensive proposals of environmental control that invoke the inexperience and possible tragedy of the myth of Phaethon.

Remember, Helios made a fundamentally flawed decision to give his son the reins, and that decision had catastrophic consequences. He did, however, give Phaethon a piece of good advice about steering the Sun chariot through the middle course of the zodiacal signs. For humanity, the best we can do between this world and the next is to admit our "own collective ignorance," remain humble, and avoid angering both the Sun god and his boss. Will this involve following the "middle course" of collective energy efficiency, environmental stewardship, and ethical choices? Certainly to do nothing is out of the question. But could we try to do too much? Will someone or some group trying to "fix" the climate repeat Phaethon's blunder? Greek mythology is replete with such stories, characters, and moral lessons.

Paradise Lost and the *Inferno*

Biblical themes permeate the Western canon, and some of them speak either directly or indirectly to the human role in weather and climate control. In *Paradise Lost* (1667), John Milton alludes to a divinely instituted shift in the Earth's axis (and thus its climate) as a consequence of the original ancestors' lapse from grace. According to Milton, while Eden was the ultimate temperate clime, watered with gentle mists, God, in anger and for punishment, rearranged the Earth and its surroundings to generate excessive heat, cold, and storms: "the Creator, calling forth by name His mightie Angels, gave them several charge."[3] The Sun was to move and shine so as to affect the Earth "with cold and heat scarce tolerable" (10.653–654); the planets were to align in sextile, square, opposition, and trine "thir influence malignant . . . to showre" (10.662); the winds were to blow from the four corners to "confound Sea, Aire, and Shoar" (10.665–666); and the thunder was to roll "with terror through the dark Aereal Hall" (10.667). The biggest change, however, resulted from tipping the axis of the Earth: "Some say he bid his Angels turne ascanse the Poles of Earth twice ten degrees and more from the Sun's Axle; they with labor push'd oblique the Centric Globe . . . to bring in change of Seasons to each Clime; else had the Spring perpetual smil'd on Earth with vernant Flours, equal in Days and Nights" (10.668–671, 677–680).

This led to massive changes in weather and climate on sea and land: "sidereal blast, Vapour, and Mist, and Exhalation hot, Corrupt and Pestilent" (10.693–695). Northern winds (Boreas, Kaikias, and Skeiron) burst "their brazen Dungeon, armd with ice and snow and haile, and stormie gust and flaw" (10.697–698), and other winds (Notus, Eurus, and the Tempest-Winds) in their season rent the woods, destroyed crops, churned the seas, and rushed forth noisily with black thunderous clouds, serving the bidding of the storm god Aeolus. But the angels had one last task: evicting "our lingring Parents" (12.638) from Eden. In this, too, Milton evokes climatic change when the blazing sword of God, "fierce as a Comet; which with torrid heat, and vapour as the *Libyan* Air adust, began to parch that temperate Clime" (12.634–636). Looking back at Paradise, "som natural tears they drop'd, but wip'd them soon; the World was all before them, where to choose thir place of rest, and Providence thir guide: They, hand in hand, with wandring steps and slow, through *Eden* took their solitarie way" (12.645–649). So you see, the wages of sin are . . . climate change.

When Dante Alighieri visited hell with Virgil in the spring of 1300, he witnessed the consequences of sin. They had left a world with "air serene" and entered "into a climate ever vex'd with storms . . . where no light shines."[4] Before

1.2 *Inferno*: "The violent, tortured in the Rain of Fire," in Dante's version of hell. (ILLUSTRATION BY GUSTAVE DORÉ, FOR *INFERNO* 14.37–39, 1861)

them, confined to the third circle, were the gluttons experiencing unique meteorological torments of eternal cold and heavy rain, hail, and snow (6.6–11), while in the seventh circle were those who had done violence to God, naked souls weeping miserably, supine, sitting, wandering, muttering, under a steady rain of "dilated flakes of fire" (14.18–27) (figure 1.2). Today we might add a new caption to Gustave Doré's illustration: Sulfurous rains fall on a wretched humanity following artificial volcano experiment gone awry; two geoengineers look on.

The heavens and "heaven" have never been strictly demarcated; in fact, they have been closely intertwined, especially when it comes to something at once as nebulous and portentous as atmospheric phenomena. Synergistic rather than conflicting interactions between the numinous and the immanent appear to be more the norm than the exception throughout history. Humans attempting to intervene in the "realm of the gods," whether through ceremonies or technologies, inevitably find themselves engaged in a complex dance with both novel and traditional steps, where stumbling and falling from grace, or at least stepping on toes, is more likely than perfect execution.

The Mandan Rainmakers

The nineteenth-century American painter George Catlin juxtaposed traditional rainmaking and Western technology in his account of the manners and customs of North American Indians. When the Mandan, who lived along the Upper Missouri River, were facing a prolonged dry spell that threatened to destroy their corn crop, the medicine men assembled in the council house, with all their mystery apparatus about them, "with an abundance of wild sage, and other aromatic herbs, with a fire prepared to burn them, that their savory odors might be sent forth to the Great Spirit."[5] On the roof of the council house were a dozen young men who took turns trying to make it rain. Each youth spent a day on the roof while the medicine doctors burned incense below and importuned the Great Spirit with songs and prayers:

> Wah-kee (the shield) was the first who ascended the wigwam at sunrise; and he stood all day, and looked foolish, as he was counting over and over his string of mystery-beads—the whole village were assembled around him, and praying for his success. Not a cloud appeared—the day was calm and hot; and at the setting of the sun, he descended from the lodge and went home—"his medicine was not good," nor can he ever be a medicine-man. (1:153)

On successive days, Om-pah (the elk) and War-rah-pa (the beaver) also failed to bring rain and were disgraced.

On the fourth morning, Wak-a-dah-ha-hee (hair of the white buffalo) took the stage, clad in his finest garb and with a shield decorated with red lightning bolts to attract the clouds and a sinewy bow with a single arrow to pierce them. Claiming greater magic than his predecessors, he addressed the assembled tribe and commanded the sky and the spirits of darkness and light to send rain. The medicine men in the lodge at his feet continued their chants.

Around noon, the steamboat *Yellow Stone*, on its first trip up the river, neared the village and fired a twenty-gun salute, which echoed throughout the valley. The Mandans, at first supposing it to be thunder, although no cloud was seen in the sky, applauded Wak-a-dah-ha-hee, who took credit for the success. Women swooned at his feet, his friends rejoiced, and his enemies scowled as the youth prepared to reap the substantial rewards due a successful rainmaker. However, the focus quickly shifted to the "thunder-boat" as it neared the village, and the hopeful rainmaker was no longer the center of attention. Later in the day, as the excitement of the boat's visit began to ebb, black clouds began to build on the horizon. Wak-a-dah-ha-hee was still on duty. In an instant, his shield was on his arm and his bow drawn. He commanded the cloud to come nearer, "that he might draw down its contents upon the heads and the corn-fields of the Mandans!" (1:156). Finally, with the black clouds lowering, he fired an arrow into the sky, exclaiming to the assembled throng, "My friends, it is done! Wak-a-dah-ha-hee's arrow has entered that black cloud, and the Mandans will be wet with the water of the skies!" (1:156–157). The ensuing deluge, which continued until midnight, saved the corn crop while proving the power and the efficacy of his medicine. It identified him as a man of great and powerful influence and entitled him to a life of honor and homage.

Catlin draws two lessons from this story. First, "when the Mandans undertake to make it rain, they never fail to succeed, for their ceremonies never stop until rain begins to fall" (1:157). Second, the Mandan rainmaker, once successful, never tries it again. His medicine is undoubted. During future droughts, he defers to younger braves seeking to prove themselves. Unlike Western, technological rainmaking, in Mandan culture the rain chooses the rainmaker.

Leavers and Takers

In his imaginative book *Ishmael* (1992), Daniel Quinn draws a basic distinction between two major streams in human culture: the Takers (the heirs of the agricultural revolution) and the Leavers (or traditional societies). As he tells it, ten thousand years ago, the Takers exempted themselves from the evolutionary process. They saw the world as having been made for them and belonging to them, so they sought to manipulate and control it. Since then, they have systematically expanded their own food resources and their population at the expense of other species. Their quest for control seemingly knows no bounds. It extends from the control of pests, both micro- and macroscopic (from bacteria to browsing deer), to the attempted control of the sky. Guided by the tacit but ubiquitous voice of Mother Culture, the assumed nurturer of Taker human societies and lifestyles, they have come to see themselves as special and superior beings who possess the knowledge of good and evil. This allows them to decide, in god-like fashion, who shall live and who shall die. The world for them is a human life-support system, a machine designed to produce and sustain human life. When the elements or other species defy him, man declares war on nature and sees it as his destiny to conquer and rule it with *complete* control:

> We'll turn the rain on and off. . . . We'll turn the oceans into farms [or carbon sinks]. We'll control the weather [and climate]—no more hurricanes, no more tornadoes, no more droughts, no more untimely frosts. We'll make the clouds release their water over the land instead of dumping it uselessly into the oceans. All the life process of this planet will be where they belong—where the gods meant them to be—in our hands. And we'll manipulate them the way a programmer manipulates a computer.[6]

Technology seems to provide the leverage to make all this possible, but, according to Quinn, Taker culture is fatally flawed in that it lacks historical perspective and the wisdom of how to live. The Takers, who act as though the world belongs to them, are the enemy of the world and are on an evolutionarily recent, unsustainable, and potentially world-shattering detour.

The older cultures, the Leavers, far from being savage, primitive, or degenerative, constitute the main stream of human evolution and trace their roots back at least 3 million years. They respect the right to life and food of all other creatures and live as members, not rulers, of the community of all life. They live close to nature in relative abundance, free from worry, in the hands of the gods, enhancing biological and cultural diversity and ecological sustainability. The Leavers, who act as though they belong to the world, allow the creatures around them a chance to fulfill their potential. In this sense, they share an evolutionary destiny.

Quinn's basic quest is to reform Taker culture by making people aware of what has been lost. He argues that people need something positive to work for, rather than something negative to work against. They need an inspiring vision more than a vision of doom, more than to be scolded, more than to be made to feel stupid and guilty:

> There's nothing fundamentally wrong with people. Given a story to enact that puts them in accord with the world, they will live in accord with the world. But given a story to enact that puts them at odds with the world, as yours does, they will live at odds with the world. Given a story to enact, in which they are the lords of the world, they will act as the lords of the world. And, given a story to enact in which the world is a foe to be conquered they will conquer it like a foe, and one day, inevitably, their foe will lie bleeding to death at their feet, as the world is now. (84)

This is the voice of Ishmael, an articulate, Bible-reading, telepathic gorilla whom Quinn uses as a transcendent messenger to humanity. Ishmael's students, in effect each reader of the book, must have "an earnest desire to save the world," must "apply in person," and must be willing to enact a life-affirming story that puts them in accord with the world. Ishmael reminds us that stopping pollution or cutting down on carbon emissions is not in itself an inspiring goal, but thinking of ourselves and the world in a new way is. By seeking to have a minimal impact on the planet, environmentalists align themselves with the values of Leaver culture. The climate engineers, however, in the name of stopping climate change, are the consummate Takers.

Science Fiction

Ultimate control of the weather and climate embodies both our wildest fantasies and our greatest fears. Fantasy often informs reality (and vice versa). NASA managers know this well, as do Trekkies. The best science fiction authors typically build from the current state of a field to construct futuristic scenarios that reveal and explore the human condition. Scientists as well often venture into flights of fancy. Although not widely documented, the fantasy–reality axis is a prominent aspect of the history of the geosciences. The chief distinction is that the fiction writers provide a moral core and compass.

An occasional whimsical story of rainmaking in the nineteenth century has given way to such a flood of science fiction that accounts of weather and climate control alone could fill a volume. The plot of the science fiction film *Voyage to the Bottom of the Sea* (1961) revolves around a heroic and unilateral engineering response to a global environmental emergency. When a swarm of meteors pierces the Van Allen radiation belt and sets it on fire, the Earth is threatened by imminent "global warming" and possible mass extinctions. With the Arctic ice cap disintegrating and Africa on fire, with world temperatures rising quickly and the end of civilization nigh, the commander of a new state-of-the-art atomic submarine (with Cadillac tail fins) proposes to extinguish the fires by launching a nuclear missile into space to cut off the burning radiation belt from the Earth. When United Nations scientists reject the plan as too risky, the commander takes unilateral action against the will of what he deems to be overcautious government representatives and elected officials. Thwarting various attempts to stop him, by saboteurs, a giant octopus, and a religious fanatic who believes it is God's will that the world end, the submarine commander fires the missile and saves the world, proving that he was right all along. The television series also featured many episodes with geophysical threats and geoengineering responses.

Other thrillers and spoofs of thrillers in recent eras had plot lines involving weather or climate control. In *Our Man Flint* (1965), super-duper secret agent Derek Flint foils an evil cabal of utopian mad scientists who are planning to take over the world through weather control. At the end of the movie *The Andromeda Strain* (1971), cloud seeding over the Pacific Ocean results in the alien "strain" being washed into the salt water, presumably killing it. *The Nitrogen Fix* (1980), by Hal Clement, depicts catastrophic global chemical and environmental changes in the not-too-distant future triggered by both extractive industry and misguided genetic engineering aimed at increasing the number and quality of nitrogen compounds. The resulting chemical reactions deplete the Earth's atmosphere of oxygen, deposit toxic and explosive compounds on the surface, and acidify the oceans. Anaerobic bacteria are the only life-forms that flourish, while humans survive only with breathing apparatus and, since most metals corrode in this harsh environment, develop a material culture based on ceramic technology. Jack Williamson's *Terraforming Earth* (2001) is based on the premise that after a devastating asteroid impact, beneficent robots will tend the human remnant, slowly terraform the Earth, and eventually reintroduce colonies of cloned humans on the planet, while Kim Stanley Robinson looks to the utopian project of terraforming the planet Mars in the not-too-distant future in his trilogy *Red Mars* (1993), *Green Mars* (1994), and *Blue Mars* (1996). In *The Case for Mars* (1996), Robert Zubrin argues that terraforming Mars for human habitation would be a relatively simple and straightforward process. Not to overlook the comedic genre, in the *Red Green Show* episode "Rain Man" (season 15, episode 297), title character Red Green sets up a homemade cloud-seeding cannon at Possum Lodge to shoot chemicals into the clouds and alleviate a drought—with hilarious unintended consequences.

In what follows, rather than overwhelming the reader with the seemingly endless themes of modern or postmodern, post-1960s science fiction, I have chosen to present some older literature that most people have not read or probably have not read recently. This literature, which is dated in many ways, yet quite relevant and enjoyable in others, strikes many of the thematic and moral chords that echo through past, recent, and current concerns about weather and climate control. I am not claiming—indeed, I think it is insupportable to claim—that science fantasy eventually finds its way into science fact. Instead, generations of readers, long before the atomic age or the space age, discovered in science fiction a more subtle kind of wish fulfillment that sets the tone but not the parameters for what might be expected in the future. The main theme here is control, but the literary genres are varied. Although some of it is tragic, much is what we might call "hard path" science fiction (with apologies to

Amory Lovins), involving massive and heroic efforts to terraform a planet or geoengineer its basic physical or biophysical systems. Such literature usually emphasizes words such as "mastery" or "domination." That is, it plays out the Baconian program involving fantasies of control. The comedic genre is well represented too, with stories that are both silly and funny. The overall effect is that no single style dominates imaginative work on weather and climate control, and some, akin to Woody Allen's movie *Melinda and Melinda* (2004), explicitly combine both tragedy and comedy.

Jules Verne and the Baltimore Gun Club

Jules Verne, the renowned French author of "scientific fiction," wrote a notable book in 1865, *De la terre à la lune*, known in English after 1873 as *From the Earth to the Moon*. In the story, when the members of the elite Baltimore Gun Club, bemoaning the end of the Civil War, find themselves lacking any urgent assignments, their president, Impey Barbicane, proposes that they build a cannon large enough to launch a projectile to the Moon. But when Barbicane's adversary places a huge wager that the project will fail and a daring volunteer elevates the mission to a "manned" flight, one man's dream turns into an international space race.

In a sequel, *Sans Dessus Dessous*, published in 1889 and appearing simultaneously in English as *The Purchase of the North Pole*, Verne revisits the possibilities of big guns, but this time with a distinct skepticism for the wonders of technology. For 2 cents an acre, a group of American investors acquires rights to the vast, incredibly lucrative but seemingly inaccessible coal and mineral deposits under the North Pole. To mine the region, they propose to melt the polar ice. Initially, the project captures the public imagination, as the backers promise that their scheme will improve the climate everywhere. They find it relatively easy to convince the public of the idea that the tilt of the Earth's axis should be eliminated (shades of John Milton). This would remove the contrasts between summer and winter, reduce the extreme stresses of heat and cold, improve health, calm the power of storms, and make the Earth a terrestrial heaven, where every day is mild and springlike. But public opinion shifts when it is revealed that the investors—members of the Baltimore Gun Club, the very same group who shot the projectile to the Moon—intend to shoot the Earth off its axis by building and firing the world's largest cannon. Initial public enthusiasm gives way to fears that if these retired Civil War artillerymen (modern-day Titans) have their way and build a kind of Archimedean lever, the tidal waves generated by the explo-

1.3 *The Purchase of the North Pole*: (*left*) building the cannon at Mount Kilimanjaro; (*center*) inside the cannon; (*right*) Fire! (ILLUSTRATION BY GEORGE ROUX, IN *THE ILLUSTRATED JULES VERNE*)

sion will kill millions of people. In secrecy and haste, the protagonists proceed with their plan, building the huge cannon in the side of Mount Kilimanjaro (figure 1.3). The scheme fails only when an error in calculation renders the massive shot ineffective. Verne concludes, "The world's inhabitants could thus sleep in peace. To modify the conditions of the Earth's movement is beyond the power of man."[7] Or is it? Perhaps he spoke too soon.

Mark Twain: Controlling the Climate and Selling It

The American humorist Mark Twain opens his book *The American Claimant* (1892) with an outrageous claim: "No weather will be found in this book. This is an attempt to pull a book through without weather. It being the first attempt of the kind in fictitious literature, it may prove a failure, but it seemed worth the while of some dare-devil person to try it, and the author was in just the mood."[8] In an opening section called "The Weather in This Book," Twain cites the undesirable and "persistent intrusions of weather" that delay both the reader and the author: "Nothing breaks up an author's progress like having to stop every few pages to fuss-up the weather." After conceding the obvious—"Weather is necessary to a narrative of human experience"—Twain seeks to keep it in its place, out of the way, "where it will not interrupt the flow of the narrative." So he promises to relegate weather to the end of his book—indeed, to its "climatic" conclusion and an appendix. With tongue still firmly in cheek, he elevates weather, or the writing about it, to the status of a "literary specialty" and points out to his discriminating readers that it ought to be "not ignorant, poor-quality amateur weather" but the "ablest weather that can be had."

Of course, talk of the weather dominates the work, just as, in the 1890s, talk of Robert Dyrenforth's experiments in Texas dominated discussions of weather control (chapter 2). In the concluding pages of *The American Claimant*, Twain gives an account of the eccentric Colonel Mulberry Sellers, the epitome of American free enterprise, who seeks to control the world's climates—and sell them—by manipulating sunspots. The colonel has just drafted a long letter explaining his scheme:

> In brief, then, I have conceived the stupendous idea of reorganizing the climates of the earth according to the desire of the populations interested. That is to say, I will furnish climates to order, for cash or negotiable paper, taking the old climates in part payment, of course, at a fair discount, where they are in condition to be repaired at small cost and let out for hire to poor and remote communities not able to afford a good climate and not caring for an expensive one for mere display. (271)

The colonel portrays himself as nothing more esoteric than a regulator and (shades of William Ruddiman) holds that climate was manipulated in prehistoric times by Paleolithic peoples—for profit!

> My studies have convinced me that the regulation of climates and the breeding of new varieties at will from the old stock is a feasible thing. Indeed I am convinced that it has been done before; done in prehistoric times by now forgotten and unrecorded civilizations. Everywhere I find hoary evidences of artificial manipulation of climates in bygone times. Take the glacial period. Was that produced by accident? Not at all; it was done for money. I have a thousand proofs of it, and will some day reveal them. (271–272)

Colonel Sellers hopes to patent a "complete and perfect" method for controlling the "stupendous energies" behind sunspots. Wielding this power, Sellers plans to reorganize the climates "for beneficent purposes. . . . At present they merely make trouble and do harm in the evoking of cyclones and other kinds of electric storms; but once under humane and intelligent control this will cease and they will become a boon to man" (272). His plans for commercialization of this technique include licensing it "to the minor countries at a reasonable figure" and providing the great empires with special rates for ordinary affairs and "fancy brands for coronations, battles and other great and particular occasions" (272). He expects to make billions of dollars with this enterprise, which requires no expensive plant, and he hopes to be operational within a few days or weeks at the longest. His first goal is improving the climate of Siberia and clinching a contract with the Russian czar, which he confidently projects will save both his honor and his credit immediately. Reminiscent of the purchase of the North Pole by the Baltimore Gun Club, the daffy colonel confides in his friend and former colleague Marse Washington Hawkins, "a stoutish, discouraged-looking man":

> I would like you to provide a proper outfit and start north as soon as I telegraph you, be it night or be it day. I wish you to take up all the country stretching away from the North Pole on all sides for many degrees south, and buy Greenland and Iceland at the best figure you can get now while they are cheap. It is my intention to move one of the tropics up there and transfer the frigid zone to the equator. I will have the entire Arctic Circle in the market as a summer resort next year, and will use the surplusage of the old climate, over and above what can be utilized on the equator, to reduce the temperature of opposition resorts. (272–273)

Sellers promises to communicate with Hawkins not by earthbound means such as letter or telegraph, but with a "kiss across the universe" using a cosmic signal sent from the surface of the Sun itself—a vast attenuation of sunbeams that will generate envy even among current proponents of solar radiation management, especially those who propose to cast a shade on the Earth using orbiting space mirrors (chapter 8). Sellers writes:

> Meantime, watch for a sign from me. Eight days from now, we shall be wide asunder; for I shall be on the border of the Pacific, and you far out on the Atlantic, approaching England. That day, if I am alive and my sublime discovery is proved and established, I will send you greeting, and my messenger shall deliver it where you are, in the solitudes of the sea; for I will waft a vast sun-spot across the disk like drifting smoke, and you will know it for my love-sign, and will say "Mulberry Sellers throws us a kiss across the universe." (273)

As he promises, Twain ends *The American Claimant* with an appendix subtitled "Weather for Use in This Book. Selected from the Best Authorities," in which he presents a rich parody of the type of dense weather writing that he fails to exclude from his own text. Here Twain contrasts the prolixity of

William Douglas O'Connor's "The Brazen Android" (1891), in which purple prose serves to invoke the purple skies of medieval London after a thunderstorm, with the terse elegance of a much greater apocalypse experienced by Noah and his family and recorded in Genesis 7:12: "It rained for forty days and forty nights."

The Wreck of the South Pole

In *The Wreck of the South Pole, or the Great Dissembler* (1899), by Charles Curtz Hahn, protagonist George Wilding finds himself shipwrecked and stranded in low southern latitudes on a continent of ice. Befriended and guided by what he takes to be mysterious spirits, Wilding makes his way south to warmer climates, to a great city inhabited by Theosophists, who, by practicing mind reading and astral projection, seek to control nature with their minds. There, the weather bureau does not predict the weather—it uses mental prowess to control it: "Their duty is to decide upon the proper kind of weather for certain seasons and days and then see that the country has it."[9] It is a land without droughts or damaging winds. There the police do not arrest criminals—they track and detain suspects who have been placed under suspicion by mind-reading surveillance.

The Great Dissembler, the most advanced Theosophist sage, has mastered a technique for keeping others from reading his mind: "I cultivated the habit of jumbling up my thoughts in the worst mess you could imagine" so no one could fasten on them (67–68). It is he who is both the chief geoengineer and the greatest general. In order to defeat the revolutionary forces threatening his city, the Great Dissembler decides to wrench the South Pole suddenly from its axis to destroy the enemy with a tidal wave and "bring back the old order of things" (72). When he executes this plan, all the climatic zones of the world will change dramatically. Cyclones, tornadoes, and earthquakes will increase in both number and intensity until the temperate latitudes merge into the tropics. With the wreck of the South Pole (a day later than planned, possibly due to the confused thinking of the Great Dissembler) comes an unexpected rift in the surrounding ice walls and unintended consequences described only as "days of terror and suffering" (74). The story ends here abruptly, with no description of the fate of the world but with the assurance that George Wilding, again stranded in a remote and icy cove but cared for and comforted by the astral bodies of his friends, will soon be reunited with them.

The Great Weather Syndicate

Weather and climate control, war, gender, and romance are juxtaposed in *The Great Weather Syndicate* (1906), by George Griffith. In the novel, Arthur Arkwright, a young British engineer, develops a machine that modifies the weather by drilling atmospheric holes to redirect the winds and clouds. This invention promises to make him the "master of the fate of the world."[10] Working through investors in the World Weather Syndicate, Arkwright sets up a chain of mountain stations equipped with "atmospheric disintegrators" that project impulses powerful enough to break up and dissipate clouds while creating partial vacuums or "holes" in the atmosphere. By coordinating its efforts, the Syndicate can determine the direction of winds and weather over any area within the range of its stations. Controlling the weather of the whole world, then, is simply a matter of multiplying the number of stations. The Syndicate "will enable us to run the world's weather and sell it out to the countries which need any particular brand at our own price" (86). For example, the Gulf Stream can be altered by this technique to benefit those willing to pay. Arkwright's love interest, Eirene, the daughter of his chief investor, introduces moral objections:

> Now I think it's wicked. You're going to upset the order of nature, you're going to make hot countries cold and cold countries hot, just so you can make profits out of them; but have you thought of all the misery and starvation and all sorts of horrors that you are going to bring upon innocent work-people who won't have a notion of what's really going on; how you will make fertile places into deserts and ruin farmers and manufacturers and all the people depending on them just because the Government of the country won't pay your price for the weather they want? No, it's just wicked. (12)

An opposing syndicate has a "pretty big idea" of its own (15). It proposes to dam the Arctic Ocean across the Bering Sea, Baffin's Bay, and Spitzbergen to stop all the icebergs from coming south and bottle up the Arctic Ocean until ice builds up there. The excess weight will then shift the axis of the Earth and cause a general redistribution of the map of the world—land, sea, and weather all included. If this evil syndicate gains control of the Earth's axis, a struggle will ensue for control of the world's weather, which can "only result in disaster to mankind" (73). Arkwright thus finds himself "at the beginning of a war for the economic control of the world," and he proposes to win "by any means within his power" (17), yet Eirene refuses to marry him until she sees how he plans to wield this power.

Eager to prove its dominance over weather, the World Weather Syndicate triggers a snowstorm in London on July 6 designed to impress the British foreign secretary. This time, the voice of Arthur's conscience is his Aunt Martha from Lancashire: "I tell him to his face that it's a sin and a shame interfering with the course of nature. For shame on thee, lad! . . . why canna' thee let the good God manage His weather in His own way? Dost'a want to bring a great city like this, and maybe all England to ruin, just to make thy own business pay?" (55).

Arthur replies that he and his investors have altruistic intentions:

> Now to be quite frank, we simply want to make money, and incidentally, increase the fertility of the world by turning deserts into paradises, for which, of course, we should expect to be paid, though not extravagantly. As the work develops we should also hope to put a stop to war . . . by just freezing the fleets of the belligerents up in their harbours, and producing such a degree of cold on any given battlefield that fighting would be impossible. (73–74)

Another female voice of conscience, Arthur's sister Clarice, worries about "all the poor people who will have to suffer" if the Syndicate engineers a frosty British winter: "[T]he people who won't be able to get work, and can't pay for wood and coal and oil, to say nothing of proper food" (78–79).

After Arkwright makes a fortune by converting formerly barren areas into arable farmland, he turns his attention to the utopian project of ending world hunger, poverty, and, especially, war. Against the world's militarists, Arkwright calls down devastating snowstorms from the heavens as a kind of meteorological Moses, freezing armies in their tracks, fogging battlefields, and locking naval vessels in ice-bound harbors. "It's weather against war, and weather will win," he tells the kaiser, after thwarting a German plot to revive the Holy Roman Empire (308).

At least in this science fantasy, techniques of weather control inaugurate a millennial reign of global peace and prosperity. The Syndicate is generally considered to be "a sort of earthly Providence" by the people in marginal lands that it helps. Eirene ultimately marries Arthur (the Controller of the Elements) so that *she* can show *him* "how to manage the climates of the world" (312). In such fiction, as later in actual proposals, the themes of precise and ultimate control of the weather and climate for economic, humanitarian, and military purposes are inextricably blended.

Earlier in their careers, some real-life twenty-first-century geoengineers worked on heroic schemes to deflect Earth-grazing asteroids. They would surely appreciate another of Griffith's novels, *The World Peril of 1910* (1907), in which astronomer Gilbert Lennard discovers a comet threatening to destroy the Earth. In the novel, American money and know-how contribute to the construction of a great cannon built into a mine shaft. The massive shot deflects the comet so that it does not strike the Earth.

A Comedic Western

The Eighth Wonder: Working for Marvels (1907), by William Wallace Cook, is a humorous "geoengineering" Western that was serialized in 1907 in the pulp magazine *Argosy*. In the badlands of North Dakota, Ira Xerxes Peck, an out-of-luck bicycle dealer, befriends a despondent but brilliant inventor, Copernicus Jones, who plans to corner the nation's electricity market by turning Horseshoe Butte, a naturally occurring iron formation, into the eighth wonder of the world, the world's largest electromagnet. It is to be Jones's Archimedean lever to move the world. "I don't think it pays, Copernicus," Peck observes timidly, "to tinker with the machinery of the universe. . . . Not unless there's money in it."[11] When the titanic magnet is turned on, everything made of iron within a 25-mile radius—tools, pumps, wagons, threshing machines, even a sheet-iron house—flies through the air and adheres to the mountain. Jones is ecstatic, as in the myth of ships imperiled by the lodestone, "ancient fables come true in modern times . . . that's what we call civilization and progress" (81).

Jones is more an inventor than a scientist, and his device actually fails to attract all the electricity from across the country. Instead, it begins to alter the seasons by deflecting the tilt of the Earth's axis. Jones takes credit for this unforeseen consequence and tries to capitalize on it by making the Northern Hemisphere permanently warmer: "We will corner the hot weather . . . and we'll make the people pay for it! . . . [W]e will select the brand of weather we want, and I will . . . hold the Earth's axis at that precise inclination" (171–175). Ever his conscience, Peck reminds Jones that "tampering with the Earth's axis, Copernicus, brings responsibilities. We must not shut our eyes to that fact" (187).

The citizens of the world respond to Peck and Jones by insisting that tinkering with the seasons is a crime against nature. The industrialists are particularly adamant, since they made much of their money in cold weather and during changes of seasons. Parroting the claims of climatic determinists, they argue the necessity of the yearly return of ice and snow to conserve the rugged character and "insistent energy" that has made the United States great, while pointing out that continuous warm weather would "sap our strength" (194): "The cold gives a zest to the blood that calls for achievement. In tropical countries the inhabitants are mostly dreamers, and excessive humidity paralyzes effort" (280).

Meanwhile, it appears that Jones did not really know what he was doing or the consequences of his actions. As Peck expresses it, "We were as two children, Copernicus and I, playing around powder with a box of matches" (197–199). Fame and fortune or infamy and prison are equal possibilities. For a ransom of $1 billion, the two geoengineers propose to stop their magnet, "leaving the seasons as we had found them." In other words, they demand an exorbitant price to maintain the status quo. Peck and Jones fend off an attack on their installation by federal troops armed only with wooden clubs (because the magnetic force has stripped them of their metal weapons), but the iron butte is finally destroyed by a cannon bombardment, since the giant electromagnet actually acts to attract the incoming shells to it! Peck and Jones survive, but Jones has seemingly learned nothing, continuing his inventive scheming under an assumed name and promising, "If anything unusual happens you'll know who should have the credit. . . . I'm off for Europe . . . to see what I can meddle with across the pond" (317–318).

The Twist in the Gulf Stream

A different genre of story tells of large-scale and catastrophic unintended consequences of tinkering in sensitive areas of the Earth's system. *The Evacuation of England: The Twist in the Gulf Stream* (1908), by Louis P. Gratacap, tells of geophysical and social dislocations caused by the collapse of the Isthmus of Panama, which diverts the Gulf Stream, causing vast climatic and social changes, including the refrigeration and depopulation of Europe.

The story begins with scientists' warnings about instabilities along the west coast of North and Central America that could result in massive geological chain reactions. Earthquakes could trigger the release of the "volcanic energy" of Panama and the West Indies, and the region could experience an "isostatic rebound"—basically a rebalancing of the Earth's crust—as it seeks a new equilibrium state. When Panama is breached (by either humans or geology), "again the waters of the two oceans will unite, and the impetuous violence of the rushing oceanic river, the Gulf Stream, that now races and boils through the Caribbean Sea, will fling its torrential waves across this divide into the Pacific."[12]

In spite of these warnings, commercial interests and the president of the United States push for the completion of the Panama Canal (actually completed in 1914 at a cost of $400 million). In the book, the excavation commences in 1909 and triggers a natural disaster. A massive series of earthquakes and tidal waves strikes Colón. The isthmus sinks, opening up a passage between the Atlantic and the Pacific that had been closed for 3 million years. Subsidence in Panama results in volcanic eruptions and the catastrophic convergence and elevation of the Caribbean islands. The Earth shudders, the poles "wobble," and the Gulf Stream, "no longer turned aside by impassible walls of land, triumphantly [sweeps] into the Pacific," opening a "new chapter in the history of the world and the history of nations" (92–99). In an understated response, President Theodore Roosevelt is quoted as saying, "It seems likely that this physical alteration may mean a change in the climate of the older portion of the earth" and an end of "the glory of England" (121–123).

With the Gulf Stream now warming the Pacific coast of the United States, Europe descends into a new ice age as the North Atlantic cools dramatically and devastating snowstorms pummel the region. Like a scene out of the film *The Day After Tomorrow* (2004), Reykjavik lies deserted. In Edinburgh, snow "fill[s] up the deep moat of the Princes' Street gardens [and] round[s] the rugged edges and wandering parapets of the Citadel" (131–136). Europe trembles "with a new apprehension" as markets panic and moral depravity sets in. London is evacuated. As the savage Scots move south, the English seek refuge in their colonies in Asia and Australia. "Heat is life, cold is death. . . . Our civilization, the civilization of Europe, has overstepped the limits of climatic permission" (187, 295). All these consequences were triggered by a macro-engineering project that went against the advice of the geologists.

Rock the Earth

World peace as a consequence of the demands of a mad scientist is the theme of *The Man Who Rocked the Earth* (1915), by Arthur Train and Robert Williams Wood. With most of the world wracked by war, a mysterious message arrives by wireless from the inventor PAX: "To all mankind—I am the dictator—of human destiny—Through the earth's rotation—I control day and night—summer and winter—I command the—cessation of hostilities and—the abolition of war upon the globe."[13]

To demonstrate his resolve and his power over the elements, PAX slows the Earth's rotation by five minutes; makes it snow in Washington, D.C., in August;

and, with a flying ring and super-powerful Lavender Ray, diverts the Mediterranean Sea into the Sahara and destroys a German siege gun as it fires on Paris. These phenomena are accompanied by geophysical marvels: strange yellow aurorae, earthquakes, tidal waves, and atmospheric disturbances. An international assembly of scientists is formed to respond, but it is designed by the diplomats to stall and fail in the hope that particular nations might gain special advantages by capturing the inventor and learning his powers.

PAX controls a source of power—atomic disintegration—that would allow the Earth to blossom "like the rose! Well-watered valleys where deserts were before. War abolished, poverty, disease!" This is reminiscent of the rhetoric, some forty years later, hyping the potential of atomic energy. Impressed, physicist Bennie Hooker sets out to find PAX and the secret behind "the greatest achievement of all time!" (111).

Meanwhile, disappointed by cease-fire violations, PAX sends his final warning to humanity, declaring that he will "shift the axis of the Earth until the North Pole shall be in the region of Strasbourg and the South Pole in New Zealand" (172). Hooker eventually finds PAX's laboratory in Labrador and witnesses his demise in an explosion near a gigantic outcrop of pitchblende: "This radioactive mountain was the fulcrum by which this modern Archimedes had moved the Earth" (216). Anticipating the founding of the League of Nations by several years, the nations of the Earth form a coalition government coordinated at The Hague and destroy all their armaments, an event that inaugurates a new age of international cooperation, peace, and never-before-experienced prosperity. In a setup to a possible sequel, Hooker is last seen exploring the solar system in his "Space-Navigating Car," powered by the Lavender Ray.

The Air Trust

The Air Trust (1915), by George Allan England, combines both geochemical and political fantasy in telling the story of a dedicated band of socialists who thwart the plans of ruthless capitalists aiming to control the world's air supply. The book is dedicated to Eugene V. Debs, "Comrade Gene, Apostle of the World's Emancipation." The author depicts scientists for hire as the willing servants of capitalists and the obedient executioners of both corporate plans and, possibly, humanity. England writes in the foreword: "I believe that, had capitalists been able to bring the seas and the atmosphere under physical control, they would long ago [have done so]."

The story begins when billionaire businessman Isaac Flint, seeking new and ever more powerful monopolies, asks:

> What is it they all must have, or do, that I can control? . . . Breathing! . . . Breath is life. Without food and drink and shelter, men can live a while. Even without water, for some days. But without *air*—they die inevitably and at once. And if I make my own, then I am the master of all life! . . . Life, air, breath—the very breath of the world in my hands—power absolutely, at last![14]

His business partner, Maxim "Tiger" Waldron, suggests "The Air Trust—A monopoly on breathing privileges! . . . Imagine that we might extract oxygen from the air. . . . [P]eople would come gasping to us, like so many fish out of water, falling over each other to buy!" (23–25).

The businessmen delegate responsibility for the details and the execution of the plan to the industrial research staff ("That's what they're for") as personified by the chemist Herzog—"a fat rubicund, spectacled man" with a keen mind, two fingers missing (from experimenting with explosives), and "character and stamina close to those of a jelly fish" (29). In the novel, the oxygen extraction plant is located at Niagara Falls and uses hydropower to run the condensers. The book includes sufficient technical details about the extraction process and the scale of the operation to suspend readers' disbelief while clearly drawing an analogy to the nitrogen fixation process developed about 1909 by the German chemist Fritz Haber and industrialized in 1913 by Carl Bosch. Benefits of commodifying the air include the sale of liquid gas refrigerants, nitrogen for fertilizer and explosives, and even ozone to "freshen and purify" the environment. But by far the most precious commodity is oxygen, the breath of life. As Flint expresses it, "We'll have the world by the wind pipe; and let the mob howl *then*, if they dare!" (69).

The plot turns around the loss of Flint's notebook, which alerts the socialist hero, Gabriel Armstrong, to the plan. He and his comrades passionately debate the need to destroy the "infernally efficient tyrants" who have taken possession of "all that science has been able to devise, or press and church and university teach, or political subservience make possible." The capitalists control "military power, and the courts and the prisons and the electric chair and the power to choke the whole world to submission, in a week!" If the socialists can destroy the Air Trust, "the great revolution will follow" to annihilate capitalism (261–262).

After working out a strategy of attack, the workers organize and, led by Armstrong, storm the plant. In a scene worthy of a Saturday matinee, they chase Flint and Waldron into one of the huge empty air tanks, as the chemist Herzog takes his own life with a vial of poison. The final scene is both ghastly and ghoulishly amusing as Waldron notices the odor of ozone and cries out, "*Flint! Flint! The oxygen is coming in!*" (325) As a huge stream of pure oxygen from a ruptured valve floods the tank, the brains of Flint and Waldron literally began to "combust":

> "*Ha! Ha! Ha!*" rang Waldron's crazy laughter. . . . All at once his cigar burst into flame. Cursing, he hurled it away, staggering back against the ladder and stood there swaying [panting, with crimson face], clutching it to hold himself from falling. . . . "Help! Help!" [Flint] screamed. "Save me—my God—save me—Let me out, let me out! A million, if you let me out! A billion—*the whole world!* . . . It's mine—I own it—*all, all mine!*" (326–327)

With a final burst of energy, "his heart flailing itself to death under the pitiless urge of the oxygen," Flint runs across the tank screaming blasphemies and slams into the opposite wall, where he falls sprawling, stone dead. Tiger Waldron attempts a final dash up the ladder to reach the door at the top of the tank. "Fifty feet he made, seventy-five, ninety"—until his overtaxed heart too bursts and he falls to his death. "And still the rushing oxygen, with which they two had hoped to dominate the world, poured [in]—senseless matter, blindly avenging itself upon the rash and evil men who impiously had sought to cage and master it!" (328).

As the plant goes up in flames, the oxygen tank explodes in a huge ball of fire. Thus the socialists foil the attempt to control the air supply of the world—and thus the world itself—and inaugurate the "Great Emancipation" of humanity from the clutches of greedy capitalists. In the words of the protagonist Armstrong, "Academic discussion becomes absurd in the face of plutocratic savagery" that seeks a "complete monopoly of the air, with an absolute suppression of all political rights." Slavery and violent revolution are the only options.

Tales of the Rainmakers

"The Rain-Maker," by Margaret Adelaide Wilson, a short story that appeared in *Scribner's Magazine* in 1917, recounts the hopes and dreams of William Converse, who operates, like the real Charles Hatfield at the time (chapter 3), by mixing and evaporating chemicals on a high tower: "The chemicals are holding the storm-centre right overhead, and the evaporation is tremendous. The rain will come this time if it holds off, the wind holds off—if only it holds off."[15]

Converse is a true believer in his rainmaking process. He came to the desert on a mission: to use his skills to atone for the death of his father, who died of thirst near this very spot some thirty years before. But Converse has much more to confront than just the desiccated sky. His wife, Linda, who thought she was marrying a prosperous entrepreneur, has become super-critical of his idealistic quest, which keeps her cooped up in a tent with a smoky stove, frying bacon and potatoes: "You've gone and thrown up a perfectly good contract in Grass Valley, a thousand sure, and more if your luck held, and you've dragged me off to this God-forsaken spot, with not a soul in thirty miles to know whether it rains or not. I want to know what you mean by it" (503–504). The high-minded Converse, like a modern-day Job, is seeking "to bring rain in the wilderness" by lifting his voice to the heavens as his father did on the night of his death. He receives no support from the vulgar, vain, and greedy Linda, though. She drives him from the tent into the night with her cutting remarks about how she no longer believes in him, and perhaps never did.

"Driven by an animal's blind desire to escape its tormentor, Converse stumble[s] down the rocky path toward the tower" (506). Devastated by her verbal assaults, he realizes that Linda has managed to shatter his faith in himself. He trips over something in the sand, and "a hot pain dart[s] through his ankle . . . it must have been a snake" (507). Pitiful and increasingly delirious, he collapses in the dry waterhole where his father met his demise. Even as he nears death, his gaze is fixed on the black and brooding sky, with its great masses of clouds sinking lower until a soft hiss, a pitter-pat of rain on the sand, informs him of "his" success: "Rain! . . . Rain in the wilderness. . . . I've not failed, after all. . . . I must find father and tell him" (509)—an impossible quest for his paralyzed body but not for his triumphant spirit.

Jingling in the Wind (1928), by Elizabeth Madox Roberts, is a stylistically complex and multivocal tone poem, "a fantasy of weather control."[16] Here we meet Jeremy the rainmaker, a man who participates in the pure sensation of nature and "gives it a point." For Jeremy, interior feelings and reflections trump the mere wetness of the rain, which is the "least significant part": "He had brought the rain into the sky. With his science and his apparatus he had engendered the rain and now, as rainman, reve of the rain, he looked about and saw his work was well done, saw his work take purpose in the clods and the parted loam."[17]

Using Jeremy's techniques, the commissioners of rural Jason County, Kentucky, in conference with the farmers, set the rain schedule for the month, but only in their own local jurisdiction. The process is so precise that if you wish for fine weather on a scheduled rain day, you merely need to cross into the next county. Much like the Kansas and Nebraska rainmakers of yore, "retorts, clouds, equations, antennae, derricks, vats, and acids" play their symbolic, if nondescript, roles in the rainmaking scenes. So does the hall of the rainmen, where licensed practitioners confer and visitors thumb charts, prod apparatus, and inscribe their names in the guest book.

Of greater rhetorical significance, however, are the debates over the morality of the technique. Half the population, led by the Reverend James Ahab Crouch ("Make the World Safe for God"), oppose rain control as a "device of the devil," blasphemous or pagan. He champions a bill in the legislature designed to crush the rainmen and preaches from his great tent how terrible it is "to subtract from the omnipotence of an omnipotent God" (179). Others, more open-minded or daring, look on it as a great benefit (25–27, 74). They plan a carnival with "a great rain display, a model rain, predicted, arranged, conducted by some rainman, controlled" (185).

Like Frank Melbourne at the Goodland County fair (chapter 3), Jeremy, known popularly as the "rain bat" for his tight-fitting black rain suit, is invited to the carnival and promises the expectant crowd a deluge by two o'clock. He works feverishly, tuning his instruments, mixing the proprietary chemicals, and conjuring up and battling with the clouds, which fight back like dragons. Finally, "out of the great rent in the beast rolled a stream of rain and the gutters were running flush, running over" (213–217). Jeremy's success is celebrated with a parade, with the rain bat riding in a convertible at the head of a motorcade as music plays, drums beat, men cheer, women swoon, and skywriters pay homage overhead. At the time the book was written, Charles Hatfield was still active in the field, and the Rock Island Railroad rainmakers persisted in memory.

N. Richard Nash's romantic comedy *The Rainmaker* (1955) is set in Three Point, Texas, "on a summer day in a time of devastating drought." Lizzie Curry's family worries more about her marriage prospects than about their dying crops and livestock. Suddenly, a charming stranger arrives, a Texas twister of a man named Bill Starbuck—*Rainmaker!*—a charlatan, but not essentially a crook, who promises, for $100, to make miracles, to bring rain. As the summer storm clouds gather overhead, lonely and plain Lizzie, too, has her love life "seeded" by the confidence man's machinations.

How to make rain? Starbuck mocks the scientific voice of the charlatan when he cries, "Sodium chloride! Pitch it up high—right up to the clouds. Electrify the cold front. Neutralize the warm front. Barometricize the tropopause. Magnetize occlusions in the sky."[18] But Starbuck, like faith healers, has his own method, "all my own," that begins with him brandishing his hickory stick and exuding confidence. After inviting himself to supper and collecting $100 in advance, Starbuck puts the family members to work for him in a test of their faith—beating on a big bass drum, painting arrows on the ground to direct the lightning away from the house, tying the farm mule's hind legs together—without allowing any questions and certainly without acting sensibly. Lizzie, who is flabbergasted by all this, admonishes her father, H. C.:

> LIZZIE: You're making a big fool of yourself! Where's your common sense?
> H. C.: Common sense? Why, that didn't do us no good—we're in trouble. Maybe we better throw our common sense away.
> LIZZIE: For Pete's sake, hang on to a little of it! (76)

Starbuck counters: "You gotta take my deal because once in your life you gotta take a chance on a con man! You gotta take my deal because there's dyin' calves that might pick up and live! Because a hundred bucks is only a hundred bucks—but rain in a dry season is a sight to behold! You gotta take my deal because it's gonna be a hot night—and the world goes crazy on a hot night—and maybe that's what a hot night is for!" H. C. responds, "Starbuck, you got you a deal!"

While the family is busy performing their rainmaking rituals, Starbuck romances Lizzie, getting her to acknowledge her own beauty. Here is where real confidence is built. But Starbuck, also known as Tornado Johnson, is wanted for selling four hundred tickets to a rain festival when it did not rain, peddling a thousand pairs of smoked eyeglasses to view an eclipse of the Sun that never happened, and selling six hundred wooden poles guaranteed to turn tornadoes into a gentle spring breeze (152). In a final confrontation with the town officers of Three Point, one of whom is sweet on Lizzie, Starbuck throws the $100 on the table and makes a dramatic escape. He returns soon thereafter, just as the drought breaks and a storm is unleashed overhead: "Rain, folks—it's gonna rain! Rain, Lizzie—for the first time in my life—rain!" (as he takes the money and races out for the second time, pausing only long enough to wave to Lizzie). "So long—beautiful!" (182).

The Rainmaker opened on Broadway at the Cort Theatre, New York City, on October 28, 1954, with Darren McGavin as Bill Starbuck and Geraldine Page as Lizzie Curry. London's *Daily Mail* called the production "a beautiful little comedy with a catch in its throat."[19] One reviewer commented that Starbuck captivated Lizzie and her family "neither to connive nor corrupt but because he must live in a glow of esteem, and what to do in that case but radiate it oneself?"[20] A 1956 film version starred Burt Lancaster and Katharine Hepburn. *New Yorker* film critic Pauline Kael observed:

> The cowtown spinster suffering from drought is Katharine Hepburn, and the man who delivers the rain is Burt Lancaster. The casting is just about perfect. Lancaster has an athletic role, in which he can also be very touching. His con man isn't a

simple trickster; he's a poet and dreamer who needs to convince people of his magical powers. Hepburn is stringy and tomboyish, believably plain yet magnetically beautiful. This is a fairy tale (the ugly duckling) dressed up as a bucolic comedy and padded out with metaphysical falsies, but it is also genuinely appealing, in a crude, good-spirited way, though N. Richard Nash, who wrote both the play and the adaptation, aims too solidly at lower-middle-class tastes. Once transformed, the heroine rejects the poet for the deputy sheriff (Wendell Corey); if there were a sequel, she might be suffering from the drought of his imagination.[21]

A musical adaptation, *110 in the Shade* (1963), played to packed houses; a remake broadcast on HBO in 1982, starring Tommy Lee Jones and Tuesday Weld, was less than memorable; and a Broadway revival featuring Woody Harrelson and Jayne Atkinson in the lead roles opened and closed with little fanfare in 1999. Still, *The Rainmaker* has perennial appeal and has been performed many times since by innumerable school and community theater groups.

Sky King and the Indian Rainmaker

Sky King, America's favorite flying cowboy, ruled the "clear blue Western skies" over the Flying Crown Ranch in Arizona (although the opening credits showed a high cirrostratus haze). With the help of his niece Penny, nephew Clipper, and private airplane, the *Songbird*, Sky King solved mysteries, rescued those in need, and fought villains—on radio from 1946 to 1954 and intermittently on television on Saturday mornings from 1951 to 1962. In June 1948, as news of cloud seeding at the General Electric Corporation reached the public (chapter 5), an episode titled "The Rainmakers Magic" aired on radio. Eight years later, in 1956, the TV episode "The Rainbird" revisited the topic, juxtaposing traditional and modern methods of weather control.

During a devastating drought, Indian dancers, medicine men, and rainmakers implore the heavens for rain. The chief and elders of the local tribe present their elderly medicine man, Tai-Lam, with an ultimatum: bring rain in two days or be replaced. Sky King, who is sympathetic, decides to help out behind the scenes by seeking advice from the local weather bureau on when and where to seed the clouds. Penny coordinates efforts, signaling Tai-Lam to begin shaking his Kachina doll and droning his pitiful rainmaking chant, while Sky King simultaneously seeds an "upper-level front" with silver iodide. A deluge follows, placing both Penny and the tribe at risk, filling the dam to its brim, and threatening to flood the valley. None of the protagonists, however, place the blame on modern weather control technology or traditional methods. The weather bureau attributes the rain to unexpected changes in a naturally occurring system. Tension returns as a second storm rapidly approaches, which could cause the dam to burst. At the risk of his life, Sky King takes off, flying into the weather front to divert it, again by cloud seeding, while Tai-Lam begins a new chant, this time to *stop* the rain. Through the mixed agency of the Kachina doll and silver iodide, all turns out well at the end of the half-hour episode. This fictional episode has its counterpart in the actual practices of the era. A. J. Liebling described a magazine clipping from 1952 titled "Old Order Changeth: Navajo Indians near Gallup, N.M. have become skeptical of—or just plain bored with—their ancient rainmaking rites. During a recent drought, they hired professional rainmakers to seed the clouds over their reservation. Result: one-and-a-half inches of rain."[22]

The futurist Arthur C. Clarke, of all people, wrote about the Zuni tribe of New Mexico, who are famous for their rain dances. At the beginning of the ceremony, just after the summer solstice, a boy representing the Fire God torches a field of dry grass. This serves as a signal for the Zuni dancers, painted with yellow mud and carrying live tortoises, to begin dancing, which continues as long as necessary, until it rains. Clarke editorialized: "That is one beauty of rain making. It always works *eventually*, though sometimes you have to wait a few weeks or months for the pay-off."[23] A cartoon contrasting traditional and scientific methods accompanied Clarke's article (figure 1.4).

1.4 Rainmaking old and new. (CARTOON BY CHARLES ADDAMS, IN CLARKE, "MAN-MADE WEATHER")

Porky Pig and Donald Duck

Commercial cloud seeding even found its way into the cartoons. The Warner Brothers Looney Tune *Porky the Rain-Maker* was shown in theaters in 1936. During a devastating drought, Porky sends his son to town with his last dollar to buy feed for the starving animals. There, next to the feed store, Dr. Quack is selling an assortment of "rain pills" for $1 from the back of his wagon. As part of his presentation, Quack hands out umbrellas to the crowd and launches a rain pill into the sky with a peashooter. The pill bursts like Dyrenforth's ordnance, and rain begins to fall immediately.

Convinced, Porky Jr. buys a box of the pills with the family's last dollar, but his angry father, in a scene reminiscent of Jack and the Magic Beanstalk, throws them on the ground. This gives rise to a series of comedic shticks. A barnyard chicken eats a lightning pill and is instantly electrified; the old gray mare eats a fog pill and is shrouded in cloud; the goose eats thunder and wind pills and all hell breaks loose. When, in the melee, one of the rain pills reaches the sky, clouds form instantly and the rains fall. As the cartoon credits roll, all ends well on the farm and everyone is happy. "That's all Folks!"

In *Walt Disney's Comics and Stories* (1953), Donald Duck, M.R.M. (Master Rain Maker), has perfected the science of rainmaking. In the opening sequence, a farmer orders 2 inches of rain on his barley field. Donald, wearing an aviator's helmet and pointing to his bag of M-3 "rain seed," offers him 2.5 inches for the same price. Donald fulfills his contract with extreme precision "to the millimeter" by seeding the farmer's X-shaped field with an X-shaped cloud he has "bulldozed" into position. The farmer and his wife are delighted, since none of the rain falls on his hay field next door: "That duck shore is a Jim Dandy! It's raining right up to the fence row! And the drops that fall on the line even have one flat side!"

Of course, Donald eventually loses his temper in every cartoon, and this one is no exception. Daisy has gone to the Idle Dandies picnic with Donald's cousin, Gladstone. Donald, jealous and angry, takes off in his cloud-seeding airplane, this time loaded with "snow starter," to gain retribution. Flying over the picnic site in Greenwood Canyon in a clear blue sky, Donald's agitation with his rival increases until he admits, "I feel mean enough now to do *anything*!" After herding some ominous rain clouds together, Donald declares, "I won't give 'em . . . anything as common as a cloudburst—I'll give 'em a *blizzard*!" In a memorable image, he pulls the control lever beyond "rain," "hail," and "snow" all the way to "blizzard," but he miscalculates and "overseeds" the clouds, turning them into a solid dome of ice.

Donald crashes his plane on the ice and parachutes down into the canyon to warn the picnickers of the danger above. The ice dome crashes down on their parked cars, but since this is a Disney cartoon, no one is injured. However, to avoid liability and preserve deniability, Donald suspends his lucrative rain business, sneaks away from the ongoing investigation, and takes an extended vacation—in Timbuktu.

Henderson the Rain King

On a more literary note, in *Henderson the Rain King* (1959), by Saul Bellow, the title character, Eugene, an introspective, earnest, and egocentric former violinist and pig farmer, seeks to find himself and escape his troubles with the modern world with a one-way ticket to Africa. Traveling cross-country on foot to visit native tribes, he unexpectedly becomes the Great White Sungo, the rain king of the Wariri, when he performs certain feats of prowess. In charge of both moisture and fertility, Henderson participates in a frenzied ceremony involving leaping, drumming, shrieking, chanting, and whipping both images of the gods and one another:

> Caught up in this madness, I fended off blows from my position on my knees, for it seemed to me that I was fighting for my life, and I yelled. Until a thunder clap was heard. And then, after a great, neighing, cold blast of wind, the clouds opened and the rain began to fall. Gouts of water like hand grenades burst all about and on me. . . . I have never seen such water.[24]

Having found at least part of himself, Henderson, significantly transformed by his experiences and eager to start anew, takes a flight back to America. In evocative passages that inspired Joni Mitchell's popular song "Both Sides Now," Bellow writes, "We are the first generation to see the clouds from both sides. What a privilege! First people dreamed upward. Now they dream both upward and downward. This is bound to change something" (280). "[Clouds are] like courts of eternal heaven. Only they aren't eternal, that's the whole thing; they are seen once and never seen again, being figures and not abiding realities" (333).

Cat's Cradle

At the urging of his older brother Bernard, Kurt Vonnegut moved to Schenectady, New York, in 1947, where he worked, unhappily, as a publicist for General

Electric—a company he once said "*was* science fiction"—in what he called "this goddamn nightmare job."[25] At a time when the air force's Project Cirrus was taking over the cloud-seeding business (chapter 5), Vonnegut published "Report on the Barnhouse Effect," a science fiction story in *Collier's* that emphasizes an inventor's moral resistance to an attempted militarization of his invention. His first novel, *Player Piano* (1952), was inspired by the mechanization he witnessed at GE and deals with the demoralizing effects of vast corporations attempting to use technology to automate everything and replace human labor with machines. The setting is Illium, a fictitious town along the Iroquois River in New York State, a dreary mill town dominated by a high-tech factory called Illium Works. In reality, Schenectady, on the Mohawk River, is the home of General Electric, while Illium is the ancient Roman name for Troy, which is also an industrial city near Schenectady in New York.

While still at GE, Vonnegut heard about the visit of H. G. Wells to the plant in the 1930s and how Irving Langmuir proposed a story idea to the famous novelist and futurist involving a new form of water (ice-nine) that would freeze at room temperature. Wells never wrote about this, but Vonnegut thought it might someday be worth pursuing. Bernard Vonnegut had, in reality, identified the hexagonal structure of silver iodide (ice-six?) as a substance that could trigger natural ice formation in clouds. Years later, Vonnegut used these ideas in his book *Cat's Cradle* (1963), where a quirky and amoral scientist named Felix Hoenikker, a loose composite of Langmuir and H-bomb scientists Stanislaw Ulam and Edward Teller, invents a water-like substance that instantly freezes everything it touches. When a tiny crystal of "ice-nine" is brought into contact with liquid water, it stimulates the molecules into arranging themselves into solid form.

Bernard obviously had a big influence on Kurt. Real-life meteorologist Craig Bohren credited *Cat's Cradle* with the "best discussion of nucleation" in print and claimed that the novel contained more information on this subject than "all the physics textbooks written since the beginning of time."[26] Indeed, Langmuir and Teller were reportedly fascinated by the theoretical possibility that a substance such as ice-nine could actually exist. In the book, Hoenikker's intent is to create a material that will be useful to armies bogged down in muddy battlefields, but the result is an unprecedented ecological disaster that destroys the world.

* * * * *

Clearly, the practice of weather control is not restricted to the West, to modern times, or to scientific practices. It has much deeper roots in world cultures and carries much deeper meaning than simply making rain or stopping it. In *Rain Making and Other Weather Vagaries* (1926), William Jackson Humphreys (1862–1949), a meteorological physicist at the U.S. Weather Bureau, classifies rainmaking into three general categories: *magical* (practices alleging personal control over secret forces of nature), *religious* (appeals to a higher power or supernatural being), and *scientific* (using natural means to alter the otherwise undisturbed course of nature). Closely following Scottish anthropologist James George Frazer's influential work *The Golden Bough: A Study in Magic and Religion* (1890), Humphreys introduces his readers to magical rainmaking practices such as bloodletting and mimicry of lightning, thunder, rain, and clouds. As did Frazer before him, Humphreys writes of ceremonies to stop the rain, involving the sympathetic magic of setting fires, heating stones, or keeping things dry. His treatment of religious rites includes appeals and supplications directed to the gods, tribal ancestors, or deceased rainmakers. In some cases, the ceremonies are intended to threaten, abuse, or annoy the powers that be. Ringing church bells in inclement weather and praying for rain were the two most common. In his writing, Humphreys tries heroically to separate myth from science and reserves "scientific rainmaking" for special treatment, but as this chapter and those to follow demonstrate, the distinction between mythological and analytical, fictional and aspirational is not so clear-cut.

Today, chemical cloud seeders have largely superseded traditional rain kings and queens, but apart from (apparently) dealing with the same topic, weather control, they hold a vastly different social status. Silver iodide flares may serve as the new fetish replacing shamanistic practices, but traditional rainmakers were and still are celebrated as central figures in their societies, while the cloud seeders are considered culturally marginal at best. If the world's cultures remain firmly rooted in myth, tradition, and storytelling, so too does the history of weather and climate control.

The hubris and folly of Phaethon, themes from Milton and Dante, and examples drawn from cultures other than our own serve to remind us of the richness and relevance of myth and storytelling. Daniel Quinn's distinction between the Takers and the Leavers, expressed through the fictional voice of Ishmael, serves further to problematize and universalize human relationships and attitudes toward the sky. Rather than standing in opposition to rationality, these stories point to fundamental relationships among nature, culture, and human solidarity that are currently not being examined in the scientistic West.

The examples of early popular sci-fi literature on weather and climate control make many of the moral points often left unsaid by scientists and engineers. Some of the stories told here are drawn from prominent authors, but most of them are probably unfamiliar. All of them, written in a variety of genres and from different angles, are relevant to later chapters. Standard histories often privilege the heroic genre. Warriors, statesmen, scientists, and lone inventors rise to face the unknown or to meet unprecedented challenges. This is particularly true in much of the history of science—but not in this book. The FIDO fog-clearing story (chapter 4) is about as close to the heroic genre as it gets.

In the fictional accounts presented here, George Griffith's *Great Weather Syndicate* fits the heroic mold, with Arthur Arkwright ending up as a managed hero. Less ruly are the heroic socialists who oppose the Air Trust. Tragedy dominates *The Wreck of the South Pole*, *The Evacuation of England*, and the short story "The Rain-Maker." Mark Twain's *American Claimant* is pure comedy, as is the geoengineering Western *The Eighth Wonder*. So, too, are the stories of Jeremy the rain bat in *Jingling in the Wind* and *Porky the Rain-Maker*, while N. Richard Nash's *Rainmaker* is a self-described romantic comedy. The *Sky King* episode is largely unclassifiable, but on balance it is indeed an adventure-farce.

The tragicomedic hybrid genre is also prevalent in this literature, from the Baltimore Gun Club's failed attempt to tip the Earth's axis for profit in Jules Verne's *The Purchase of the North Pole* to PAX and his Lavender Ray in *The Man Who Rocked the Earth*, and Kurt Vonnegut's Felix Hoenikker and the practitioners of the absurd human-centered philosophy of Bokononism in *Cat's Cradle*. Even Donald Duck, as "Master Rain Maker," strikes out in anger and slinks away in shame to avoid blame. There are ample opportunities in this type of analysis to reward additional scholars with literary interests—if we can only break out of our narrative ruts. There are no classical heroes here. It is the tragicomic—the voices of Verne, Vonnegut, and even Donald Duck—that seems to come closest to the actual tone of most of the checkered history of weather and climate control.

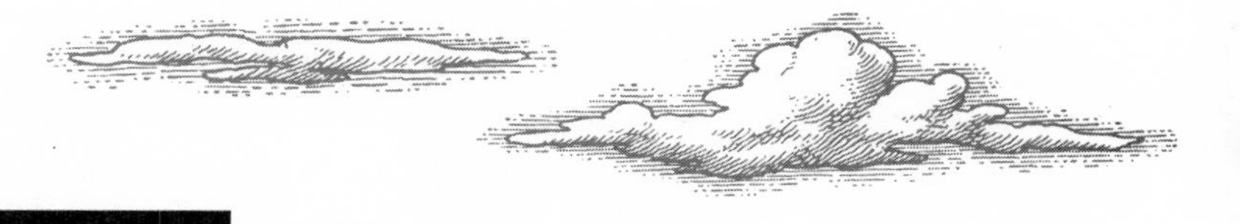

2

RAIN MAKERS

It is not generally known . . . that the question of causing rain by artificial means is no new one.

—ROBERT DECOURCY WARD, "ARTIFICIAL RAIN"

THE quest to control nature, including the sky, is deeply rooted in the history of Western science. In the dedication to *The Great Instauration* (1620), Sir Francis Bacon (1561–1626) encouraged his "wisest and most learned" patron, James I, to regenerate and restore the sciences. Bacon's program involved "collecting and perfecting" natural and experimental histories to ground philosophy and the sciences "on the solid foundation of experience of every kind."[1] His wide-ranging catalog of particular histories included aerial and oceanic topics that are relevant here: lightning, wind, clouds, showers, snow, fog, floods, heat, drought, ebb and flow of the sea. The goal was to replace Aristotelian natural philosophy, stimulate rapid progress in science, improve the human condition through technology, and eventually control nature.

Bacon's philosophy identified three fundamental states of nature: (1) the liberty of nature, (2) the bonds of nature, and (3) things artificial. In the first category, nature is, well, "natural"—free and unconstrained. The second category comprises mistakes and monstrosities resulting from motions that are violently forced or impeded. The third category involves art and technology—mechanisms and experiments constraining nature to operate under human control. Thus gentle rains falling from the sky may water a garden naturally; rainmaking, which seeks to bond and bend natural processes, is a violent or forced act, a monstrosity; and designed irrigation systems, employed by many agriculturalists, constitute artifice. To cite another example of the three states, a shade tree and a gentle breeze may provide some respite on a hot day; towing icebergs to lower latitudes or turning the blue sky milky white with sulfate aerosols to attenuate sunbeams, however, would be violent acts involving forced motions and would constitute errors of potentially monstrous proportions; and the design of building ventilation and cooling systems, subject to individual choice, is clearly within the realm of artifice. As a third example, the eruption of a volcano is considered a force of nature; making an artificial volcano or otherwise tinkering with an existing one would certainly be a mistake; but deflecting lava flows around a village is an artificial but useful thing to do.

In *New Atlantis* (1624), the scientists of Solomon's House practice both observation and manipulation of the weather: "We have high towers . . . for the view of divers meteors—as winds, rain, snow, hail, and some of the fiery meteors also. And upon them in some places are dwellings of hermits, whom we visit sometimes and instruct what to observe . . . and engines for multiplying and enforcing of winds to set also on divers motions."[2] In great experimental spaces, researchers imitate and demonstrate natural meteors such as snow, hail, rain, thunder, and lightning and "some artificial rains of bodies and not of water" (400). Three so-called mystery men are in charge of expanding the repertoire of practices not yet brought into the arts, and three pioneers or miners try new experiments "such as themselves think good" (410); that is, they manipulate nature without further review or oversight, a task requiring perfect virtue and judgment by the experimentalists.

Bacon was conversant with a venerable humanistic tradition that divided history into three parts—ancient, medieval, and modern—but his valuation of the three eras was asymmetric. He granted grudging respect to the ancients, denigrated the Middle Ages, and elevated modern accomplishments to equal or soon-to-be-greater status than those of antiquity. For Bacon, the rise of modern science was due to "the true method of experience . . . commencing . . . with experience duly ordered and digested, not bungling or erratic, and from it educing axioms, and from established axioms, again new experiments."[3] "New discoveries," Bacon argued, "must be sought from the light of nature, not fetched back out of the darkness of antiquity" (154). He elaborated at length on his new method, calling for researchers to work together and making the important point that the sciences were about to enter a period of great fertility. Bacon's communitarian campaign was taken over by innumerable practitioners in the seventeenth century. His greatest legacy, without doubt, was institutional, in that his outlook was absorbed by the Royal Society of London and by many other scientific societies.

Scientific Revolutions "*de l'air*"

The "scientific revolution," although subject to intense historiographic debate, is a term that commonly refers to the transformation of thought about nature through which the authority of ancient texts was replaced by the "mechanical philosophy" and methodology of modern science. Most, but not all, historians see it as a series of events in the sixteenth and seventeenth centuries or, more narrowly, from 1543 (*De Revolutionibus* of Copernicus) to 1687 (*Principia* of Newton). The standard accounts privilege astronomy, physics, and medicine, but also in this era natural philosophers turned away from the traditional practice of preparing commentaries on Aristotle's *Meteorologica* and instead began focusing on new techniques for describing, measuring, and weighing the atmosphere. Behind this turn was the hope that somehow quantification might lead to understanding and trigger a cascade of new capabilities, including prediction and control. Beginning with the Accademia del Cimento in Florence, the scientific societies of Europe attempted to make histories of the weather and promoted the collection, compilation, dissemination, and discussion of meteorological observations from remote locations and over widespread areas of the globe. Adherents of the new mechanical and chemical philosophy insisted that all atmospheric phenomena could be reduced to their component processes and could be explained by an emerging body of natural laws. They developed new instruments—thermometers, barometers, hygrometers, and calibrated rain gauges—for observing and quantifying aspects of the atmosphere. New practices and perspectives meant that henceforth no atmospheric process, however seemingly insignificant, would be left unrecorded. As a result, a culture of measurement emerged, linked to a new meteorological science of planetary proportions. This "descent, with variation," of viable meteorological instruments, so proudly traced by scientists and historians, is only one aspect of the story, since many techniques resulted in dead ends—in extinct or forgotten practices. The lack of uniform standards and global and temporal coverage, however, remained a continuing challenge.[4]

In 1949 one of the early champions of the idea of a scientific revolution, the historian Herbert Butterfield, wrote the following:

> Since the Scientific Revolution overturned the authority in science not only of the middle ages but of the ancient world—since it ended not only in the eclipse of scholastic philosophy but in the destruction of Aristotelian physics—it outshines everything since the rise of Christianity and reduces the Renaissance and Reformation to the realm of mere episodes, mere internal displacements, within the system of medieval Christendom. Since it changed the character of habitual mental operations even in the conduct of the non-material sciences, while transforming the whole diagram of the physical universe and the very texture of human life itself, it looms so large as the real origin both of the modern world and of the modern mentality that our customary periodization of European history has become an anachronism and an encumbrance.[5]

More recently, a prominent feminist scholar, Carolyn Merchant, saw the same events as a disaster of unmitigated proportions: "The removal of animistic, organic assumptions about the cosmos constituted the death of nature—the most far-reaching effect of the Scientific Revolution."[6] She argured that because scientists had redefined nature as a system of dead, inert particles moved by external rather than inherent forces, their endorsement of the reductionistic framework of the mechanical philosophy legitimized nature's manipulation and progressive destruction. Power over nature was fully compatible with the values of scientists' ultimate supporters—governments—especially the military establishment, commodifiers, and other ideologues and opportunists of various stripes. Others wonder if there have been many scientific revolutions, or perhaps none at all![7]

Most historians agree that since the seventeenth century, scientists have attempted to complete the Baconian program, elevating the attainment of natural knowledge to the sine qua non of human achievement, and then wielding this knowledge to gain power over and control of nature for the stated purpose of improving the human condition, however narrowly defined, but often falling short of this goal. This program, the opening wedge of a revolution articulated in different ways by Galileo, Descartes, and others, was more than a new set of techniques in the laboratory or the field. It was a revolution in thought that placed humanity at the conceptual and willful center of the universe, redefined our relationship with the natural world, elevated the scientific method to the pinnacle of truth recently vacated by the church fathers, and dealt a blow to apocalyptic thinking. As the Enlightenment eroded belief in divine providence as a moving force in history, the historiographic void was filled by the notion of progress, a secular notion based on the development and application of human reason to the challenges of understanding, prediction, and ultimately, control.

Great Fires and Artificial Volcanoes

In the closing decades of the eighteenth century in Europe, and slightly later in Russia and the United States, serious attempts were made to broaden the geographic coverage of weather observations, standardize their collection, and publish the results. Individual observers in particular locales dutifully tended to their journals while networks of cooperative observers gradually extended the meteorological frontiers. No one, however, had yet proposed a serious scientific-based program of weather control. James Pollard Espy (1785–1860) was a leading meteorologist of his day, the first to be employed by the U.S. government in this capacity. Born into a farm family in Washington County, Pennsylvania, and educated at Transylvania University in Kentucky, he worked as a frontier schoolmaster and lawyer until he moved to Philadelphia in 1817. There he supported himself by teaching mathematics and classics part time at the Franklin Institute while devoting his free time to meteorological research. From 1834 to 1838, he served as the chairman of the Joint Committee on Meteorology of the Franklin Institute and the American Philosophical Society. He won the latter's Magellenic Prize in 1836 for his theory of hail. Working with the scientific societies of Philadelphia, Espy gained the support of Pennsylvania's legislature to equip weather observers in each county in the state with barometers, thermometers, and other standard instruments to provide a larger, synoptic view of the weather, especially the passage of storms. He also maintained a national network of correspondents and volunteer observers. During this period, he invented a "nephelescope," an early cloud chamber, which he used in his popular lectures and, in his technical work, to calculate the amount of heat released by condensing water vapor.

Espy moved to Washington, D.C., in 1842. In his first government appointment, as professor of mathematics in the navy, he developed a ventilator for ships and expanded his network of meteorological correspondents. He also held a joint appointment as the "national meteorologist" in the U.S. Army Medical Department, a position that boosted his storm studies by providing him access to the meteorological reports of the army post surgeons. From 1847 to 1857, his salary was provided by annual appropriations from Congress. With Joseph Henry, he established the Smithsonian meteorological system of observers and experimented with telegraphic weather reports. Several of his major reports on meteorology appeared as U.S. Senate executive documents.[8]

Espy viewed the atmosphere as a giant heat engine. According to his thermal theory of storms, all atmospheric disturbances, including thunderstorms, hurricanes, and winter storms, are driven by steam power. Heated by the Sun, a column of moist air rises, allowing the surrounding air to rush in. As the heated air ascends, it cools and its moisture condenses, releasing its latent heat (this is the "steam power") and producing rain, hail, or snow. Espy emphasized, correctly, the importance of knowing the quantity of vapor in the air, "for it is from the latent caloric [or heat] contained in the vapor that all the force of the wind in storms is derived. It is only when the dew-point is high that there is sufficient steam power in the air to produce a violent storm; for *all storms are produced by steam power*."[9] His theory was well received by many scientists of his time, including a committee of the French Academy of Sciences chaired by François Arago. The convective theory is now an accepted part of meteorology, and for this discovery Espy is well regarded in the history of science.

Espy strayed from the scientific mainstream when he promoted his idea that significant rains of commercial importance for agriculture and navigation could be generated by cutting and burning vast tracts of forest. He believed the heat and smoke from these fires would create huge columns of hot air, producing clouds and triggering precipitation, much like the effects of volcanic eruptions. He listed five scientific reasons why setting large fires should produce rain: (1) experiments showed that expanding air cools dramatically, and (2) under certain conditions of high humidity forms both a visible cloud and significant amounts of precipitation; (3) chemical principles indicated that the "caloric of elasticity" (a venerable term for latent heat) released in the condensation of this vapor is immense, equal to about 20,000 tons of anthracite coal burned on each square mile of cloud extent. Espy's convective theory further held that (4) this release of heat would keep the cloud buoyant, lower the barometer, and "cause the air to rush inward on all sides toward the center of the cloud and upward in the middle, thus continuing the process of condensation of vapor, formation of cloud, and generation of rain."[10] Espy derived his final point empirically by collecting observations and testimonials to the effect that (5) air does indeed rush inward on all sides toward the center of the region where a great rain is falling and upward into the cloud.

Espy explained that three things can prevent rains from accompanying great fires: (1) winds, (2) excessive moisture, and (3) stability of the upper levels of the atmosphere. He released small balloons and tracked their flight in order to get a sense of the winds, and he used a hygrometer to measure atmospheric moisture and estimate its changes with height. Stability was more of a problem, for as he observed, in the present state of science, the levity of an upper stratum of air could not always be known. Correspondents, friends, and even a congressman laughed at Espy when they heard of his proposal to make rain, but he assured them that science was on his side. He even ventured a prediction of how the experiments might turn out in favorable conditions and felt there was no disgrace in desiring to see a great experiment made. He anticipated that his labors would be crowned with success.

In 1838 Espy petitioned the U.S. Senate to reward him in proportion to his ability to make rain by burning woodlands. James Buchanan (D-Pennsylvania) apologized to his colleagues for the "strange petition" he was about to present, but assured them that it came from "a very respectable and scientific man" with excellent references and credentials:

> The petitioner . . . says that he has discovered a means of making it rain in a tract of country at a period of time when there would be no rain without the use of his process. Mr. Espy proposed to make the experiment at his own expense; and he proposed that Congress should pass an act engaging to reward him with a certain sum if he succeeded in making it rain in a tract of country ten miles square; a still higher sum if he produced rain in a tract of country one thousand square miles; a still higher sum if he produced rain in a tract of five thousand square miles; and, lastly, to give him a still greater compensation if he should cause the Ohio river to be navigable all summer from Pittsburgh to the Mississippi.[11]

Buchanan supported the petition based on Espy's scientific reputation, but "scarcely knew himself what to say about it." Senator John J. Crittenden (W-Kentucky) "doubted very much, whether, even if this thing was possible, it would be a good policy to encourage the measure." He thought that no mortal should have the power that Espy professed to have and no one could take the Ohio River under his special protection:

> Why, sir, he might enshroud us in continual clouds, and, indeed, falsify the promise that the earth should be no more submerged. And if he possesses the power of causing rain, he may also possess the power of withholding it; and, in his pleasure, instead of giving us a navigable river, may present us with rock and shoals and sandbars. He thought that this would be too dangerous a power to entrust to any individual . . . unless . . . we had some very summary process of manufacturing sunshine. (39)

The senators, obviously enjoying the discussion, pointed out that no citizen should be empowered to hoard up the clouds and vapors or to dispense them at will. Buchanan's motion failed, and Espy's petition lay on the table. That year, and for several years following, Espy looked closer to home, seeking, but failing to receive, government support for rainmaking. "Magnificent Humbug" opined the *Genesee Farmer*. According to the *Boston Quarterly Review*, "The public at large think of him as a sort of madman, who fancies that he can produce artificial rain."[12]

Espy's magnum opus, *The Philosophy of Storms* (1841), includes a long section titled "Artificial Rains," in which he compiled testimonies of rainfalls accompanying volcanic eruptions and large fires: "The documents which I have collected on this subject, if they do not prove that the experiment will succeed, do at least prove that it ought to be tried."[13] Espy concluded that if a large body of air is forced to ascend in a column, a large self-sustaining cloud will be generated and cause more air to rise up into it to form more cloud and rain. He argued that this was certainly the case in volcanic eruptions and should also be the case for great fires. He cited the mysterious connection between volcanoes and rain as noted by the famous geographer and explorer Alexander von Humboldt, who observed that sometimes during a volcanic eruption a dry season changed into a rainy one. Thus he argued that the rainmaking effects of a giant forest fire should mimic those of a volcanic eruption.

Espy scoured the literature for supporting evidence. He cited Martin Dobrizhoffer, an Austrian Jesuit evangelist in South America who wrote that he witnessed the tribes of the Abipones in Paraguay producing rain (in an admittedly very rainy climate) by setting fire to the plains. He also cited the practice of American Indians burning the prairies to produce rain, and he called for his correspondents to send in reports and testimonies of similar instances supporting his theory. An observer in Louisiana wrote that a conflagration in the long grass in the prairies of that state was soon followed by rain.

In 1845 Espy issued a circular letter "To the Friends of Science" with specific details of his rainmaking plan. He proposed a massive experiment along the Alleghany Mountains (a region quite familiar to him): "Let forty acres . . . be fired every seven days through the summer in each of the counties of McKean, Clearfield, Cambria, and Somerset, in Pennsylvania; Alleghany in Maryland; and Hardy, Pendleton, Bath, Alleghany, and Montgomery, in Virginia." Espy anticipated the effects of upper-air wind shear by recommending that woodlots several miles apart be fired, "so that the up-moving column of air which shall be formed over them may have a wide base, and thus may ascend to a considerable height before it may be leaned out of perpendicular by any wind which may exist at the time."[14]

He also proposed an even larger, continental-scale project that involved simultaneously firing masses of timber in the amount of 40 acres every 20 miles, every seven days, along a line of 600 or 700 miles in the western United States along the Rocky Mountains. Espy predicted that the *probable* outcome of this managed system would be regular, gentle, and steady rains sweeping across the entire country like clockwork for the benefit of farmers and navigators. Here is how Espy explained his plan:

> A rain of great length, north and south will commence near or on the line of fires; this rain will travel eastward; it will not break up till it reaches far into the Atlantic Ocean; it will rain over the whole country east of the place of beginning; it will rain only a short time in any one place; it will not rain again until the next seventh day; it will rain enough and not too much in any one place; it will not be attended with violent wind, neither on land or on the Atlantic Ocean; there will be no hail nor tornadoes at the time of the general rain nor intermediate; there will be no destructive floods, nor will the waters ever become very low; there will be no more oppressive heats nor injurious colds; the farmers and the mariners will always know before the rains when they will commence and when they will terminate; all epidemic diseases originating from floods and subsequent droughts will cease; the proceeds of agriculture will greatly increase, and the health and happiness of the citizens will be much promoted. (51)

Espy presented the testimonies of eyewitnesses who saw both clouds and rain produced by fires. The good citizens of Coudersport, Pennsylvania, including attorneys, judges, and ministers, attested that both clouds and rain were produced by the burning of a fallow field in July 1844. Similar phenomena had attended a prairie fire in Indiana the previous summer. Surveyor George Mackay claimed to have stimulated convective showers in Florida by cutting and burning "exceedingly inflammable" saw grass: "We often fired the saw-grass marshes afterward; and whenever there was no wind stirring, we were sure to get a shower."[15] Apparently, a number of farmers in Florida were in the habit of setting grass fires to produce rain when they planted their corn. A forest fire in Isle Royale, Michigan, in 1846 produced similar results, as did extensive forest fires in Nova Scotia and, apparently, coal burning in the industrial city of Manchester, England.

Perhaps the most striking eyewitness testimony of "steam power" of the atmosphere being kindled by a great fire was sent to Espy by the Reverend J. D. Williamsom, who was hiking with a companion on a mountain summit near Keene, New Hampshire, in July 1856: "The weather was excessively hot. Not a cloud was to be seen, nor was there a breath of wind stirring. Looking to the southeast at a distance of some five or six miles, I saw a fire just kindled in a fallow of some acres in extent. The column of smoke ascended perpendicularly and unbroken."

Williamsom, who was familiar with Espy's theory, remarked to a companion that the fire should soon produce rain unless disturbed by upper currents:

> Up went the column strait as an arrow, and anon it began to expand at the top and assume the appearance of cloud. This cloud, with its base stationary, expanded upward, and swelled as if a huge engine was below with its valve open for the escape of steam. . . . Soon the rain began to descend . . . [and the cloud] sailed off in an eastern direction, pouring down torrents of rain. . . . I have ever regarded [this event] as a perfect and undeniable demonstration of the truth of [your] theory, and I can no more doubt it than I can doubt the evidence of my senses.[16]

For his work in mapping and forecasting and for his tireless promotion of rainmaking, Espy earned the derisive sobriquet "the Storm King."

Eliza Leslie's "Rain King"

The year Espy moved to Washington, the popular magazine writer Eliza Leslie published a short story in *Godey's Lady's Book* called "The Rain King, or, A Glance at the Next Century," a fanciful account of rainmaking a century in the future, in 1942. In the story, Espy's great-great-grand-nephew, the new Rain King, offers weather on demand for the Philadelphia area. Various factions vie for the weather they desire. Scores of alfalfa farmers and three hundred washerwomen petition the Rain King for fine weather forever, while corn growers, cabmen, and umbrella makers want consistent rains. Fair-weather and foul-weather factions apply in equal numbers until the balance is tipped by a late request from a high-society matron desperately seeking a hard rain to muddy the roads and prevent a visit by her country-bumpkin cousins.

When the artificial rains come, they satisfy no one and raise widespread suspicions. The Rain King, suddenly unpopular because he lacks the miraculous power to please everybody, takes a steamboat to China, where he studies magic in anticipation of returning someday with new offerings. "Natural rains had never occasioned anything worse than submissive regret to those who suffered inconvenience from them, and were always received more in sorrow than in anger," Leslie wrote. "But these artificial rains were taken more in anger than in sorrow, by all who did not want them."[17]

Leslie's short, humorous fantasy revealed a dramatic and instantaneous change in public attitudes "precipitated" by artificial weather control. Although Leslie was no meteorologist, her tale "showed a far better grasp of weather's human dimensions and of the pitfalls of weather control than anything Espy ever wrote."[18] Since then, however, the intractable human dimensions of weather and climate control have taken a backseat to the technical schemes of optimistic rain kings and climate engineers with relatively simple ideas, or at least angles.

Espy received honorable mention in 1843 in Nathaniel Hawthorne's "Hall of Fantasy"—a marketplace of wild ideas that most of us visit at least once but some dreamers occupy permanently; a marketplace seemingly perfectly suited to the millennial ideas of rain kings and climate engineers. Here the statues of the rulers and demigods of imagination—Homer, Dante, Milton, Goethe—are memorialized in stone, while those of more limited and ephemeral fame are made of wood. Plato's Idea looms over all. Here are social reformers, abolitionists, and Second Adventist "Father [William] Miller himself!" Civil and social engineers propound ideas of "cities to be built, as if by magic, in the heart of pathless forests; and of streets to be laid out, where now the sea was tossing; and of mighty rivers to be stayed in their courses, in order to turn the machinery of a cotton-mill."[19] "Upon my word," exclaimed Hawthorne, "it is dangerous to listen to such dreamers as these! Their madness is contagious" (204). Here are inventors of fantastic machines aimed to "reduce day dreams to practice": models of a railroad through the air, a tunnel under the sea, distilling machines for capturing heat from moonshine and for condensing morning mist into square blocks of granite, and a lens for making sunshine out of a lady's smile. "Professor Espy was here," reminiscent of Aeolus, the god of the winds, "with a tremendous storm in a gum-elastic bag" (206). The "inmates of the hall," it is said (remember that all pass through here on occasion), take up permanent residence by throwing themselves into "the current of a theory," oblivious to the "landmarks of fact" passing along the stream's bank.

Cannon and Bells

Charles Le Maout (1805–1887), a pharmacist and mine assayer in Saint-Brieuc, near the coast of Brittany, was a dedicated pacifist. One of his powerful arguments in favor of peace went far beyond typical arguments invoking the carnage, desolation, and miseries of war. He thought that war, especially cannonading but also the ringing of bells, destroyed the fragile equilibrium of the aerial elements and was responsible for undesirable atmospheric perturbations of all kinds, including rain, hail, thunder, lightning, harsh winters, and possibly airborne epidemic diseases. He wrote:

> To have a proper idea of the fragility of the atmosphere in which we are destined to live, like fish in the depths of the sea, we ought to imagine ourselves inhabiting a crystal palace which, on the firing of a cannon, would be shattered to atoms over our heads. . . . As soon as the cannon cease firing or the bells cease sounding, when the sky is cloudy or overcast, the weather clears up and the blue sky and sunshine appear. . . . I am not thus wrong to say that God creates fine weather and man turns it foul.[20]

During the memorable siege of Sevastopol (1854–1855), which he observed, Le Maout said "all of nature was affected" by the cannonading, which he claimed caused a widespread outbreak of whooping cough. He convinced Marshal Jean-Baptiste Philibert Vaillant, the scientifically minded French minister of war who had instituted telegraphic weather reports, to order his artillery officers to record the weather on days when cannon were being fired. The results were inconclusive, and Vaillant, unimpressed by the outcome, disavowed the theory in the *Journal officiel de l'Empire*, concluding, "The famous influence of cannon is illusory."[21]

Disappointed but undaunted, Le Maout collected his own statistics to show that the weather in years with peace was more salubrious than in those with war. He advised keeping both the guns and the church bells of Europe and the Mediterranean silent, both in war and during celebrations, since their concussions disrupted the natural course of the winds and produced clouds and condensation at immense distances:

> Man has two powerful agents at his disposal [guns and church bells], for influencing the atmosphere. He can, if he pleases . . . govern the aerial phenomena; and (were all human disturbance to cease on the surface of the globe) the air, in obedience to the laws of attraction, would probably return to a state of repose, as does the surface of the sea when not agitated by storms.[22]

Conversely, he argued that selective cannonading and bell ringing during times of drought might provide relief for agriculture. Le Maout was convinced that he had presented the most powerful argument for the establishment of universal peace and urged his readers to propagate and popularize this doctrine for the sake of humanity. Waxing poetic, he wrote:

> Nature prepares the storms and tempests; man makes them explode.
> God makes good weather; man makes it bad.
> He who sows with gunpowder will reap the storm.[23]

War and the Weather

In America, the enthusiasm for "harvesting the storm" with gunpowder and other explosives was just beginning. During the Civil War, some observers began to suspect that the smoke and concussion of artillery fire generated rain. After all, didn't it tend to rain a day, or two . . . or three . . . following most battles? The heavy fighting at Gettysburg on the first three days of July 1863 under fair skies was followed by torrential downpours on July 4 that lasted all day and into the night, resulting in roads knee-deep in mud and water that hampered the Confederate retreat. Skeptics hastened to point out that the connection between war and the weather was an ancient one—and a shaky one.

In Plutarch's "Caius Marius" (75 C.E.), "it is observed, indeed, that extraordinary rains generally fall after great battles; whether it be, that some deity chooses to wash and purify the earth with water from above, or whether the blood and corruption, by the moist and heavy vapors they emit, thicken the air, which is likely to be altered by the smallest cause."[24] According to William Jackson Humphreys, Plutarch's first option was a matter of belief, not science, while his second option was not significant, since only about 0.01 inch of rain would fall over a square mile if ten thousand soldiers, assuming they were nothing but blood and sweat, "were wholly evaporated and then all condensed back."[25] Humphreys posed a plausible explanation for the apparently high correlation between rains and battles. He noted that plans were usually made and battles fought in good weather, so that after the battle in the temperate regions of Europe or North America rain will often occur in accordance with the natural three- to five-day periodicity for such events. Perhaps generals simply preferred to fight under fair skies, with rainy days therefore tending naturally to follow. Perhaps it would tend to rain several days after doing most anything!

In 1871 Chicago civil engineer and retired Civil War general Edward Powers published his book *War and the Weather, or, The Artificial Production of Rain*, in which he reviewed the weather following selected battles and contended that rain followed artillery engagements—usually within several days. Powers found a "perfect explanation" for this in the theory of oceanographer Matthew Fontaine Maury, who maintained that there were two great atmospheric currents, the equatorial and the polar, flowing aloft in nearly opposite directions. Powers argued that the concussion of battle caused these higher strata to mix and release their moisture. He envisioned stimulating rainfall on demand through the agency of loud noises, perhaps by detonating explosive charges carried aloft by kites or balloons. In times of drought, when the ground was bone-dry, he envisioned tapping into the elevated rivers of air that carried abundant moisture from the Pacific Ocean. Analogous to drilling for groundwater, aerial explosions would merely release the moisture that was already up there, traveling overhead. Seven decades later, this "river of air" would be called the jet stream and would be deemed important not for its moisture, since it is absolutely desiccated, but for its dynamic effects on high-flying aircraft and on surface weather.

When critics pointed out that loud concussions, if effective, should cause it to rain immediately, not hours or days later, Powers fell back on his two-current theory: "The center of the atmospheric disturbance caused by a battle should remain in the vicinity of the battlefield while the two currents are mixing together and initiating the process that leads to rain—a process which, it is plain, must require time in reaching a state of effective action."[26] However deficient in meteorological details, Powers's theory was appealing to desperate farmers, like those in New England at the time, since it directed their hopeful gaze aloft, away from their parched fields and devastated crops. Powers reminded them that there is an ocean of moisture derived not from surface evaporation but from the Pacific Ocean and just waiting to be tapped. However, one observer noted that no effect on the weather had been perceived in the Rocky Mountains after years of blasting for mining and road-building operations.[27]

Powers sought support for his theory from the U.S. Army Signal Office weather service and through his representative, Charles Farwell (R-Illinois), who championed this cause for the next two decades. After reviewing Powers's theory and his proposal to fire three hundred cannon arranged in a circle a mile across, the House Committee on Agriculture concluded in a report that the government should act unilaterally on this issue of great significance and support Powers's field experiments: "We have the powder, and we have the guns, and the men to serve them, and we ought not to leave to other nations and to after-ages the task of solving the great question as to whether the control of the weather is not, to a useful extent, within the reach of man."[28] In another proposal, Powers suggested employing the siege guns at the Rock Island Arsenal in Illinois for rainmaking experiments at a cost, per rainstorm, of $21,000, an amount he claimed was much less expensive than the cost of irrigation or the loss of crops due to lack of rain, but an amount that could outfit more than a score of family farms. The proposals were not funded.

Powers finally found an ally in Daniel Ruggles of Fredericksburg, Virginia. Ruggles was a West Point graduate, a former general in the Confederate Army, and the owner of a ranch in Rio Bravo, Texas, who received a patent in 1880 "for producing rain fall . . . by conveying and exploding torpedoes or other explosive agents within the cloud realm."[29] Ruggles's "invention" consisted, in brief, of a balloon carrying torpedoes and cartridges charged with such explosives as nitroglycerine, dynamite, gun cotton, gunpowder, or fulminates, and connecting the balloon with an electrical apparatus for exploding the ordnance.

Like his predecessor Espy, Ruggles made surprising claims to have "invented a method for condensing clouds in the atmospheric realm, and for precipitating rainfall from rain-clouds, to prevent drought, to stimulate and sustain vegetation, to equalize rainfall and waterflow, and by combining the available scientific inventions of the age, to guard against pestilence and famine, and to prevent, or to alleviate them where prevailing."[30] He claimed that the concussions and vibrations of the explosions would, under the proper conditions, consolidate the "diffused mists" passing overhead into rainfalls. His scheme favored the remote detonation of the explosives using timed fuses or electric wires, but for more precision (and much greater risk) he also imagined aeronauts bombing the clouds with torpedoes attached to parachutes. Promising scientific rigor (still a challenge today in rainmaking), he proposed to select clouds on which to experiment in conformity with "well-defined meteorological data," which he listed as "barometric tension, thermometer and its changes, hygrometer, anemometer, anemoscope . . . , elevation, average rainfall, river stages, and magneto-electric condition of the atmosphere" (10).

Arguing that if God had not wanted us to manipulate the clouds, he would not have placed them so clearly in our line of vision, Ruggles promised "to *appropriate the atmospheric laws of cloud-land*, in sunshine and in storm, and direct them, so far as may be practicable, within the sphere of the great industrial interests and energies of man" (12). Dazzled by his own genius, the scope of the undertaking, and the prospects for "untold advancement," he exclaimed, "The field is broad—very broad; as deep as it is broad—it is very deep!" (12).

Ruggles claimed (as did every generation afterward) that he was taking the next step technologically, in this case by ascending above the Earth's surface into the atmospheric realms with balloon probes and human aeronauts using the latest chemical explosives and electrical devices, all under the banner of advanced engineering and meteorological science unknown to Espy:

> The gigantic stride of the engineer through the cloud-capped mountains, and with miraculous force rendering asunder the foundations of old ocean's bed; the modern "Prometheus," magneto-electric lightning, had not then been enchained; the leviathan "steam" had not then been bound to the billowy ocean's foam; aerial navigation sat with clipped wings in the portals of the temple of science; the grand triumphs in chemical philosophy in the development of explosives; in the condensation of the elements of light in the photographic art; the development of mines of vast extent and fabulous wealth; the unfolded banner of meteorological

> science—no, none of these grand revelations of occult science were available to him. They had then [in Espy's day] scarcely dawned upon the horizon of the human mind. (13)

Wrapping up his argument, which was by now a secular sermon, with themes borrowed from the march of progress and the pulpit, Ruggles claimed that his technique might alleviate human suffering both in the United States and around the world:

> The conformation of our continent, crowned with lofty mountain ranges, its great bounding rivers, its broad fertile plains, and its boundless forests—all swept by the rain-clouds of surrounding oceans—all, all give assurance that a combination of skill and industry will materially protect our soil from impending drought, and from those visitations of desolating famine so often chronicled in the eastern world. . . . [If this plan works,] no other scheme of philanthropy known to man—save that embodied in the Christian dispensation—transcends it! (17–18)

Describing his scheme as an "advanced step" in the science of "meteorological engineering," Ruggles appealed, unsuccessfully, to the U.S. Senate Committee on Agriculture for $10,000 in support of his rainmaking experiments. A Civil War veteran who had witnessed major battles with no rain at all wrote in a letter to *Scientific American* that if cannon explosions in a battle do not cause rain, Ruggles's patent balloon will not do it either.[31] An editorial writer opined, "We do not think the invention is worth a cent or the patent either."[32]

A Perfect Imitation of Battle

Robert St. George Dyrenforth (1844–1910), a controversial and flamboyant patent lawyer from Washington, D.C., was certain that rain could be caused by explosions in midair. He read whatever he could about rainmaking, including Le Maout's pamphlets from France and the second edition of Powers's book, published in 1890, and consulted with meteorologist John P. Finley of the Signal Service and many others. During a severe and prolonged western drought, Charles Farwell, now a U.S. senator, succeeded in obtaining appropriations of $9,000 for the support of a new series of field experiments on rainmaking by concussion. He recommended that Secretary of Agriculture Jeremiah Rusk be placed in charge of the project. The newly formed U.S. Weather Bureau, also under Rusk's supervision, was quite skeptical of rainmaking by concussion, and the chief of the Division of Forestry, Bernhard E. Fernow, thought that the entire enterprise was under-conceptualized, with no reasonable expectation that the experiments would be effective. Nevertheless, Rusk chose Dyrenforth as the lead investigator and special agent of the government.

Dyrenforth was born in Chicago and received his education in Germany, at Prussian military academies, at the Polytechnic School in Karlsruhe, and at the University of Heidelberg, where he was awarded a doctorate in mechanical engineering in 1869. He served as a war correspondent during the Austro-Prussian War of 1861 and, during the Civil War, attained the rank of major in the Union Army, but later he claimed he was a "general." After studying law at Colombian College in Washington, D.C., he worked as an attorney for the Patent Office and in private practice. It was said that Dyrenforth was boastful of his accomplishments, even alleged ones, and was extremely demanding of both his family and his subordinates.[33]

Dyrenforth decided that the best rainmaking policy would be to attack the atmosphere on multiple fronts with balloons, kites, dynamite, mortars, smoke bombs, and even fireworks. His primary idea was to stimulate condensation of moisture or deflection and mixing of opposing moist and cold air currents by concussion, using whatever explosive devices were available to him. In this, he was firmly following trails blazed by Powers and Ruggles. Dyrenforth theorized that as secondary effects, the explosions would generate shock, pressure, and heat, creating a powerful upward current in the form of an eddy or a whirlpool and inward- and upward-rushing streams of air in line with Espy's convective theory. The explosions should also generate electrical charges that would polarize the Earth and sky, generate a magnetic field, and possibly enhance the condensation of moisture—a theory reminiscent of the one articulated by the American chemist Robert Hare in the 1830s. Following a line of reasoning attributed to the Scottish physicist and meteorologist John Aitken. Dyrenforth expected smoke from the gunpowder to provide nuclei for the agglomeration of suspended particles of moisture.

Another idea was that balloons inflated with one part oxygen and two parts hydrogen and detonated aloft with an electric spark would supposedly form a small amount of liquid water in the process, thereby seeding the clouds with sympathetic nuclei for the aggregation of more water. Critics pointed out that producing hydrogen and oxygen gases in the field was slow and required bulky and expensive equipment and supplies. Moreover, since a large exploding balloon could be expected to produce no more than 6 ounces of water, it would probably be more efficient to fly a pint of water into a cloud on a balloon or kite and just release it. Dyrenforth was persistent, however, since he favored a secondary effect of this technique: the loud bang produced by the exploding balloon. He noted that a small bubble of oxy-hydrogen produces a report like a "horse-pistol," and recalled an occasion years before, when physicist Joseph Henry had detonated 50 cubic feet of the mixture in a buried vessel, and the explosion tore a hole in the ground 18 feet in diameter. Dyrenforth's experiments with rackarock (an explosive widely used in coal mining) and a 10-foot oxy-hydrogen balloon on his country estate in Mount Pleasant, near the current National Zoological Park in Washington, D.C., were witnessed by Secretary of the Smithsonian Samuel P. Langley, John Wesley Powell of the U.S. Geological Survey, patent-holder Daniel Ruggles, and other luminaries. Although Dyrenforth did not generate rain that day, he did induce a letter of protest to the secretary of agriculture from his neighbor William J. Rhees, chief clerk of the Smithsonian, who claimed the blasts disturbed his fine herd of Jersey cows, shook his farmhouse, and alarmed his family. Beyond the neighborhood upset, this exercise in backyard bombing, with smoke billowing and flaming oxy-hydrogen balloons falling from the sky, was dramatic, it attracted a large crowd of onlookers, and it was fun. It amounted to the government's declaration of war on both drought and boredom.[34]

Nelson Morris, a prominent Chicago meatpacker who was said to own the largest herd of Black Angus cattle in the world, offered his "C" Ranch near Midland, Texas, as a site for the field trials. The ranch was located at 32°12'N, 102°20'W, at an elevation of about 3,000 feet, in dry, hilly country just off the right-of-way of the Texas and Pacific Railroad. Morris sweetened the offer with free room and board for Dyrenforth's team and payment of all local expenses. The site is located on Ranch Road 1788, not far from the New Mexico towns of Alamogordo, Socorro, and Roswell, if you get my drift and are looking for a day trip.

The advance party left Washington, D.C., on July 3, 1891, by train, carrying suitcases, mortars, and 2 tons of cast-iron borings furnished by the navy for making hydrogen. The full account is in Dyrenforth's final report, but as recorded more humorously by the *Farm Implement News* of Chicago, the party consisted of half a dozen special scientists, "all of whom know a great deal, some of them having become bald-headed in their earnest search for theoretical knowledge."[35] Myers and Castellar were the balloonists; Rosell, the chemist; Curtis, the meteorologist; and Draper, the electrician. In St. Louis, they added 8 tons of sulfuric acid in drums, 5 additional tons of cast iron, 1 ton of chloride of potash, and 0.5 ton of manganese oxide, along with casks, balloons, and other supplies. Once they got to Texas, the railroad provided them free passage to Midland, where they arrived on August 5. Waiting for them on the siding was a block of pure tin rolled into thin sheets for making electrical kites and six kegs of blasting powder donated by a local coal mine.

The arrival of "Generals" Dyrenforth, Powers, and Ruggles coincided with a summer dry spell but also, conveniently, with the traditional (and commonly known) onset of the rainy or monsoon season, when winds from the Gulf of Mexico and the Gulf of California typically bring showers and thunderstorms to the high plains. Dyrenforth, broad-shouldered and brash, wearing cavalry boots and a campaign pith helmet, gave a public address to dramatize the situation and heighten the sense of accomplishment if his rainmaking experiments happened to succeed. He emphasized the barrenness and extreme aridity of the region, the heat, harsh Sun, cloudless sky, dry south wind, alkaline soil, and feeling that his skin was turning into parchment. Although even as he spoke the winds were beginning to loft moisture across Texas from the south, he pointed out the current local dry conditions. He was sure that, by his technique, Midland's wells and lakes would fill up, its fine soils would produce a bountiful crop, and there would be little left to desire—if the region could only be supplied with water.

At the "C" Ranch, a front line of attack was established with sixty makeshift mortars constructed from 6-inch well pipe. His crew tamped charges of dynamite into prairie dog holes, placed them on flat stones, and draped mesquite bushes with rackarock. The electrical kites and the oxy-hydrogen balloons formed the second and third lines of battle. The contract meteorologist, George E. Curtis, deployed barometers, thermometers, sling psychrometers, and an anemometer, but curiously no rain gauges and no electrical measuring apparatus.[36] The first rain fell on August 13, before any experiments had been made, but the *Chicago Herald* reported this event as "heavy rain at the ranch in response to the party's efforts." This on a day when the U.S. Weather Bureau was predicting rain and showers in the state, mainly east of Midland. Still, Dyrenforth assumed that he had had some effect in causing the nimbus clouds to drop their loads on Midland and claimed that he had caused a "very heavy rain" of about an inch that day.

The reporter on assignment from the *Farm Implement News* had a different perspective. The headline read: "News from the Rainmakers. They fail on account of the dry weather and because their apparatus won't work. The elements play them false." Of their efforts he wrote,

> Their kites fail to fly . . . , their hydrogen tanks and their balloons leak, and even the clouds fail to cooperate. . . . When the gas generating furnace caught fire, eventually a cowboy roped the blazing furnace and dragged it into a stock tank and extinguished the flame. . . . And when a cumulus cloud happens to pass their way, rain often falls before they can make their explosive apparatus work.[37]

Beginning on the evening of August 17, a massive barrage of aerial and ground explosions echoed throughout the night. At dawn, the skies were clear, but twelve hours later rain began to fall in the area. Dyrenforth took immediate credit for this, even though the ranch received only a trace amount of rain. About a week later, on August 25, a day after the weather observer, Curtis, had departed, Dyrenforth declared the weather "settled and dry," this according to the opinion of the ranch hands. The team set off a barrage of explosions all day, ending at eleven o'clock that night with Dyrenforth commenting that the "atmosphere at that time [was] very clear, and as dry as I have ever observed it." But seemingly, the concussions had done their job, for "at about 3 o'clock on the following morning, August 26, I was awakened by violent thunder, which was accompanied by vivid lightning, and a heavy rainstorm was seen to the north—that is, in the direction toward which the surface wind had steadily blown during the firing, and hence the direction in which the shocks of the explosions were chiefly carried."[38]

Professor Alexander McFarlane of the University of Texas had a different perspective: "The trial of Friday, August 25, was a crucial test, and resulted not only in demonstrating what every person who has any sound knowledge of physics knows, that it is impossible to produce rain by making a great noise, but also that even the explosion of a twelve-foot balloon inside a black rain cloud does not bring down a shower."[39] Dyrenforth left the next day for Washington with an inconclusive set of results, but clearly thinking and claiming that he had made a difference. He instructed his expedition, under the direction of John T. Ellis, to carry on in El Paso at the invitation of the mayor, as long as expenses were paid.

In need of munitions, Ellis contracted with the Consolidated Fireworks Company of North America in New York City for six dozen bombshell salutes, each weighing 21 pounds. He also bought 2,000 cubic feet of oxygen and 1,000 pounds of dynamite. The city of El Paso paid the $477 bill for equipment and shipping. The team conducted experiments in September about 1.5 miles north of the city center, on a ridge about 5,700 feet above sea level.

On September 18, a team of twenty-three artillerists fired at the sky all day long, in what one observer called "a beautiful imitation of a battle." Many prominent witnesses were in attendance for the event, including the mayor, the local weather bureau observer, curious citizens, and dignitaries from Mexico. They assembled on the ridgeline with their buggies and parasols to watch Ellis inflate his hydrogen balloon and ascend into the heavens (figure 2.1). Most brought their lunches and were treated to an all-day fireworks display. Witnesses reported seeing clouds and lightning flashes downwind at sunset (not at all unusual in this

2.1 Weather engineers and onlookers in El Paso, Texas, watching the inflation of the balloon in which John T. Ellis is to make his ascent. (*HARPER'S WEEKLY*, OCTOBER 10, 1891, 772)

climate). Ellis reported hopefully, "Soon after midnight rain had begun to fall *within a few miles of El Paso*, to the south and southeast."[40] Remember, however, that they were experimenting *north* of the city and the prevailing winds were from the south. In other words, whether the Ellis team was responsible or not, they were willing to take credit for any rain that fell anywhere in the region. The final bill presented to the city for one thundershower during the "rainy" season was $1,300.

From there, the team proceeded by invitation to Corpus Christi and San Diego, Texas, which were reportedly "suffering a severe drought"; according to Ellis, when the group arrived and before they could set up their equipment, "a heavy rain had set in from off the Gulf of Mexico and the weather continued stormy for several days." Still, they decided to bombard the rain-swept skies. Although many shells were detonated with no apparent effect, Ellis reported, selectively, that one explosion, in heavier clouds than usual, "was immediately followed by a downpour which lasted for several minutes and soaked the [observing] party to the skin before they could enter a carriage" (33).

According to long-term climatological records, West Texas was well watered in 1891, with rainfall up to an inch more than average. Or could it be that the amount was enhanced and the statistics skewed by the Dyrenforth team's purported successes? Lieutenant S. Allen Dyer, second in command of the expedition, concluded from his experiences that "rain can be produced by artificial means . . . and 'rainmaking' will prove a practicable and most valuable success when the conditions are favorable for rain" (41). Eugene Fairchild, an expedition member, testified: "I am convinced that the experiments have been entirely successful, and furthermore that the scheme is practicable" (53). But how practical is it to have more than twenty artillerists staying in a town at a cost of more than $1,500 just for materials? Nevertheless, some prominent citizens of San Diego said they were "astonished" at the results and were sure that the rain was a direct result of the experiment. Judge James O. Luby sent Dyrenforth his congratulations: "What was my surprise, after retiring for the night, to hear the patter [of rain] on the shingles; I then knew that, in the language of the festive cowboy, you had 'got a cinch on Old Pluvius,' and that the 'Powers' that be, go there with the limpid *aqua pura*" (55).

Nevertheless, Dyrenforth, who had spent $17,000 for the three experiments ($9,000 from the government and $8,000 from local sources and assistance in kind), sounded a note of uncertainty in his official report: "The few experiments which have been made, do not furnish sufficient data from which to form definite conclusions, or evidence upon which to uphold or condemn the theories of the artificial production or increase of rainfall by concussion" (57). Still, he ventured the following three positive "inferences":

> First, that when a moist cloud is present, which if undisturbed, would pass away without precipitating its moisture, the jarring of the cloud by concussions will cause the particles of moisture in suspension to agglomerate and fall in greater or less quantity, according to the degree of moistness of the air in and beneath the cloud.
>
> Second, that by taking advantage of those periods which frequently occur in droughts, and in most if not in all sections of the U.S. where precipitation is insufficient for vegetation, and during which atmospheric conditions favor rainfall, without there being actual rain, precipitation may be caused by concussion.
>
> Third, that under the most unfavorable conditions for precipitation . . . storm conditions may be generated and rain be induced, there being, however, a wasteful expenditure of both time and material in overcoming unfavorable conditions. (58)

To paraphrase all this, if you go to a dry area during a typically rainy season and conduct entertaining and impressive demonstrations but do not take any careful measurements, you can usually convince the eyewitnesses of your efficacy and, in turn, claim credit for any rain that does fall nearby.

The media had a field day with Dyrenforth's experiments. The *Nation* criticized the government for wasting tax dollars, observing that the effect of the explosion of a 10-foot hydrogen balloon on aerial currents would be less than "the effect of the jump of one vigorous flea upon a thousand-ton steamship running at a speed of twenty knots."[41] *Scientific American* pointed out that after the rainmakers had telegraphed from Texas to all parts of the country announcing the wonderful success of their bombs, it was discovered that the meteorological records for that locality had indicated probabilities for rain for a day or two in advance of the firing, and that the rain would have fallen all the same without any burning of powder or sending up of balloons. The article was accompanied by an illustration of traditional rainmaking in India and the cutting remark that there "seems to be little doubt that the swinging of a Hindoo head downward is just as effective for producing rain as the making of loud noises."[42] The *Farm Implement News* published a satirical cartoon of Dyrenforth and his team in action (figure 2.2).

F. W. Clarke's humorous "An Ode to Pluviculture; or, The Rhyme of the Rain Machine," published in *Life* in 1891, was undoubtedly inspired by Dyrenforth's experiments. In the poem, the hapless farmer, Jeremy Jonathan Joseph Jones, seeks to break a drought using

> cannon, and mortars, and lots of shells,
> And dynamite by the ton;
> With a gas balloon and a chime of bells
> And various other mystic spells
> To overcloud the sun.

His third shot into a cloudless sky "brought a heavy dew"; his fourth, tornadoes, "thunder, rain and hail." Jeremy drowned in the ensuing flood, and his farm is now a lake. All efforts to stop the deluge were in vain,

> Until the Bureau at Washington stirred,
> And stopped the storm with a single word,
> By just predicting—Rain![43]

Curtis, the meteorologist on the Dyrenforth expedition, ended his official report on a sour note: "These experiments have not afforded any scientific standing to the theory that rain-storms can be produced by concussions."[44] He thought

2.2 Robert St. George Dyrenforth claimed success after his federally funded rainmaking mission to Texas in 1891. After receiving a telegram from the weather bureau saying "Rainstorm approaching," Dyrenforth orders his assistants to speed up: "Hurry up the inflation, touch off the bombs, send up the kites, let go the rackarock; here's a telegram announcing a storm. If we don't hurry, it will be on us before we raise our racket." (CARTOON BY H. MAYER, IN *FARM IMPLEMENT NEWS*, SEPTEMBER 1891, 25)

it had only encouraged the "charlatans and sharpers" who were busily engaged in defrauding the farmers of the semiarid states by contracting to produce rain and by selling rights to use their various methods. But Senator Farwell, who had supported the experiments, was very upbeat in an interview with the *New York World*: "For twenty years I have had no doubt rain could be produced in that way, and quite expected the experiments to be successful. . . . When Prof. Dyrenforth makes his official report of these experiments, I expect that [the government will appropriate] $1 million, may be, or $500 thousand any way, for rainmaking."[45] Dyrenforth ultimately claimed victory and was actually reappointed as government rainmaker in 1892 to continue the work in San Antonio, Texas, with a grant of $10,000, although he spent less than half of that. He distanced himself from all the press coverage and hoopla, but claimed in his official report that his practical skills, combined with his use of special explosives "to keep the weather in an unsettled condition," could cause or at least enhance precipitation—when conditions were favorable! Not everyone was convinced, however.[46]

In 1891 Lucien I. Blake, professor of physics and electrical engineering at Kansas State Agricultural College, reviewed the Dyrenforth experiments and criticized the working assumption that concussion alone could make it rain. Blake noted that the effect of "air quakes" (basically energy from sound waves) should be immediate, yet Dyrenforth reported rain hours or days after the explosions. Perhaps, argued Blake, the smoke and particles from the explosions had a greater effect than the concussions. He pointed out that scientists had recently discovered that moisture does not condense in dust-free air but only in the presence of dust nuclei, or "Aitken nuclei." Blake further observed that every hailstone had a bit of dust in it and pointed to his own experimental seeding results with powders of carbon, silica, sulfur, and common salt that precipitated the moisture in a condensation chamber, and on burning sulfur and gunpowder to produce heavy, visible clouds of vapor.[47]

A year later, Blake proposed a field test to produce rain in the free atmosphere by raising, at intervals of about half a mile, a number of relatively inexpensive tethered balloons, each lifting a 30-pound smoldering ball of turpentine mixed with sawdust, straw, or paper pulp. These would generate a considerable smoke screen and might produce the right type of nuclei in the proper (but not excessive) concentrations needed for rain. Although he had insufficient funds for the field test, he claimed that his reasoning was based on sound laboratory experiments and would be much cheaper than Dyrenforth's elaborate explosive techniques.[48]

Observers from afar also commented on the explosive American rainmaking attempts. In *Transactions of the Epidemiological Society of London* for 1892, Sir William Moore noted that a rainmaker in New York had exploded 200 pounds of dynamite carried aloft by a balloon over the Croton Aqueduct and was immediately rewarded with a heavy downpour. He thought it "quite possible" to produce rain, since in his understanding clouds were "masses of minute vesicles" in an aeriform state. Their liquefaction could be caused by an explosion and the resulting compression that forces the moisture to coalesce, become larger drops, and fall as rain. Contrary to Powers, Ruggles, and Dyrenforth, all of whom maintained that concussive explosions could intervene directly in copious streams of invisible high-altitude moisture, Moore held that the amount of rain produced artificially would be insufficient unless clouds were already present, an unlikely situation during droughts in tropical lands. Instead, he recommended that governments invest in irrigation systems.[49]

One of the more colorful ideas for bringing down the rain at the time came from G. H. Bell of New York in 1880. He proposed building a series of hollow towers 1,500 feet high—one set of towers to blow saturated air up to cooler air and have the moisture condensed into rain, the other set to suck in rain clouds and store them for use as needed. The inventor considered that the same system could be used to prevent rain by reversing the blower so that the descending air might "annihilate" the clouds.[50]

Other explosive ideas were in the air as well. A weather patent to destroy or disrupt tornadoes was filed by J. B. Atwater in 1887. His device consisted of dynamite charges with blasting caps installed on poles and situated a mile or so southwest of a settlement. A tornado crossing the elevated minefield was supposed to detonate the explosives with its high winds and flying debris, hopefully disrupting its circulation and protecting the town. With the likelihood of a given area being visited by tornadoes rare and their recurrence even more rare, the installation of minefields, even elevated ones, never caught on—fortunately so for the generation of children then playing in the fields.[51]

The most improbable invention, however, belongs to Laurice Leroy Brown of Patmos, Kansas, who filed a patent application in 1892 for an "automatic transporter and exploder for explosives aiding rain-fall" (figure 2.3). The device was basically a large tower (*A*) with a sloping wire (*B*) connected to a battery (*C*) on which an operator can hang a stick of dynamite (*D*) on a pulley (*E*) and have it roll along a track until it completes an electrical circuit through a wire (*F*) and point (*G*) at the end of the track (*H*). The completed electric circuit was intended to ignite the dynamite and set off shock waves to stimulate rainfall, according to the ideas published by Edward Powers. Although erecting, and especially operating, such a device would certainly be a welcome diversion on the Kansas plains, possible design flaws include the danger to the operator of climbing a high metal tower with sticks of dynamite during an electrical storm and the apparent certainty that the first detonation of explosives at the end of the track would completely destroy the apparatus at the base of the sloping wire.[52]

* * * * *

In the nineteenth century, the scientific rain kings—James Espy, Charles Le Maout, Edward Powers, Daniel Ruggles, and Robert Dyrenforth—were altruistic monomaniacs who based their vision of a prosperous and healthy world order on the ultimate control of a single weather variable: precipitation. Grasping at scientific straws while posing as masters of an esoteric aerial realm, they

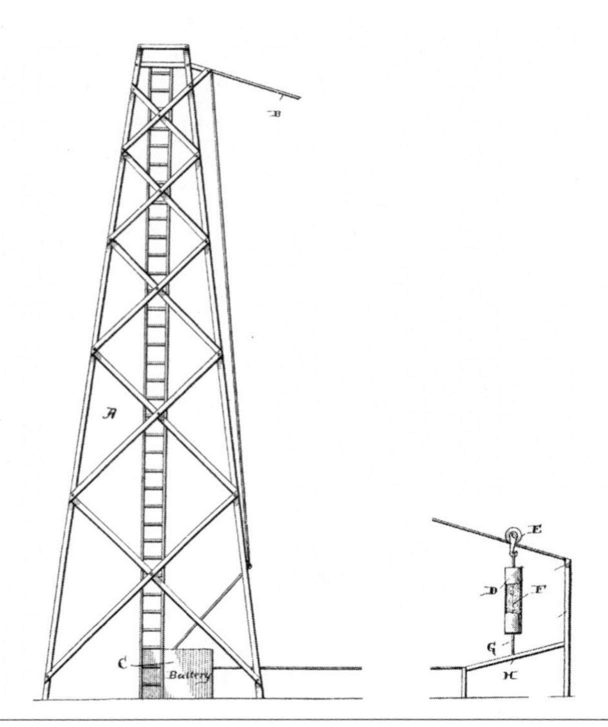

2.3 Tower and dynamite detonator proposed by Laurice Leroy Brown. Aside from the danger of climbing a high metal tower when storms are building, the dynamite (*D*) sliding down the sloping wire (*B*) would completely destroy that part of the apparatus. (ADAPTED FROM U.S. PATENT APPLICATION 473,820, APRIL 26, 1892)

appealed to the public's sense of the possible and, for funding, to the government's general lack of good sense. They wrote speculative books, brandished patents, and tinkered with their gadgets and toys, many of them incendiary or explosive, like children with firecrackers on the Fourth of July. It would be unfair to call them charlatans, since they explained their technical principles, experimented in the open (often with military surplus equipment), and avoided direct or deceptive marketing techniques. Yet there was often more hoopla than actual theory, and in lieu of results, their efforts produced perhaps less promise than hype.

Of course, things are different now, if only much larger in scale. Twenty-first-century climate engineers behave as, well, altruistic monomaniacs who base their vision of a prosperous and healthy world order on the ultimate control of a single climate variable: either solar radiation or carbon dioxide (chapter 8). Yes, things are truly different now. No longer do "climate kings" grasp at scientific straws while posing as masters of an esoteric aerial realm; nor do they appeal to the public's sense of the possible and, for funding, to the government's general lack of good sense. Or do they? There is no flood of speculative books, patents, articles, and gadgets regarding geoengineering. Or is there? Surely the "boys with their military toys" syndrome has long since passed. Or has it?

3

RAIN FAKERS

Among the many people who "live by their wits" there is a class who prey upon others subtly yet publicly. Their impelling motives, cupidity and desire for notoriety are stimulated by their vanity, and their rudder is hypocrisy. Although it is their business to live at the expense of others, it is not as parasites or fawning dependents; rather, they make dupes of their patrons, and they do this by pretending to possess knowledge or skill of a high order in some professional line. Their victims become their prey through sheer credulity and the predatory class [is known as] charlatans.

—DANIEL HERING, *FOIBLES AND FALLACIES OF SCIENCE*

"IT is not in human nature to suffer from a prolonged or repeated evil without seeking for a remedy"[1]—so wrote Daniel Hering in 1924 regarding weather control. In the struggle of the agriculturalist against hail and drought, that "remedy" was to seek new techniques for altering the weather. When the rainmaker mixed his proprietary chemicals and a sprinkle of rain touched the parched prairie, it was hard to dissuade the relieved farmers from believing that they had witnessed a miracle. Hering called this charlatanism an "old, familiar form of delusion"—*post hoc, ergo propter hoc*—and a weather control, "vagary." After the hail cannons were discharged with a mighty roar and the storm clouds dissipated, "it [was] hard . . . to convince the relieved grape growers that the cannons [had] not shot the storm away" (249).

The hoopla and hype of Robert Dyrenforth and his team could well be considered a form of charlatanism, except that they made some attempt, modest as it was, to explain their assumptions and they conducted their affairs without extensive marketing efforts. Like James Espy before him, Dyrenforth fits better into the sincere but deluded category of those who became overly enthusiastic about a single technique or theory. The hail shooters and the rainmakers who mixed secret chemicals, however, preyed on misguided hope and gullibility.

At War with the Clouds

Over the years, two basic approaches have prevailed concerning what to do when severe weather threatens: ceremonial and militaristic. Sacrifices, prayers, and the ringing of consecrated storm bells were favored by most until about 1750; since then, military assaults on the clouds have predominated. In ancient Greece, the official "hail wardens" of Cleonae were appointed at public expense to watch for hail and then signal the farmers to offer blood sacrifices to protect their fields: a lamb, a chicken, or even a poor man drawing blood from his finger was deemed sufficient. But woe to the negligent hail watcher if the signal was not given in time to offer the sacrifices and the crops were subsequently flattened. He himself might be beaten down by the angry farmers. The Roman philosopher Seneca mocked this practice as one of the "silly theories of our Stoic friends."[2] In Norse tradition, making a loud racket during storms was said to frighten away the demons of the storm. This was also a widespread practice among early and medieval Christians. A passage in the Bible about the "prince of the power of the air" convinced Saint Jerome that there were devils around when storms were about. Witches, too, were accused of causing bad weather. The *Compendium Maleficarum* (1626) contained an illustration of a witch riding a goat in the storm clouds. Throughout the Middle Ages, processions, often involving entire villages, were held in times of storm.[3]

Church bells were inscribed, consecrated, and even baptized. In his *Meteorological Essays* (1855), the noted French scientist and politician François Arago cited a number of traditional prayers that were recited during the installation of a new church bell, including the following: "Bless this bell, and whenever it rings may it drive far off the malign influences of evil spirits, whirlwinds, thunderbolts, and the devastations which they cause."[4] As well as calling the faithful to prayer and assembly and warning the community of invaders, the peals of the church bell were thought to agitate the air, disperse sulfurous exhalations, protect against thunder and lightning, and disperse hail and wind. The German playwright and lyric poet Friedrich Schiller placed as the motto of his famous "Song of the Bell" the Latin inscription customarily adorning many church bells: *Vivos voco; Mortuos plango; Fulgura frango* (I call the living; I mourn the dead; I break the lightning).[5] In Austria, it was traditional to ring "thunder bells" or blow on huge "weather horns" while herdsmen set up a terrific howl and women rattled

3.1 Medieval hail archers. (OLAUS MAGNUS, *HISTORIA DE GENTIBUS SEPTENTRIONALIBUS*, 1555)

chains and beat milk pails to scare away the destructive spirit of the storm. But is it dangerous to ring church bells during thunderstorms? Because a large number of bell ringers had been struck dead by lightning, Archduchess Maria Theresa of Austria banned the practice in 1750. The French government followed suit in 1786, but noted in its decision that the demons were still suspected of throwing lightning at churches. Still, bell ringers were well advised to avoid any proximity to or contact with a wet rope connected to a large metal object in a high tower during electrical storms.

Confronting the storm with displays of military might was also a venerable practice (figure 3.1). The mythical King Salmoneus of Elis, who traced his lineage to Aeolean roots, was an arrogant man who imitated thunder by dragging bronze kettles behind his chariot and hurled blazing torches at the sky to imitate lightning. It was his impious wish to mimic the thunder of Zeus as it rolled across the vault of heaven. Indeed, he declared that he actually *was* Zeus and designated himself the recipient of sacrificial offerings. Zeus punished this ridiculous behavior by striking him dead with a thunderbolt and destroying his capital city of Salmonia. His mistake of playing god brought down the wrath of heaven against him, but also triggered the annihilation of both the unjust and the just in his kingdom. In this case, imitation was not rewarded as the sincerest form of flattery. In the fifth century B.C.E., Artaxerxes I of Persia was said to have planted two special swords in the ground with the points uppermost to drive away clouds, hail, and thunderstorms.[6] In France in the eighth century, the populace erected long poles in the fields to do the trick. The poles were not anticipations of Benjamin Franklin's lightning rods, but were festooned with pieces of paper covered with magic inscriptions to protect against storms, a practice that the emperor Charlemagne regarded as superstitious. In the farm communities of central Europe, it was traditional to ignite gigantic heaps of straw and brushwood in advance of an approaching storm. The main effect of this was likely not meteorological, but it did foster a sense of shared risk and community engagement. Of course, the burning pyres contributed to the awesome spectacle of flashing lightning and pealing thunder.

In more recent times, according to Arago, nautical men generally believed that the noise of artillery dissipated thunderstorms and that waterspouts could be disrupted by the firing of cannon (figure 3.2). He mentioned the case of the Comte d'Estrées, who in 1680 fired on storms off the coast of South America and dissipated them, reportedly to the amazement of the Spanish witnesses. In 1711, however, a furious French naval bombardment in the harbor of Rio de Janeiro was followed by a tremendous thunderstorm (216).

3.2 Naval vessel firing its guns at a triple waterspout. (ESPY, *THE PHILOSOPHY OF STORMS*)

The practice of firing storm cannon apparently spread from sea to land. An entry on *orage* by Louis de Jaucourt in the famous French *Encyclopédie* of 1750 states that the dissipation of storms by the noise of cannon "does not seem out of all probability" and may be worth the cost of an experiment. By 1769 a retired French naval officer, the Marquis de Chevriers, had set up his battery in France to fight against strong hail and damaging storms. Ever the empiricist, Arago examined the weather records of the Paris Observatory, where, within earshot, regular gun practice took place for more than twenty years at a nearby fort. He found no effect of the cannonading on dissipating the clouds (214–218). By the middle of the nineteenth century, however, the opposite opinion—that the concussions of great explosions might make it rain—had garnered renewed public attention, but certainly not acceptance, through the work of Charles Le Maout, Edward Powers, Daniel Ruggles, and Robert Dyrenforth.

Hagelschiessen

For centuries, farmers in Austria shot consecrated guns at storms in attempts to dispel them. Some guns were loaded with nails, ostensibly to kill the witches riding in the clouds; others were fired with powder alone through open empty barrels to make a great noise—perhaps, some said, to disrupt the electrical balance of the storm. In 1896 Albert Stiger, a vine grower in southeastern Austria and burgomaster of Windisch-Feistritz, revived the ancient tradition of *hagelschiessen* (hail shooting)—basically declaring "war on the clouds" by firing cannon when storms threatened.[7] Faced with mounting losses from summer hailstorms that threatened his grapes, he attempted to disrupt, with mortar fire, the "calm before the storm," or what he observed as a strange stillness in the air moments before the onset of heavy summer precipitation.

Stiger gained notoriety on his first attempt. A gentle rain in his valley reportedly accompanied his shooting on June 4, 1896, with damaging hail falling elsewhere. He experienced a very militant summer, shooting at the clouds on forty different occasions. His hail cannon were constructed from 12-inch iron mortars (or pipes) and were loaded with a quarter pound of black powder; but some of them burst upon firing. Their replacements were made of steel with funnel-shaped chimneys taken from the smokestacks of worn-out railway engines, which the state provided to Stiger and others free of charge. The devices resembled megaphones pointed vertically and were installed on strong bases made of oak, some with wheels for towing. Later models had a steel ring welded inside

3.3 International Congress on Hail Shooting, 1901

the barrel to act as rifling, giving the discharged gases and smoke a distinct rotation and a whistling sound, said to be effective in agitating the air (figure 3.3).

Stiger erected lines of small huts overlooking his valley, with the funnels of the hail cannon protruding through their roofs. They were spaced about half a mile apart and were located along ridgelines. These "hail forts" were staffed by a small army of officers, artillerists, and signalmen who systematically fired at the clouds. The huts served the dual function of getting the shooters a bit closer to the clouds and keeping them and their powder dry so that firing could proceed even in the pouring rain.

A number of sentries occupied mountain watchtowers during hail season. Their assignment was to sound a warning so that the artillerists could break up the ominous calm before a gathering storm. The forward sentry was the town's telegraph operator, who kept a magnetic dip needle in his office. When the instrument behaved erratically, its agitation was taken as an indication of the presence of great electrical tension in the air. Messages from nearby towns might also warn of advancing storms. The telegraph operator spread the warning locally by first hoisting a red flag, which alerted carriage operators and other drivers to keep a tight rein on their horses in anticipation of the coming barrage. Then he fired a warning shot, which signaled the men to run to their posts to begin their fusillade of up to two shots per minute and up to a hundred shots per station per storm.

Although the efficacy of the system was never proved, the kaiser had a favorable opinion of it, and the technique spread to nearby countries. Some guns were sold in northern Italy by 1900, and some insurance companies decided to offer lower rates to growers within earshot of the hail cannon. Some provincial governments provided funds so that towns could appoint a general officer, instruct the artillerists, test and operate the cannon, and stockpile powder provided by the military. It was an exciting day in the neighborhood when the hail cannon started roaring. According to one commentator, the discharge was impressive: "From the mouth of the cannon issues a mass of heated gas, smoke, and smoke rings, propelled violently against the lowering cloud . . . like puffs of a locomotive, but with far greater energy of propulsion . . . a veritable gas attack in the realm of the aeronaut."[8] Even though no ammunition was involved, it was said that the power of the shot could kill small birds.

In 1907 the American meteorologist Cleveland Abbe, who had been publishing critiques of hail shooting since the turn of the century, reported the demise of the practice in Italy. A special commission of the Accademia dei Lincei in Rome had just issued a report that concluded, after testing more than two hundred cannon and other explosive devices through the course of five summer seasons, that there was no rational basis for expecting the noise, smoke, heat, or grand vortex rings to have any significant effect on enormous hail-generating clouds that extended over 30,000 feet in height. The study indicated that the vortex rings issuing from the hail cannon reached no higher than about 300 feet above the surface and had no influence on the storm clouds. The commissioners recommended that the Italian government no longer encourage "such expensive and useless work."[9] Although official support waned, the practice lingered, for hope springs eternal, and on occasion the clouds did disperse following a bombardment. Given the enormous sense of relief felt by the grape growers, it was hard to convince them that their artillery had not shot the storm away.

Contemporary hail shooters still make noise in farming communities on the Great Plains of the United States. In the film *Owning the Weather* (2009), Mike Jones and his crew discharged a radio-controlled stovepipe-shaped cannon nestled inside a corral padded with bales of straw. They claimed that the cannon's whistling "sonic boom disrupts the formation of hail" and lessens the chances of its formation. Jones was aware of the checkered history of this practice, but claimed that a revival was under way because of new technologies and "new understanding of the physics involved."[10] More-serious scientists were of the opinion that hail shooting gave a bad name to weather modification practices. In 1926 William Jackson Humphreys denigrated the practice in the epigraph of his book: "Trying to avert or destroy the hailstorm whether by scare or by prayer, by shooting or electrocuting, has been one of our fatuous follies from the earliest times down to the very present."[11]

Hurricane Cannon

William Suddards Franklin (1863–1930), a physics professor first at Lehigh University and later at the Massachusetts Institute of Technology, thought he understood atmospheric instability and how to use it for weather control. In 1901 he proposed to do something about hurricanes before they made landfall by exploding charges of gunpowder to initiate convection and thus dissipate a storm's source of energy before it could intensify. For Franklin, it was just an idea: "Please don't think that I have the machinery all designed and constructed to put this idea into effect. In fact I have made no experiments and do not know if the plan is at all practical."[12]

Franklin speculated about controlling the weather by using small amounts of judiciously placed energy. Just as an unstable brick chimney might collapse in a gust of wind, so, in an unstable atmosphere, it might be possible to trigger storms by exploding 5 to 10 tons of powder. Using the domino effect as a metaphor, he pointed out how turning a "number of grasshoppers" loose in a room full of dominos would surely result in their collapse.[13] Franklin was convinced that the atmosphere also responded to what he called "impetuous processes," such as a single spark causing a raging fire or the movement of a single insect setting off a storm:

> Imagine a warm layer of air near the ground overlaid with cold air. Such a condition of the atmosphere is unstable, and any disturbance, however minute, may conceivably start a general collapse. Thus a grasshopper in Idaho might conceivably initiate a storm movement, which would sweep across the continent and destroy New York City, or a fly in Arizona might initiate a storm movement, which would sweep out harmlessly into the Gulf of Mexico. These results are different surely, and the grasshopper and the fly may be of entirely unheard-of varieties, more minute and insignificant than anything assignable. Infinitesimal differences in the earlier stages of an impetuous process may, therefore, lead to finite differences in the final trend of the process.[14]

Such minute disturbances "may be the determining factor" in what Franklin called "atmospheric collapse," triggering the time and place of severe "domino storms" downstream.[15] Note that triggering unstable equilibrium processes is not the same as the butterfly effect, or more properly the Lorenz attractor, of Edward Lorenz's chaos theory. The set of nonlinear, three-dimensional, and deterministic solutions to the Lorenz oscillator, when graphed, resembles a butterfly. I once asked Ed if a butterfly could actually affect the Earth's general circulation, especially given viscosity. He smiled and said, "Perhaps if the butterfly was as big as the Rocky Mountains." I call this the "Mothra effect."

Franklin thought that the ultimate goal of meteorology was to devise a means for controlling storm movements by the suitable expenditure of energy at the critical time and place. Whether or not it could ever actually be realized, Franklin concluded, this was a "legitimate conception to say the least," well worth the attention of meteorologists. He praised the smoke-ring cannon of Burgomaster Stiger as a possible means for controlling all kinds of storm movements and thought it might hold the key to weather control. After laying out the mathematics of the forecast problem and the need for some future computer to solve thousands of equations simultaneously, Franklin proposed a more prosaic method of protecting Florida, by "touching off" the local energy of the atmosphere when a hurricane approached. Imagine a line of overgrown hail cannon along the coast consisting of twenty or thirty very large open steel cones, 15 to 20 feet in diameter at the base, 40 to 50 feet in diameter at the top, and each 100 feet high, with a ton or more of gunpowder per cone to be exploded. This would drive the air in the cone (60,000 cubic feet of it) upward as a kind of giant "smoke ring" that would start a rising column of air, thus stealing this energy from the approaching hurricane. Although such a project could cost several million dollars, according to Franklin, the people of southern Florida would benefit if they funded it.[16]

Abbe thought that Franklin's suggestions were "not the best that science has to offer." He pointed out that neither the concussions of cannonading nor Stiger's special vortex ring cannon had ever been proved to be effective. Abbe concluded, "The importance of unstable equilibrium in the atmosphere is a matter that has been so thoroughly investigated since the days of Espy that Professor Franklin has only to study the modern literature of meteorology and the mechanics of whirlwinds in order to realize the folly of his argumentation."[17] Abbe wanted experimental trial, not peasant-like faith. More than seventy years later, Ross Hoffman would again propose hurricane control using a distorted understanding of chaos theory as his guide (chapter 7).

Kansas and Nebraska Rainmakers

In the 1880s and 1890s, intermittent drought conditions in Kansas and Nebraska, some regionally severe, combined with economic turmoil and crop failures to encourage farmers to seek the services of rainmakers. According to climatological records, the Midwest received nearly normal rainfall in 1891. Kansas and nearby states, however, experienced a summer dry spell (but not a full-blown drought) that was threatening to stunt the crops. The farmers, seeking to be proactive, contacted rainmaker Frank Melbourne—known variously as "the Australian," "the Irish Rainmaker," or "the Ohio Rain Wizard"—who promised a soaking areal rain for $500. "Let every farmer who is able act promptly and contribute to this fund," advised the *Goodland News*, "and we will give to Goodland and Sherman county a valid boom such as they have never enjoyed before."[18] Plans were made for the rainmaker to be the star attraction of the county fair, along with horse racing, public speeches, and a grand evening ball. The governor and members of the State Board of Agriculture were invited as special guests, and the Rock Island Railroad announced reduced fares for all. Those opposed to the effort cited the hubris of meddling in the "Lord's business" and the dangers of unintended consequences such as setting off a tornado "that would blow the town from the face of the earth" (310).

The rainmaker arrived amid great fanfare, with his proprietary chemicals and rain machine. But he arrived slightly damp, since a period of unsettled weather had just begun and a light rain was already falling. Determined to collect his fee by wringing even more moisture out of the clouds, Melbourne (perhaps a model for Jeremy the "rain bat") proceeded to the fairgrounds, where he mixed his chemicals in solitude on the second floor of a mysterious shed, especially erected to his specifications. The shed was cordoned off by a 20-foot rope perimeter patrolled by Melbourne's brother, who remained on the ground floor as a sort of bodyguard and bouncer. The general public could do little more than gaze at the shed, hoping to catch a glimpse of the "cloud making substances" escaping through a hole in the roof. Melbourne built up anticipation by releasing reports from neighboring towns announcing major rainfalls downwind, for which he took full credit. Ultimately, however, he failed to deliver on his contract. His excuse, which many accepted, was that conditions were not right for rainmaking; the relatively cool nights and strong winds had rendered his chemicals and rain machine ineffective. Before leaving town for far-off engagements, never to return, Melbourne lined his pockets by selling his secret formula and copies of his rainmaking machines to local entrepreneurs. Soon three new enterprises—the Inter-State Artificial Rain Company, the Swisher Rain Company, and the Goodland Artificial Rain Company—were sending "rain-making squads" throughout the region.[19]

In 1894 the *American Meteorological Journal* reported that entrepreneurial rainmakers had succeeded in convincing a number of people, and even some paying clients, that they could, for a price and with the proper chemicals, draw down moisture from the arid skies. The *Kansas City Star* reported that the rainmakers possessed good timing, for they often commenced their experiments just as rain was due, convincing the gullible onlookers that their success was no coincidence. It did not hurt that, according to climatological records, rainfall was near normal in the region in 1893 and 1894.[20]

In those years, the Rock Island Railroad Company maintained a popular rainmaking department and hauled a special car along the tracks with an agent who claimed not to be producing rain but to be assisting nature in the task by supplying certain missing (but unnamed) elements to the atmosphere through concussions, gaseous mixtures, and electrical discharges. By 1894 the railroad had developed ten such rainmaking outfits, frequently deploying three units at a time to operate in tandem. Clinton B. Jewell, the railroad's chief dispatcher, offered his rainmaking services free of charge. His mobile rainmaking car, inspired by Dyrenforth's experiments and outfitted at company expense with what Jewell claimed were the secret chemicals and apparatus of Melbourne, rode the rails as a kind of traveling fireworks and vaudeville show, detonating dynamite, launching exploding balloons and rockets, and dispensing foul-smelling volatile gases charged with electricity, the last said to chill the air to enhance condensation. He promised to deliver "Kansas Weather" to his clients across the Midwest.[21]

Jewell gave reporters a tour of his car and a briefing on his procedures. He said his gas formula used "metallic sodium, ammonia, black oxide of manganese, caustic potash, and aluminum," these mixed with an "alloy known as murium," an imaginary radical thought to be an active agent in hydrochloric acid. These materials were both toxic and potentially explosive. When rain was to be produced, Jewell parked his car on a side track and filled an 800-gallon tank on the roof with water. Inside the car's laboratory was a wide shelf laden with bottles of chemicals and various sorts of apparatus. Under the shelf were large locked boxes, which were never opened in the presence of visitors. A second shelf supported a twenty-four-cell battery connected with wires to a very large jar. Another set of wires ran to the "rain machine proper," which consisted of six large jars grouped by twos in which the gas was made and from which it was released from the car through three pipes. Other pipes, bottles, and vessels completed the scene, making the car look like a small chemistry laboratory. Jewell explained that no force was necessary to send the warm, lighter-than-air, bluish gas into the sky: "When the rainmaking machine is in operation, 1,500 feet of gas escapes from each of the three pipes each hour. The warm gas ascends steadily over the span of four hours to an altitude of between 4 and 8,000 feet." After several hours, the gas inexplicably "turns cold instantly and drops with a rush, creating a vacuum, into which the moisture contained in the air rushes, forming clouds, and they form the storm center." Seeking a way to "make rainfall almost instantly," Jewell said he was working on an apparatus to send his gases up in liquid form enclosed within a shell, which, when it burst, would release the liquid, spreading it in all directions, instantly forming a large volume of cold gas. Jewell and his colleagues gladly took credit for any rainfall, near or far, that coincided with their operations. In at least one case, however, a hailstorm came up in Belleville, Kansas, that broke windows and outraged the locals, who threatened to sue for damages. Nevertheless, Jewell claimed that his trials frequently produced between 0.5 inch and 6 inches of rain, "each time contrary to the predictions made by the weather service."[22]

One widely publicized appearance of a rainmaker at a fair in Dodge City, Kansas, described a test of the liquid gas bombs:

> Shortly before noon, a special train pulled in bearing the rainmaking contraption on a flatcar. The apparatus was described as a monster mortar, "a sort of cross between a cannon of exceptionally large caliber and a giant slingshot." The workmen spent hours preparing the equipment for the demonstration. Thousands of people milled around the car, asking questions and offering advice. When the contraption finally was ready, an official of the railroad company quieted the crowd. He said that no one knew whether the apparatus would produce results. He pointed out that the company had the interests of the people at heart and was willing to spend its own money in an effort to produce rain for the district's crops. Chemical bombs were placed in the cannon and thrown into the air by the slingshot. A dozen or more bombs were discharged, emitting a cloud of yellowish smoke.[23]

Reportedly, the crowd was satisfied with the demonstration and fully expected to be drenched soon by a downpour. But nothing happened. The lasting result was equivalent to that of a good fireworks display—memorable but evanescent.

"An Unfortunate Rain-maker"

Harper's Weekly published a spoof of Kansas rainmaking in 1893 with its tale of "an unfortunate rain-maker," the fictitious Mr. Schermerhorn Montgomery, of

Hankinside, Kansas. Sooner or later, it was inevitable that something would go wrong. Montgomery got into legal trouble by causing a flood when he claimed to have made rain: "It did not seem possible that a man could go about carrying, as it were, thunder storms in one pocket and long steady rains in another, and not fall into some sort of a complication with common folks who do not have even a heavy dew in the whole house."[24]

Montgomery advertised that he made it rain only at night and on Sundays. He also claimed responsibility for cool northwesterly winds in the summer, but never charged for them. "I throw in a wind with each rain ordered," he explained, "the same way you get a baked potato when you order a chop. Fogs, frosts, cloudy days, and aurora borealis extra. Earthquakes should be spoken for two days in advance of the time needed" (735).

One morning after a particularly heavy rain, Montgomery set out to collect $1 from every farmer in the county for his services, but he met with considerable opposition. The first farmer somehow "knew that warn't no artificial rain," the second "reckoned it was a naterel thunder-storm," and the third demanded proof that the shower was a Montgomery special. At a public meeting, Montgomery addressed the skeptical farmers:

> "I produced that rain myself," said he. "It came, like all of my rains, in the night, when your hired man can't be put to any practical use. I saw the country needed rain, and I went out last night while you slept and made it. Consequently today your fields rejoice and your grateful cattle low their mellow thanksgivings from pastures revivified and gladdened by my beneficent rain." (735)

Following this oration, a corn farmer rose and asked Montgomery if he was absolutely certain that it was his rain. "Every drop of it," answered Montgomery. "Then," replied the guileless farmer, "you are responsible for the ten acres of my corn which the storm washed away. I shall sue you for damages" (735). And he did, to the tune of $400.

Adding editorially that "the science of rain-making is in its infancy" (which it always seems to be), *Harper's* noted that the business of artificial rainmaking (or, for that matter, hurricane diversion or climate engineering) would always be vulnerable to lawsuits that would be impossible to prevent and devastating to the enterprise: "A rain-maker, without his umbrella, standing in the middle of a vast Kansas prairie watching his rain pour down in torrents, and his patrons' crops ride gaily past on the hurrying flood and [with] no way to stop it, must be a most melancholy spectacle" (735). It seemed that Montgomery the rainmaker had not figured out a way to turn the rain off!

Charles Hatfield, the "Moisture Accelerator"

Charles Mallory Hatfield (ca. 1875–1958), who ran his proprietary operations mainly in the western states, garnered both widespread fame and quite some notoriety in the opening decades of the new century. Hatfield was born in Kansas and moved with his family to California as a youth, later working as a sewing machine salesman and eventually city manager of the Home Sewing Machine Company of Los Angeles. In 1898 he began to study meteorology; *Elementary Meteorology*, by William Morris Davis, was his favorite text, which he heavily annotated, and his favorite chapter, undoubtedly, was the one on the causes and distribution of rainfall.

Hatfield turned to rainmaking in 1902, trying his first experiments on his father's ranch in Bonsell, near San Diego. There he climbed a windmill and stirred and heated some chemicals in a metal pan, watching and waiting as the vapors rose into the sky. When a heavy rainstorm followed, it convinced him that his technique worked. He got into professional rainmaking on a bet, by claiming that he could produce 18 inches of rain in Los Angeles in the winter and spring of 1904/1905. Thirty prominent businessmen signed up to offer him $1,000 if he could accomplish this by May 1; the goal was exceeded a month early. Not that Hatfield had "done" anything. The long-term average rainfall in Los Angeles is 15 inches a year, more at higher elevations, and has ranged over the years from as little as 4 inches to more than 38 inches. Hatfield was lucky that year. The previous year's rainfall total had been a meager 8.7 inches; in 1904/1905, the year of his wager, it was 19.5 inches; and the following year, without Hatfield's involvement, it was 18.2 inches.

What Hatfield had "done" was erect a high tower near Esperanza Sanitarium in the San Gabriel Mountains above Pasadena and mix his noisome but ultimately harmless chemicals diligently throughout the winter. He believed that his technique worked best during the winter rainy season and at an altitude above 3,000 feet, two facts that he likely learned from Davis. When a reporter from the *Los Angeles Examiner* caught up with Hatfield in March, he described his theory as "a beautiful one":

> When it comes to my knowledge that there is a moisture-laden atmosphere hovering, say, over the Pacific, I immediately begin to attract that atmosphere with the assistance of my chemicals, basing my efforts on the scientific principle of cohesion. I do not fight Nature as Dyrenforth, Jewell and several others have done by means of dynamite bombs and other explosives. I woo her by means of this subtle attraction.[25]

His primary apparatus consisted of galvanized evaporation pans containing chemicals and water to be absorbed by the atmosphere, "where the fluid begins to work to attract and accelerate moisture." He also used a standard weather bureau rain gauge to document his results. His first tower was 14 feet square and 12 feet high, with a small opening underneath to create an updraft and thus assist the evaporation. Working with his brother Paul, Charles said he stayed up most of the night, with Paul coming on duty from four to eight o'clock in the morning. Then Charles would work again until six in the evening, sleep for three hours, and get ready for the next night. One of the brothers was constantly on watch. They had devised several alarms "for the detection of unannounced visitors during the night," and they kept a "small arsenal" inside their tent. Charles told the reporter, "I can assure you anyone who is looking for trouble will find it. I devote some time to hunting in the mountains." Hatfield said his technique was much more subtle and less noisy and flamboyant than those of his predecessors, but that he charged much more. He claimed that he never wanted to apply to Washington for a patent, "for that would mean the publication of information and rain-producers would spring up like mushrooms all over the country" (as they did after 1947). When asked about those who were skeptical of his methods, Hatfield quickly added, "Censure and ridicule are the first tributes paid to scientific enlightenment by prejudiced ignorance" (8).

Willis L. Moore, chief of the U.S. Weather Bureau, called Hatfield's method "fake rainmaking" and pointed out that widespread and "excessive" rains were prevalent throughout the West that winter:

> It is, therefore, apparent that the rainfall which was supposed to have been caused by the liberation of a few chemicals of infinitesimal power was simply the result of general atmospheric conditions that prevailed over a large area. It is hoped that the people of southern California will not be misled in this matter and give undue importance to experiments that doubtless have no value. The processes which operate to produce rain over large areas are of such magnitude that the effects upon them of the puny efforts of man are inappreciable.[26]

By operating in the climatologically established rainy seasons (usually in midwinter in California), by consulting U.S. Weather Bureau forecasts, by taking contracts in drought-stricken regions on the chance that conditions would improve, and by claiming success for any nearby shower, Hatfield was able to operate at a substantial profit. Billing himself in newspaper ads as a "moisture accelerator," he built his tall, mysterious towers, usually in the hills and often near a lake, and equipped them with large shallow pans in which he patiently mixed and evaporated proprietary chemicals—until it rained. He used the "no rain, no pay technique," with a clause in the contracts to cover his daily expenses in case of failure. Cynics said he was just betting his time against the expected fee that it would rain somewhere in the region during the contracted period. Hatfield's claims extended over an area that was about 100 miles in radius, which increased his chances of apparent success a hundredfold, compared, for example, with a circle merely 10 miles in radius. The careful reader will note that *any* rainmaking technique, traditional or technological, will be followed by rain in a large enough designated area if the practitioner is sufficiently persistent. It may take weeks or months, but it *will* rain—eventually, somewhere, and sooner if the technique is practiced during the rainy season. If you extend the spatial dimension to cover the globe, it is raining very hard somewhere on the Earth right *now*; and if you wait long enough, it will rain where you are. Hatfield also fielded requests to suppress the rain. The following appeal, published in the local newspapers, was addressed to him concerning the weather in Pasadena in January 1905 for the Tournament of Roses Parade: "Great moistener if you will listen now, And make this vow: Oh, please, kind sir, don't let it rain on Monday!"[27]

Hatfield plied his trade along the West Coast and into Canada and Mexico. In the summer of 1906, following a drought in the Canadian Yukon and after his initial success in Los Angeles, the provincial governors became an "easy mark" for Hatfield's self-promoting efforts. They awarded him a $5,000 contract for "meteorological experiments on the Dome," the mountain peak near Dawson. The largest mining concerns raised an additional $5,000 by private subscription. According to the contract, should Hatfield fail to produce sufficient rain to satisfy a board of seven evaluators, he was to receive only his cost of transportation and shipping to and from the Klondike and maintenance for himself and an assistant.[28]

These arrangements generated concerns in the Canadian Parliament a continent away in Ottawa. The Honorable George E. Foster, of North Toronto, was the most vocal: "Suppose that man Hatfield gets his apparatus to work and tinkers with the vast and delicate atmosphere of the universe; is it not possible that he may pull out a plug or slip a cog, and this machinery of the universe once started agoing wrong may go on to the complete submersion of this continent?"[29] And what if damage is done across international borders?

> If this government starts Mr. Hatfield shooting up into the sky, discharging his wondrous and mysterious combination of chemicals into the atmosphere and interferes with the vast chain of atmospherical mechanism to which the United

> States has some claim as well as ourselves, what about the Monroe Doctrine? . . . International complications, international conflagrations may take place, and for aught we know we may be involved in a tremendous bill for damages. (562)

Foster thought that the weight of scientific opinion was not in favor of Hatfield: "I believe the United States [Weather] Bureau . . . and they give it as their scientific opinion that he is an unmitigated fake and that anybody who has truck with Mr. Hatfield is very close to being bereft of good common sense. But that is only [the opinion of] a weather bureau. What is a weather bureau compared with the Yukon council and the Dominion government?" (563). The parliament ultimately decided that rainmaking was indeed the business of the local Yukon council, and Hatfield found a way to claim "success" for his efforts.

Hatfield was once described as "smiling, buoyant and fast talking, with a strong chin, large nose, high forehead and light blue, twinkling eyes . . . a quietly dressed, slender man of middle height with square shoulders, who is crowding forty" (figure 3.4).[30] By another description, he was "a man on a mission . . . ,

3.4 Closely guarded to keep the inventor's secret. Charles Mallory Hatfield's rainmaking plant on the shore of Chappice Lake, Alberta: "A deck surmounted by an open tank containing chemicals." The inset shows Hatfield. ("THE RAIN-MAKER: FIGHTING DROUGHT WITH CHEMICALS," *ILLUSTRATED LONDON NEWS*, FEBRUARY 4, 1922)

wiry, bordering on downright skinny . . . the greyhound narrowness of his face . . . exaggerated by a long, aquiline nose . . . yet . . . possessed of a quiet charisma, a patina of self-confidence that belied his unimpressive physiognomy. On occasion, when he was in full flow, his piercing blue eyes could take on the glaze of the evangelist."[31]

Hatfield is remembered largely because his rainmaking activities in January 1916 coincided with a severe flood in San Diego. According to city water department records, more than 28 inches of rain fell that month, the Morena Reservoir overflowed, and the Lower Otay Dam burst, sending a wall of water into downtown San Diego that killed dozens of people, left many others homeless, and destroyed all but 2 of the city's 112 bridges. Seeking to avoid lawsuits, the city of San Diego denied its connection to Hatfield, who had a vague contract for rain enhancement, and never paid him the $10,000 he claimed was due to him. Hatfield pursued the suit against the city for two decades before it was finally dismissed, without payment, in 1938.[32]

But Hatfield was not ready to cease his practice, and his services were sought across the country. In 1920 he took a contract in Washington State under the sponsorship of the Commercial Club of Ephrata. Hundreds of curiosity seekers gazed from afar at his strange tower on the shore of Moses Lake, from which mysterious gases were said to emanate. Nothing happened immediately, but soon after his departure the skies opened up, releasing a deluge. Skeptics saw no connection between the cloudburst and Hatfield's earlier efforts, but the miracle man claimed the rain as his own, bearing his private brand—although he did admit that it arrived somewhat late. The *Seattle Post-Intelligencer* reported: "The wonder worker himself must admit that his process is somewhat crude and unfinished when his storms wander all over the state, washing out orchards and bursting canals. Possibly some legislation may be necessary to compel the rainmaker to hog-tie his storms in the future."[33] He was back in Washington State a year later at $3,000 an inch, and collected $4,000 an inch from the United Agricultural Association of Alberta, Canada, until, after 2 inches of rain fell, he was asked to "turn off the faucets."[34]

In 1922 he took his equipment to drought-stricken Naples, Italy. American papers reported that after the rains came, he was received as a hero, and the Italian government tried, unsuccessfully, to offer him 1 million lire for his secret. Two years later, the authoritative *Monthly Weather Review* informed its technically oriented readers that the rainmaker had failed in California and had folded his tower and silently left the Bakersfield area after falling well short of his goal of producing 1.5 inches of rain in a month.[35]

Hatfield and the U.S. Weather Bureau had never been on good terms, although he knew the local officials personally and was a heavy user of weather bureau data, maps, and forecasts. He typically took contracts in areas that had experienced lower than normal precipitation and worked in seasons when rainfall might be expected to occur. This combination ensured that the local citizenry was desperate for rain, increased the chances of getting a contract, raised the price, and bettered the odds that average or above-average rainfall—for which Hatfield could take credit—was just around the corner. In 1918 Ford Ashman Carpenter, the weather bureau station manager in San Diego and Los Angeles, looked back on several decades of attempted rainmaking in southern California. Without naming names (but clearly alluding to Hatfield), Carpenter recalled that the rainmaker "possessed a limited education" and lacked the ability to differentiate cause from effect. Using a system of "no rain, no pay" but still always collecting his expenses in advance, the rainmaker typically operated in the rainier months of January and February, after a dry autumn. Carpenter concluded that by far the most important feature of the rainmaker's work consisted of playing on the credulity of the people: "It is therefore a psychological rather than a meteorological problem, for the fundamental factors are those of the mind and not of matter."[36] It is in this sense that Hatfield served as the model for Starbuck in the Broadway play *The Rainmaker* (1955). He was even invited to its Los Angeles premiere.

Betting on the Weather

In the early 1950s, more than $2 million in legal claims were filed against New York City by upstate residents for purported damage caused by the cloud-seeding efforts of Dr. Wallace E. Howell over the Catskill Mountains reservoirs. Although the lawsuits were eventually dismissed because of technicalities, an elderly raconteur and bon vivant, Colonel John R. Stingo, who often referred to himself as "the Honest Rainmaker," was astonished that men of science at that time were becoming targets of damage suits and hard feelings, when decades earlier his own rain-inducing efforts had generated nothing but good feelings for all involved. The noted *New Yorker* columnist A. J. Liebling caught up with Stingo (whose name means literally "strong brew") at a series of Manhattan watering holes and heard his creatively embellished, colorful, improbable, and possibly misleading stories of how in yesteryear he had lived by his wits and bet with the odds (but never with his own money) on prizefights, on the horses (when betting at the track was outlawed), and, of course, on the weather. For him, rainmaking was a confidence game and not at all a scientific endeavor. Liebling described Stingo as neatly dressed, short of stature, lively, and quick of wit, with the air of an old military man; his habitual expression "that of a stud-poker player with one ace showing who wants to give the impression that he has another in the hole";[37] his typical regimen a series of golden gin fizzes with egg yolk in the morning, hard liquor at midday, and then beer and wine for the duration.

Stingo, a weaver of tall tales, claimed that as a youth in 1908 he had witnessed a memorable but ultimately futile rainmaking extravaganza in the Lower San Joaquin Valley. To save his wheat crop, Captain James McKittrick had invited Egypt's leading rainmaker, (the fictional) "Sudi Witte Pasha," and his entourage of twenty-two professors and holy men, and assorted cantors, priests, bell ringers, soothsayers, dancing girls, chefs, servants, and bodyguards, to his (also fictional) 212,000-acre estate, "Rancho del McKittrick." Their rainmaking technique consisted of chants, prayers, ablutions, and dancing, lots of dancing, over the course of three days. On the fourth day, the pasha and his crew scattered ground-up kofu beans from ancient Persia in the fields and hosted a feast for three hundred guests, an "Orgy in Imploration for Rain," that lasted into the fifth day. After several more days of waiting, with no rain in sight, the formerly jolly Captain McKittrick, who was out about $200,000 in expenses, decided to ship the pasha and his entourage to the nearest railroad station, thence "to the outgoing Pelican Express for Phoenix, Fort Worth and New Orleans," and finally by steamer "from the Crescent City through the Straits of Gibraltar to the palm-waving beaches of dear old Cairo" (12–13). Stingo was impressed by the pasha's show but judged his timing unfortunate in that the Fates did not deliver normal rainfall that week. He took away from the experience the impression that an American market might require a show with less exoticism and more displays of cold science and impressive paraphernalia, perhaps with a spiritual note.

The colonel then related his early efforts out west in 1912, when he was the front man or setup man for the rainmaking show of "Professor Joseph Canfield Hatfield" (again a made-up name). To clinch a deal, Stingo (at that time working under his given name, James A. Macdonald) would warm up a crowd of farmers with a version of the following speech:

> Rain—its abundance, its paucity—meant Life and Death to the Ancients, for from the lands and flocks, herds, the fish of the sea, the birds of the air, the deer and mountain goat they found sustenance and energized their being. All the elements depend upon the Fall of Rain, ample but not in ruinous overplus, for very existence. Through human history the plentitude of Rain or its lack constituted the difference between Life and Death, the Joy of Rain or existence and misery. (11)

Next onstage was the director of ordnance, Dr. George Ambrosius Immanuel Morrison Sykes, who would explain how barraging the clouds with cannon and a rapid-fire Gatling gun would wring out their moisture. Then J. C. Hatfield himself, not a particularly eloquent or convincing man but a true believer in his techniques, would mumble something about how Marco Polo had returned from Cathay with an explosive yellow powder and stories of its use as a rainmaking device in ancient China. Finally, it was up to Macdonald to close the deal (or set the hook) by getting the local farm officials to sign a contract for "detonationary services" with the Hatfield Rain Precipitation Corporation, at $10,000 an inch of rain for up to 3 inches and $80,000 for a full 4 inches. Stingo confessed to Liebling, "All first-class Boob Traps must contain a real smart Ace-In-The-Hole," and the rainmaking company's consisted of converting weather bureau tables, charts, and rainfall averages into a set of betting odds, a kind of pari-mutuel handicapping that the company estimated to be 55 percent in its favor. He then described how 5,000 people had turned out on a hot, dry afternoon to watch the team set off the ordnance show, how Hatfield climbed the mountain like an Old Testament prophet, how Sykes's spouse gathered the faithful together to pray for rain, how the guns roared and the smoke billowed, how a storm came up at midnight and drenched the valley, how the rainmakers took credit for this, and how the local populace subsequently prospered, praising Hatfield's powers and paying the rainmakers the $80,000. Macdonald's cut was $22,000. The team repeated the show, successfully, in Oregon the next summer, but business tapered off in subsequent years after the farm guild bought its own cannon (16–19).

We know from newspaper accounts that one of Stingo's stories was based in fact. In 1930, in the depths of the Great Depression and Prohibition, Stingo was working the crowds at Belmont Park, living by his wits, and writing a horse-racing column for the *New York Enquirer*, an unpaid position but one that kept him in circulation. During a rainy week in early September, he spotted his old comrade in rainmaking, Sykes—now a "minister of Zoroastrianism," a flat-earther, and another Wizard of Oz—who was in the process of trying to convince the track officials (and other "solvent boobs") that his California-based U.S. Weather Control Bureau could *prevent* rain during the racing season and save the track from bankruptcy (29–35). Sykes, who used neologisms freely, claimed that he controlled the weather through "dynurgy, xurgy, psychurgy . . . isogonic force, quantumie . . . Bolecular energy, freenurgy—especially freenurgy . . . and thermurgy."[38]

Soon Stingo and Sykes were back in business as equal partners: Stingo was to run the office, calculate the odds of rain in September, do most of the talking, and close the deal; Sykes would act as the incomprehensible true zealot and set up and operate his mysterious rain machine (which could be run in reverse, he said, to drive away the rain). The deal included a 10 percent cut for Mrs. Sykes, not for leading the prayers, as she had done in 1912, but for improving public relations by mixing pitchers of "Pisco Punch," a legendary smooth and fruity drink featuring Peruvian pisco brandy, made popular during the California gold rush.

Stingo sweet-talked the millionaire sponsors of this project, Joseph E. Widener, president of the Westchester Racing Association, and his female associate, the "long, lissome, and lucreferous" Mrs. Harriman, whose critical weaknesses included belief in the occult and "an inquiring mind." Then he made his pitch: "We, the U.S. Weather Control Bureau . . . agree to induct, maintain and operate the Iodine Silvery-Spray and Gamma-Ray-Radio system" for a payment of $2,500 for each of two dry Saturdays and $1,000 for each intervening weekday, with a forfeiture of $2,000 for every day it rained, the track to pay for the buildings and labor. The rain suppressors stood to make $10,000 if every day was clear and to lose $4,000 if it rained every day. They calculated the climatological odds as 0.7499 in their favor.[39]

Soon Sykes began installing his equipment. The "Vibratory Units and the Chemicalized Respository" were located in an abandoned clubhouse and the "Detonary Compound" in a one-room, five-sided, windowless shed at the other end of the track. Sprouting from the roof of the clubhouse were two rods of shiny steel; inside were an old Ford Model-T engine serving as the Vibrator, an impressive mass of wires, batteries, long rows of gaudy glass jars (probably filled with colored water), and a washtub full of evil-smelling chemicals. The shed looked like an ornate "cabalistic" pentagram covered with a spiderweb of wires. Padlocked doors and security guards confronted the curious. As described at the time by *New Yorker* racing columnist Audax Minor, "On the roof was a big five pointed star strung with radio aerial wire and festooned with ornaments from discarded brass beds and springs from box mattresses. The star always faces the way the wind blows" (41). Over the roof of the shed Sykes had constructed a small platform where he could stand to "direct magnetic impulses" and conduct the show.

The two buildings stood about a mile apart and were linked, as described by Stingo, by the "Ethereal Conduit upon which traveled with the speed of Light augmented 30,000-fold the initiatory Pulsations the Vibrator, and thence, via the antennae, to the natural Air Waves and channeled Coaxial Appendixtum" (42). The press played along for a while, publishing accounts of the mysterious equipment and reporting Sykes's claim that he had "one of the most powerful radio installations in the Western Hemisphere," which led local residents to wonder if static electricity generated by the equipment was interfering with their radio reception. A visit from representatives of the Federal Communications

Commission revealed that to run his contraption, Sykes had "borrowed" some 32,400 kilowatts of electric power daily from the grid without the knowledge of Consolidated Edison; this resulted in local power failures and an official warning to Sykes not to do that again.

The weather remained fair and dry that week, save for a light mist on the final Saturday, and the Weather Control Bureau netted $8,000, with additional income derived from side contracts and bets, some with local mobsters. To celebrate, Mrs. Sykes threw a party for several hundred people at the Hotel Imperial in Manhattan, replete with chamber music, dancing long into the night, and endless bowls of Pisco Punch. All was well, but it did not end well. When reporters accused Sykes of just being lucky, he announced, perhaps after one too many drinks, that he would prove his power by throwing his machinery into reverse to produce torrents of rain between 2:30 and 4:30 on the next Monday afternoon. The odds were heavily against him; Stingo likened it to "breaking up a full house to draw for four of a kind." It did not rain that afternoon, and a "deluge of derision" broke over Sykes and the Weather Control Bureau, resulting in the loss of pending future contracts at Belmont and Churchill Downs (47–48).[40] Thus, according to Stingo, the rainmaker's art was eclipsed, not to be revived for a score of years, next time not with old Ford motors, radios, and secret chemicals, but with airplanes, dry ice, and silver iodide (chapter 5).

Seeding the American West

Irving P. Krick (1906–1996) was a talented, charismatic "rainmaker" in both the business sense and the meteorological sense. The term that adheres most readily to him is "maverick." Krick was a child prodigy on piano as well as a student of physics. After completing his doctorate in meteorology at Caltech, he helped establish the university's Department of Meteorology, but he lacked a strong theoretical background. The program he developed emphasized the training of applied meteorologists, especially for the rapidly growing airline industry. Krick himself spent most of his time developing the Krick Weather Service, of Pasadena, California, using department space and weather bureau equipment. He specialized in speculative ultra-long-range forecasts, which the U.S. Weather Bureau considered doubtful. Krick gained a moment of fame by forecasting calm winds for the set of *Gone with the Wind* the night the burning of Atlanta was filmed. His forecasts were based on so-called analog methods using data from historic maps, which he codified for use with a simple slide-rule gadget. The forecasts were also tailored to be just what the client wanted to hear. Filmmakers favored rainfall, since they saved money on the days it rained and they did not shoot outside, so Krick provided them with "wetter" forecasts than normal. The hydropower company Edison Electric, on the contrary, preferred dry forecasts that favored water conservation in its dams, and Krick was glad to oblige by predicting clear skies for the company whenever he could.[41]

Krick's use of analog forecasting techniques almost led to disaster in World War II, when, serving in the U.S. Army Air Force as one of six principal Allied forecasters tasked with predicting the date of the D-day invasion of Europe, he urged that the invasion proceed on June 5, when winds in the English Channel would have swamped the Allied invasion force. Ever the self-promoter, Krick later tried to take credit for the actual June 6 forecast, but more-balanced accounts indicate that the undertaking was truly a group effort, with Krick again playing the role of a maverick.[42] The Norwegian meteorologist Sverre Petterssen (1898–1974), who was centrally involved, later expressed his opinion of the situation:

> I knew Krick very well. In 1934 he had spent about two weeks with me in Bergen, [Norway] and in 1935, [when I was] a visiting professor at California Institute of Technology, I had worked with him for a period of four months. I considered him a very able, intuitive forecaster who could rise to considerable heights if he would dig deeper into the theoretical background of weather prediction. . . . However, wisely or unwisely, Krick took a liking to industrial applications and offered his services first to the film industry and later to any industry, anywhere. Krick's main protector at Caltech was its President, Dr. Robert A. Millikan, who had organized U.S. weather efforts in the First World War. Millikan was a top level science advisor and confidant of General [H. H.] Arnold, the Commanding General of the U.S. Army Air Corps. . . . I knew that Krick, after a brief service in the U.S. Navy, had transferred to the Army Air Corps and that his long-range forecasting system had some kind of official sanction there. General Arnold saw to it that many of the senior air weather officers were sent to Krick to study his techniques. . . . I had little confidence in any system of mechanical selection of analogs and I thought it . . . would not be difficult to look up the true meaning of the word "quackery" and then ignore the forecasts altogether.[43]

Krick returned to Caltech after the war, but, according to its former president Lee A. DuBridge, "everybody around the campus, and other meteorologists and other scientists around the country, said that Krick was a fake." DuBridge wanted Krick's department to deemphasize long-range forecasting and proprietary methods and to focus on "a real study of the physics of the atmosphere." He ordered a review of Krick's work by scientific elites Vannevar Bush, Karl Compton, and Warren Weaver, who concluded that Krick "claims to do things that he can't do [and] claims to have done things he didn't do." About the same time, the weather bureau accused Krick of using its equipment on loan to Caltech for commercial gain. In 1948 DuBridge discontinued the meteorology program at Caltech and accepted Krick's resignation, but Krick already had a new job, as a commercial rainmaker.[44]

Soon after the General Electric Corporation's cloud-seeding experiments, Krick requested and received a set of GE reports on weather research. He followed with interest the saga of Project Cirrus and Irving Langmuir's claims for the silver iodide generator in New Mexico (chapter 5). Krick visited GE in February 1950, seeking advice on the latest cloud-seeding technologies, but the visit was tense. He had been marketing his rainmaking projects out west by representing himself as having an unofficial relationship with GE, while dropping the names of Langmuir, Vincent Schaefer, and Bernard Vonnegut with his customers as if they were his close colleagues. When the science editor of the *San Francisco Chronicle* contacted GE in conjunction with a story he was planning on Krick's rainmaking claims, he was informed that there was "no connection whatsoever" between Krick and GE, other than supplying him with background material as was done for many others.[45]

During the western drought of the early 1950s, Krick began cloud-seeding operations for large agricultural concerns. His clients included wheat farmers, ranchers, and stream-flow-enhancement projects on the Salt River in Arizona and the Columbia River in the Pacific Northwest. In the latter project, Krick was credited by the Bureau of Reclamation with an 83 percent enhancement of the river flow, but the weather bureau considered this claim meaningless and sought to discredit him whenever possible. At the height of its operations, Krick's company was conducting seeding operations covering 130 million acres of western lands, in all the areas where Charles Hatfield had operated (figure 3.5).[46] By 1954 cloud seeding had been attempted in about thirty nations, including Australia, Canada, France, South Africa, Spain, Peru, and Israel, and that year a total of fifty-seven commercial cloud-seeding projects were under way in twenty-five states and territories, including Hawaii and Puerto Rico.[47]

Later, Krick snagged a contract to run silver iodide generators in Squaw Valley for the 1960 Winter Olympics and claimed that the deep snow pack was in part the result of his efforts. He remained active in ultra-long-range forecasting, expressing his belief in an orderly universe (and atmosphere) and the sorting out of its regularities through the use of analog methods and digital computers. Echoing a common sentiment about long-range prediction at the time (chapter 7),

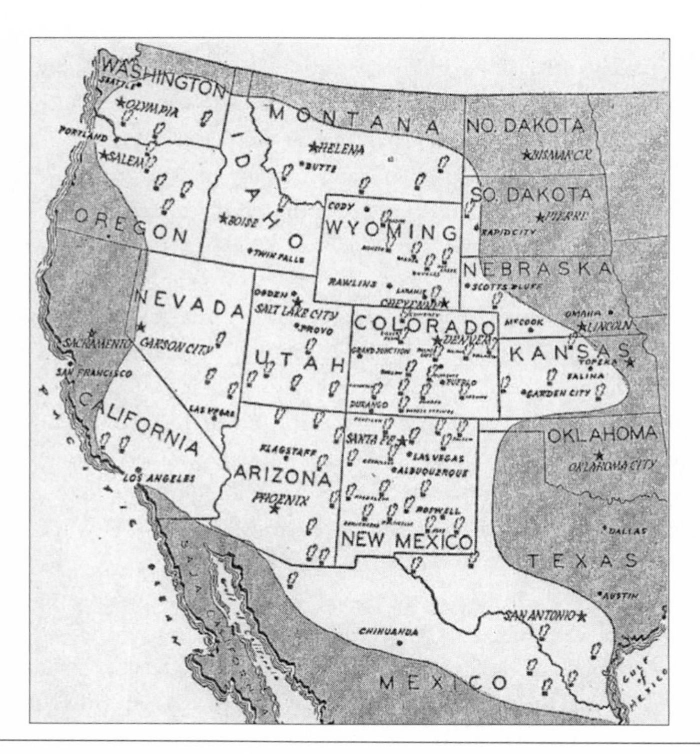

3.5 Irving Krick's generators for cloud-seeding operations in seventeen western states and Mexico. (WILLARD HASELBACH, "'RAIN MAKER OF THE ROCKIES': HISTORY'S BIGGEST WEATHER EXPERIMENT UNDERWAY," *DENVER POST*, APRIL 22, 1951, 17A)

Krick proclaimed, "Give me enough time, men, and electronic computers and I'll tell you the Newfoundland weather for 200 years from now."[48] He also claimed, echoing the surety and determinism of the famous mathematicians Gottfried Leibniz and Pierre-Simon Laplace, "If we had precise information back to the Ice Age, we could pinpoint the weather at 3:10 P.M. on March 11, 3004 in Tokyo"—a prediction no one would be around to verify.[49] Although he said that he had voted for Richard Nixon in the 1960 presidential election, Krick offered a free (and lucky) long-range forecast for the inauguration of President John F. Kennedy: "fair, cold, and dry." Controversy continued to follow Krick, however. He resigned from the American Meteorological Society (AMS) after being accused of making unsubstantiated claims for his forecasting methods and for violating the society's code of professional ethics, but he rejoined in 1985, because, he said, he had outlived most of his enemies. His necrology in the *Bulletin of the American Meteorological Society* constitutes a study in understatement.[50]

Deadly Orgone

In 1951 a near fatal experiment with radium led Wilhelm Reich (1897–1957), an eccentric Austrian-born physician and practicing psychoanalyst, to conclude that he had discovered a new type of proprietary energy he called "deadly orgone," which, in its material form, he claimed, appeared in the air as black toxic specks that could calm the winds, cause tree leaves to droop, silence the birds and insects, and even sicken humans. The following year, Reich invented a "cloudbuster," a cluster of hollow pipes resembling a Gatling gun, to attract and remove the deadly orgone from the atmosphere. Running water through the tubes served to rinse out and drain off the accumulated toxins. Or so he claimed.

Reich, who had worked with Sigmund Freud on human sexuality in the 1920s, moved to Germany in 1930 and joined the Communist Party, seeking to combine social theory and personal liberation from sexual taboos. When the Nazis came to power, Reich was forced into an itinerant life in a number of Scandinavian countries, where he experimented, using basic electrical equipment, on what he termed "bioelectric energy." His experiments led him to believe that he had discovered a fundamental motive power of the universe, which he first called "bions" and later "orgone energy." He postulated that this energy permeated all life and was also present in the atmosphere. Moving from Norway to the United States in 1939, Reich lectured on the psychological aspects of orgone energy and devised a simple device he named the Orgone Energy Accumulator to demonstrate his theories on both healthy and cancerous tissue. In the late 1940s, accused of fraud and suspected of conducting a sex racket, Reich moved his operation to a remote location in Rangely, Maine, to an estate he called Organon. It was here that he discovered "deadly orgone."[51]

Reich claimed to be able to prevent or produce rain wherever he pointed his cloudbuster; he even devised a smaller-scale medical device that he pointed at his patients! After all, isn't it either raining or not raining all the time? And aren't patients either mostly healthy or unhealthy? An eyewitness to a demonstration in Maine in 1953 reported: "The strangest looking clouds you ever saw began to form soon after they got the thing rolling."[52] Maintaining Reich's legacy, a dedicated band of enthusiasts is currently clearing the air of "chemtrails," with homemade cloudbusters constructed from copper pipe, quartz crystals, and metal filings. They are "repairing the sky."[53] They do so at the risk of their health, however, since plans published on the Internet do not include a drain for the deadly orgone. Use of this device will be followed by rain or clearing skies—your choice.

Provaqua

If Charles Hatfield were active today, he might be working for Earthwise Technologies, trying to peddle the company's ion rain project. Unsung heroes often emerge, however, to expose the charlatans and to contest unsupportable claims. Richard "Heatwave" Berler, a television weatherman in Laredo, Texas, deserves to receive a journalism award for using moments stolen from his nightly weathercast to confront the charlatans and reveal the madness. In late November 2003, in response to an unsolicited proposal, the Webb County Commissioners Court issued a contract to Earthwise Technologies for rainmaking in the vicinity of Laredo. The project, called Provaqua, involved building four large ion-generating rain towers spanning the Rio Grande watershed at a cost of up to $5 million. Webb County taxpayers were asked to pay $1.2 million, with the balance coming from Mexico.

Earthwise, a sole proprietorship operating out of Dallas, Texas, was promoting an unproven Russian technology known as IOLA (ionization of the local areas). Three years earlier, the company—or, more accurately, Steven Howard, its president and sole employee—made an unsuccessful bid to install up to twenty-five "ionization platforms" in the Houston–Galveston area, a heavily populated region and, according to the Environmental Protection Agency, a non-attainment area for air pollution. For a fee of $25 million a year, he offered to clear the region's air of particulate matter and reduce concentrations of ozone near the ground. According to Howard, the company's patented IOLA technology would create an ascending "convection chimney" to draw in polluted air and disperse pollutants more rapidly and at greater heights than occurs naturally. Much like a giant home air purifier, Howard explained, the devices would help precipitate heavier particles and could mitigate the formation of ground-level ozone.

The Laredo project claimed to be able to harness and redirect natural atmospheric energy processes in the Earth's hydrological cycle. According to Howard, clouds were not necessary to produce rain. Ions floating up from the tall electrified towers that his company proposed to erect would cling to humidity in the air, generating clouds and producing a slow, gentle rain. The ions would also attract new "aerial rivers of moisture" from the Gulf of Mexico and would disperse pollution and freshen and purify the air. In a presentation to the commissioners court, Howard further explained that IOLA "changes the electrical charge of water vapor, thereby speeding up the natural velocity of condensation." Earthwise offered to generate a 15 to 20 percent minimum increase in measurable rainfall, with a maximum 300 percent increase.[54] Local TV channel KGNS interviewed the excited Webb County chief of staff, Raul Casso, who explained that society had wanted to create rain for centuries and that he believed it was now possible: "Making it rain . . . has always been one of man's age-old aspirations. . . . [Y]ou have [dowsing] forks and diviners, and rain gods and all sorts of things that people have done to try to evoke rain; but you can't do it—until now."[55] Heatwave Berler, however, smelled a rat.

In the closing moments of one of his evening weathercasts, Heatwave humbly expressed his concerns about the project, saying that he was not arguing that it was impossible for the project to work, just that there was no evidence of it working. He interviewed Casso, asking him if the county commissioners had sought the opinion of any scientists before making the decision to spend $1.2 million of the taxpayers' money on the project. Casso initially listed the various civic groups they had talked to, but eventually admitted that no, they had not asked any scientists. Heatwave's questions generated a list of explanations from Howard (doing business as Earthwise Technologies), all of which Heatwave systematically debunked.

Heatwave, now fully engaged with the issue, used his weathercast to express his concerns about the lack of peer-reviewed articles and improper documentation provided to the commissioners. He also found it interesting that Howard, like Dyrenforth and Hatfield long before him, was trying to make it rain during the naturally occurring rainy season. Earthwise was claiming experimental successes based on only one year of precipitation measurements, a timeframe that Heatwave stressed was much too short when dealing with weather, especially when rainfall amounts in different years and in different locales can vary by as much as an order of magnitude. He drew the analogy to tossing a coin once and then concluding that all coin tosses would have the same outcome. When Heatwave discovered that similar projects elsewhere had been terminated due to lack of evidence, Earthwise Technologies responded that there was a lot of research and articles on the methodology, but that unfortunately it was all in Russian and had not been translated. This puzzled Heatwave, since the American Meteorological Society and the World Meteorological Organization, to name only two organizations, had a long history of cooperation with Russian meteorologists and issued reports and abstracts in translation to overcome language barriers. The absurdity of the situation spurred a spoof advertisement for "Dud Light" on the local radio, the gist of which was that for only $5 million you can get a machine that magically makes rain, with instructions in Russian and with the guarantee that it has failed to work everywhere else it was tried.[56]

The fiasco ended in a dramatic Webb County Commissioners Court meeting in December 2003 during which Judge Louis Bruni aggressively and embarrassingly supported funding. He was voted down by the county commissioners because of the overwhelming rejection of the project from their voter constituency, largely brought about by Heatwave's investigations. A local magazine, *Laredos*, summarized the mood of the meeting: "While the early minutes of the meeting were glossed by a thin patina of civility, the proceedings quickly degenerated into a side show of blatant disdain, sarcasm, chicanery, the rearing of ugly heads, a couple of juggling acts, patronizing platitudes, and for some on the sidelines of county government, incredulity that public leaders conducted county business in this manner."[57] Humble Heatwave Berler had stood up to and defeated the rainmakers, saving the county and the region millions of dollars and further embarrassment.

* * * * *

Hail shooting to protect a crop and rainmaking in times of drought are usually considered to be desperate acts by desperate people. But there are other dimensions, both cultural and psychological. One is the solidarity of a community trying to do something, anything, to augment Providence. Another is the sheer entertainment value of a traveling rainmaker's entourage coming to town with its mysteries, loud fireworks, and showmanship. Many times, people do both: pray *and* hire a rainmaker. Charles Hatfield undoubtedly turned a profit by working with the moist air masses provided by nature and predicted by the weather bureau. John Stingo and George Sykes combined climatology, handicapping, and complicated apparatus in executing their confidence game. They, like Clinton Jewell and others, kept their secret techniques under close wraps. Others, like Frank Melbourne, made their money by selling their secrets as a kind of franchise operation to the highest bidders.

Common traits of successful charlatans include seeking financial gain by taking credit for natural rains. Little to no capital and no business training are needed. A sense of ethical responsibility or long-term engagement with a community may be detrimental. Use of the latest technologies, juxtaposed in odd and mysterious ways with claims of esoteric knowledge, and recitation of a scientific mantra also seem to help.

Practicing meteorologists were uniform in their criticism of rainmaking and hail shooting. In 1895 meteorologist Alexander McAdie wrote: "Rainmakers of our time bang and thrash the air, hoping to cause rain by concussion. They may well be compared to impatient children hammering on reservoirs in a vain effort to make the water flow."[58] Weather forecaster Ford Carpenter's examination into the methods of the rainmaker revealed "a disregard of physical laws," with no proof or prospect of success;[59] and Cornell University president David Starr Jordan ridiculed rainmakers when he called their attempts to grow rich without risk or effort "the art of pluviculture,"[60] a practice that William Humphreys defined as "the growing and marketing of rain-making schemes, a never-failing drought crop."[61]

Are there charlatans out there now? Certainly there are huge commercial interests, similar to Irving Krick's, hoping to profit from the scientific and social angst surrounding looming water shortages, damaging storms, and climate change. The Provaqua project in Laredo is one obvious example. Massive ocean iron fertilization schemes to cash in on carbon credits also come to mind (chapter 8). Weather control is currently being practiced on five continents in some forty-seven countries, through some 150 experimental and operational programs. To what effect? In 2002 the Texas Department of Agriculture provided funding of $2.4 million for rainmaking activities. Throughout the American West, agricultural, water conservation, and hydropower interests are conducting routine weather modification operations that cover about one-third of the total area. They are not sure if their efforts are effective, *but they are afraid to stop*!

In 2003 the National Academy of Sciences issued the report *Critical Issues in Weather Modification Research*. The study cited looming social and environmental challenges such as water shortages and drought, property damage and loss of life from severe storms as justifications for investing in major new national and international programs in weather modification research—in essence, finding engineering solutions for nature's shortfalls and wrath. Although the report acknowledged that there was no "convincing scientific proof of efficacy of intentional weather modification efforts," its authors believed that there should be "a renewed commitment" in the field. The fact is, weather modification has never been shown to work in a reliable and controllable way, and the report admitted as much: "Evaluation methodologies vary but in general do not provide convincing scientific evidence for either success or failure."[62] This has been true throughout history, and it remains true today.

During the 2008 Summer Olympics, China spent more money on rainmaking and rain suppression than any other nation—but with no verifiable results. The country has developed a cadre of peasant artillerists, supported by a high-tech weather central, who stand ready to bombard every passing cloud with chemical agents assumed to either dry it out or make it precipitate. Note the use of cannon. In every era, weather and climate controllers employ the latest techniques: explosives, proprietary chemicals, electrical and magnetic devices. Aviation was added in the early twentieth century, as were radar and rocketry by mid-century. Since then, every new technology of any meteorological relevance has been proposed or actually tried in the controversial quest for weather control.[63] With so much invested and so little to show for it, perhaps there are more charlatans out there than we might imagine.

4

FOGGY THINKING

Fog is a cloud that is earth bound.

—ALEXANDER MCADIE, "THE CONTROL OF FOG"

FOR most of human history, at least until 1944, people were at a loss to know what to do about the fogs and vapors obscuring their view. Natural fog, seen from afar, is quite beautiful as it pools in the river valleys or burns off on a sunny morning, but those enshrouded by it may not fully welcome the whiteout conditions it brings. Of course, such obscuration can be a good thing, as in Virgil's *Aeneid* when Venus cloaks her son and his companion in a thick fog to protect them on their journey, or when, following a massive artillery barrage in World War I, the fog, "mute but masterful . . . countermanded all battle orders, and the roar of a thousand batteries gave way to stillness."[1] Sometimes fog is used as a theatrical curtain. Shakespeare employs the weather to reveal Hamlet's mental state when he apprehends the sky filled with "a foul and pestilent congregation of vapors." Coleridge's ill-fated albatross first appears to the Ancient Mariner out of an ice fog. Then there is London or pea soup fog, mixed with the smoke of millions of chimneys, Sir Arthur Conan Doyle's "dun-coloured veil," sometimes yellow, sometimes brown, composed of an unhealthy mixture of smoke and vapor.[2] Actual, as opposed to literary, fogs were deemed unhealthy and undesirable, capable of interrupting or suspending normal activities such as shipping or aviation.

In 1899 Cleveland Abbe described a local fog dispeller suitable for use on ships to assist navigation, or perhaps to increase precipitation. It was called the Tugrin fog dispeller. In foggy weather, a pipe 3 inches in diameter with a musket-shaped flange at the end was used by the navigating officer to direct a powerful stream of warm air from the engines to "blow a hole right through the fog," causing it to fall as raindrops and providing forward visibility of several hundred feet, sufficient to avoid a collision.[3] Abbe further suggested that if the pipe was aimed vertically, it could be used to condense and precipitate fog moisture—for example, for agricultural uses along the California coast. According to meteorologist Alexander McAdie, in March 1929 a murky smoke-fog, the densest and most persistent in twenty years, settled down over New York City, forcing transatlantic liners to lie at anchor. Commerce was suspended and commuters were stranded for several days. With the rise of commercial and military aviation, efforts to dispel fog were driven largely by the desires (and actual needs) of pilots to overcome the vulnerabilities and limitations that fog imposed. In the second quarter of the twentieth century, electrical, chemical, and physical methods of fog dissipation included the electrified sand trials of L. Francis Warren and his associates, the experiments with chemical sprays of Henry G. Houghton, and the operational FIDO fog burners of World War II. All these projects were relevant to aviation safety, and all were of interest to the military.

Electrical Methods

From the time of Benjamin Franklin, the role of atmospheric electricity in meteorological processes, including its suspected role in stimulating precipitation and its possible role in clearing fogs, was under active investigation. In the early nineteenth century, chorographer John Williams proposed a scheme to dehumidify the British climate by electrifying it. For personal, political, and vaguely scientific reasons, he argued that climatic change in England became noticeable around 1770, with the spring and summer months becoming cloudier, wetter, and colder and the winters milder. Williams attributed this shift to human "change effected on the surface of our Island," due to the cutting of forests, digging of canals, and enclosing of lands—all of which had combined to increase the amount of moisture released into the atmosphere and caused adverse effects on human health and agriculture. These physical changes, he claimed, were themselves due to political and economic changes, including the American Revolution, the inflated price of grain, and heavy taxation on labor and agriculture. It was a view that sprang from the author's personal malaise and a generally unsettled mood in Britain. Williams argued that the newly "ungenial seasons" might be ameliorated by building electrical mills, two per county, with giant rotating cylinders to diffuse excess electrical fluid into the surrounding air. He imagined that the newly electrified air would then act to dissolve fogs and dissipate rain clouds. The electrical mills were never built, and the British, as ever, are still damning their damp and cloudy climate and discussing their "peculiar weather," with no ready answers as to what, if anything, is wrong with it or how, if at all possible, to fix it.[4]

In the 1830s, the American chemist Robert Hare, a professor at the University of Pennsylvania, promoted an electrical theory of storms. He imagined that the atmosphere behaved like a charged Leyden jar with two electrical oceans of opposite charge: the celestial and the terrestrial. Clouds acted as the mediators between the two, suspended like pith balls in a static electrical field. When the electrical balance was disturbed, the atmosphere behaved in a way that counteracted gravity. The net result was a local diminution of pressure, inducing inward- and upward-rushing currents of air that resulted in rain, hail, thunder, lightning, and, in extreme cases, tornadoes. Hare argued strenuously that he had discovered a new electrical "discharge by convection" in the atmosphere, which formed the motive power of storms and was to be considered the complement of the famous electrical discharge by conduction discovered by Franklin in lightning strokes.[5]

In 1884 British physicist Oliver Lodge demonstrated that smoke and dust can be precipitated by the discharge of a static electric machine. He then asked, "Why should not natural precipitation be assisted artificially?"[6] In his largest-scale experiment, he cleared a smoke-filled room and discovered that electrical charges encourage the coalescence of infinitesimally small cloud droplets into "Scotch mist or fine rain." He opined that clearing London fogs and abating industrial or urban smoke might be a "difficult but perhaps not impossible task," equivalent to such other noble quests as navigating the Arctic Ocean, exploring the Antarctic continent, scaling Mount Everest, and conquering tropical diseases. Lodge regarded the future prospects with hope and felt that the control of the atmosphere "will be tackled either now or by posterity" (34). But applying this technique outdoors was another matter, and he admitted the propensity of physicists "to rush in where meteorologists fear to tread!" Thus the stage was set for the cloud modifiers to add electricity to their tool kit as they attempted to make rain and dissipate fogs.

Electrified Sand

On the basis of Lodge's theory, a U.S. patent was awarded in 1918 to John Graeme Balsillie of Melbourne, Australia, for a "process and apparatus for causing precipitation by coalescence of aqueous particles contained in the atmosphere."[7] Balsillie claimed to be able to ionize a volume of air and switch the polarity of the electrical charges in the clouds "by means of suitable ray emanations," making them more attractive to one another and thus producing artificial rain. His apparatus, complete with a schematic diagram, consisted of an array of tethered balloons or kites linked to an electrical power supply on the ground. His patent claimed that Röntgen rays from a tube carried aloft and beamed to reflect off a metallic-coated balloon would ionize the surrounding air. In an age in which mysterious X-rays could penetrate flesh to reveal bone, the development of a rainmaking ray gun might be just around the corner. Balsillie's balloons were charged to 320,000 volts—or at least he said they should be—and the ionization, he claimed, would extend outward for a good 200 to 300 feet from each balloon or kite—to be flown in formation (more or less) during a brewing storm. There is no evidence, however, that this patent was anything more than the inventor's flight of fancy—except for its influence on L. Francis Warren and his associates.

Round 1: Dayton, Ohio

"Fliers Bring Rain with Electric Sand," the *New York Times* headline announced on February 12, 1923. The story itself, however, was quite underwhelming. Between 1921 and 1923, field trials conducted in Dayton, Ohio, at McCook Field seemed to show that electrified sand could dissipate clouds and might someday both dispel fog and generate artificial rain. The demonstrations were the brainchild of Luke Francis Warren (fl. 1930), a self-styled and self-taught independent inventor and dreamer who frequently misstated his credentials as "Dr. Warren of Harvard University." Credibility and financial support came from Wilder D. Bancroft (1867–1953), a well-ensconced but controversial chemistry professor at Cornell University. Technical assistance came from Emory Leon Chaffee (1885–1975), a Harvard University electrophysicist, and the U.S. Army Air Service provided aircraft facilities (and a patina of respectability). Although the hope of making rain and driving mists from cities, harbors, and flying fields was great, the hype was even greater. Little is known about Warren, save for a few press clippings, but his story can be told through documents in the Bancroft Papers at Cornell.

Wilder Bancroft, grandson of the famous historian and statesman George Bancroft, was expected to do great things. He studied physical chemistry with Wilhelm Ostwald in Leipzig and J. H. van't Hoff in Amsterdam before joining the faculty of Cornell University in 1895. Bancroft was seemingly more adept at writing than at chemistry. He attracted students with his genteel style and wit more than with his laboratory technique, while he dedicated his considerable writing skills to the new *Journal of Physical Chemistry*, which he edited for thirty-seven years. During the Great War, Bancroft served in the Chemical Warfare Service and wrote its history; after the war, he chaired the Division of Chemistry of the National Research Council. Back at Cornell, Bancroft worked on colloid chemistry, the chemical physics of finely divided matter in suspension—for example, in such complex fluids as ink, wine, milk, smoke, and fog. Thinking about fog, specifically fog dissipation, brought Bancroft into the controversial field of weather control. If, in laboratory tests, electric fields precipitated smoke and fog, why would they not do so in nature?

At the time, Bancroft was under fire from critics for his lack of clarity in organic chemistry and for having missed most of the new physical implications of quantum mechanics. He was busy trying to keep his struggling journal afloat, more by diplomacy and fund-raising than by the influx of new ideas. The marketing of ideas was important to Bancroft. He once opined, "Since the greatest discoveries are likely to be ones for which the world is least ready . . . the greatest scientific men should really be super-salesmen." On weather control, however, he chose to stand on the sidelines as an investor and cheerleader and allowed his associate Warren to take the point position as advocate and business "rainmaker," if not super-salesman. As the airplane was opening up a new era in weather control, Bancroft wrote to Warren in 1920, "[i]t would probably be absolutely prohibitive in cost to produce rain by spraying clouds from beneath; but it is quite possible that you can get satisfactory results by spraying from above."[8]

To get his ideas off the ground, Warren lobbied in Washington, D.C., lunching and dining on Bancroft's dime, with "leading men of the air force." Initially, the military offered merely to take electrical measurements at its flying fields. General Electric was interested in providing the electrical equipment. Major William Blair, who had led the meteorological efforts of the U.S. Army Signal Corps during the war, offered the use of an airplane. The lobbying possibilities were endless. Warren wrote to Bancroft that he had to move quickly, or "I shall be forced to go through the entertainment and visit stunts with a 'new bunch of guys', but as I like them all, and have a soft spot in my make-up for all mankind, I do not apprehend serious trouble, but only inconvenience, as there will be days here when I can do little more than spend denario [mainly Bancroft's] and kick up my heels away from home." Detained in Washington over a weekend, Warren ended his letter to Bancroft with a list of his possible activities, including "attending the aviators ball at Langley Field, playing in the parks with the kids on Sunday, or flirting with the hat girls at the restaurants."[9]

His lobbying efforts eventually paid off, though, and the army provided funds for an initial field test. In the summer of 1921, Warren contracted for electrical work to be done by the physicist Chaffee at Harvard's Cruft High Tension Electrical Laboratory. Chaffee examined the theoretical basis for charging small particles with high voltage, built a generator that would run off an aircraft motor, and designed the best way to disperse the sand, which he determined was through an electrically energized nozzle and the prop backwash.[10] Warren arrived in Dayton on September 7 and began to install equipment on the aircraft. The army paid to bring Chaffee out at $25 a day plus expenses, but Warren chose to stay off the payroll to protect his business rights, since he was then in the process of applying for multiple international patents. The U.S. Army Air Service provided him with two planes, pilots and observers, a car and driver, a stenographer, and a coordinating officer—Major T. H. Bane. Not all was going smoothly, however. Warren had fallen behind on payments to his creditors and, as usual, was writing to Bancroft seeking financial aid "to help me out of this mess." His plan was to "go above detached clouds and try to cause precipitation in the form of trailing rain. We should be able to pull this stunt off within ten days, I hope."[11]

In the "stunts" (they can hardly be called experiments), a La Pere plane flying above the cloud tops sprinkled sand charged to approximately 10,000 volts by an on-board wind-driven generator. The electrified sand was dispensed through musket-shaped nozzles (figure 4.1) and further scattered across the clouds by the action of the airplane's propeller. Sometimes, but only sometimes, these aerial "attacks" opened clearings in fair-weather cumulus clouds or dissipated them completely. Although the stated goal of the project was to clear airport fogs and generate rain, no tests were conducted on low-level stratus or nimbus clouds. Other than a dramatic exhibition of the prowess of aviators (it was known at the time that the backwash from propellers alone could bust up clouds by mixing them with surrounding drier air), nobody knew why the electrified sand technique should work.

Alluding in vague terms to small-scale smoke-clearing demonstrations under laboratory conditions, Warren offered up some technical mumbo jumbo about the effect of electrified sand particles accelerating the "free electrons in a mass of air." He told the press and his patrons (but never published) his theory that "each electron attaches itself to a certain number of molecules and so forms a gas ion, upon which moisture condenses, thereby making a cloud particle."[12] He claimed that his technique produced "a so-called trigger action, forcing the elec-

4.1 Fog dispersal apparatus: sand being discharged through nozzles that are carrying a potential of 10,000 volts. (NATIONAL ARCHIVES PHOTO B8241)

trical charge in the cloud to change from a static to a kinetic state that will rapidly spread or flash over the whole clouded area from the spraying of only a few pounds of dust over a small part of a highly charged storm movement and force precipitation when the wet bulb conditions are favorable over the dry section" (3). By changing the polarity of his generator, he said, he could reverse the process and produce "a large hole, in a fraction of a minute . . . through the entire cloud from top to bottom" (3). Of course this is gobbley-gook, akin to the unsupported technical claims invoked by the charlatan rain fakers. When asked why he was intercepting only fair-weather clouds, Warren cited the absence of suitable fog in Dayton and the danger of flying through rain clouds, since all were "highly electrified and it was not deemed safe to deal with them with high voltage until measures were taken to guard against possible accidents to the pilots and planes" (3).

Aviation pioneer Orville Wright, who worked at McCook Field, witnessed one of the test flights through his office window and sent a telegram to the *New York Times*. He testified that he saw aviators cut to pieces three cumulus clouds in ten minutes, but saw no rain fall: "Having little knowledge of meteorology and the other sciences involved in the experiment, I do not wish to be understood as expressing any opinion as to the practical value of the experiment nor of the possibilities that may develop from them."[13] Navy commander Karl F. Smith was also watching. His memo, "Dr. Warren—Rainmaker," noted that as an observer he was "not gullible and had remained skeptical," but seeing a cloud split in two by the technique was "absolutely uncanny."[14] The military applications were obvious. Smith envisioned special-purpose "clearing ships" for enhancing aerial navigation by dissipating fog or for cutting holes in clouds for bombing operations while keeping the main cloud bank intact for cover. Although he was not fully convinced, he thought Warren's technique "so important" that, after a few more trials, the U.S. Navy Bureau of Aeronautics should either present it to the Patent Office or purchase outright Warren's rights and retain them for military purposes. Had Warren agreed, he could have cashed in on his invention then and there. Instead, he reserved his rights, immediately formed the A. R. Company (for "Artificial Rain"), and issued Bancroft 1,000 shares of stock at $5 a share. He also filed a U.S. patent application for "Condensing, Coalescing, and Precipitating Atmospheric Moisture."[15]

When the story was initially reported in the newspapers, cartoonists immediately got to work. One set of panels published in the *New York World* fantasized about using the technique for raining out Sunday baseball games, ruining a rival's new hat, fighting fires, disrupting parades, and selling umbrellas (figure 4.2).

In March 1923, one month after the initial publicity, U.S. Weather Bureau librarian and widely read weather popularizer Charles Fitzhugh Talman reported that meteorologists remained unconvinced by the Dayton tests.[16] In a weather bureau press release, William Jackson Humphreys contrasted the puny efforts of the rainmakers with the enormous scale of the atmosphere and called their techniques "entirely futile." He compared the techniques of Bancroft and Warren with those of earlier rain kings: "The idea of the college professor and his aviator friends out in Cleveland, to sprinkle electrically charged sand on a cloud while above it in an airplane, is picturesque and plausible," he noted, "but won't work in commercial quantities."[17] Given the enormous forces at work in the atmosphere, Humphreys warned farmers in arid regions not to pay out their good money for so-called rainmaking devices: "Wet weather a la carte—the dream of meteorologists, farmers, and umbrella salesmen for a good many years—is still an empty mirage."[18] In response, Bancroft wrote: "No use arguing with Weather Bureau. Prefer to wait for results and let them do the explaining."[19] A cartoonist captured the tension between the new high-tech possibilities and domestic farm life, with the grizzled older man representing both worlds (figure 4.3).

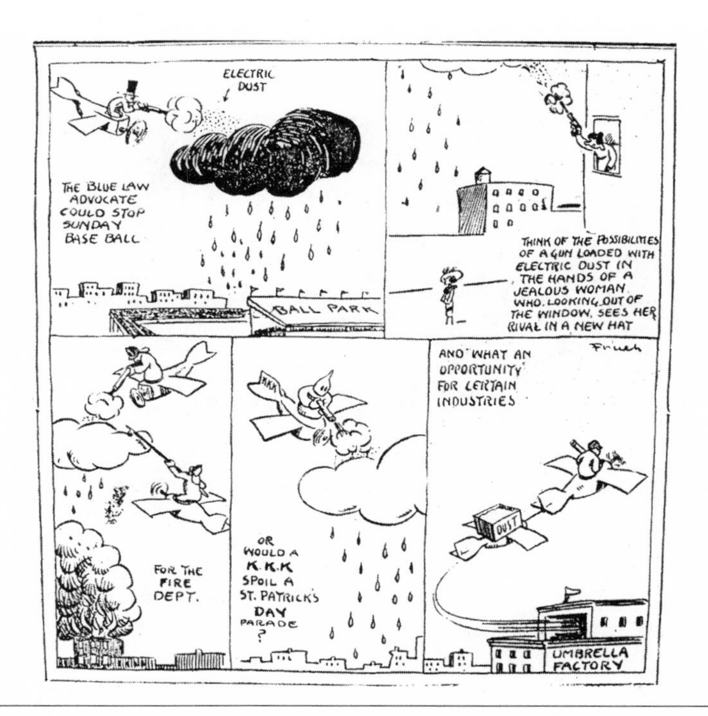

4.2 "Rain to Order": lampoon of possible applications of rainmaking using electrified sand. (CARTOON BY AL FRUEH, IN *NEW YORK WORLD*, FEBRUARY 15, 1923; BANCROFT PAPERS)

The May 1923 issue of *Popular Science Monthly* described the Warren–Bancroft demonstrations and hyped the story: "Think of it! Rain when you want it. Sunshine when you want it. Los Angeles weather in Pittsburgh and April showers for the arid deserts of the West. Man in control of the heavens—to turn them on or shut them off as he wishes."[20] In an illustration, an electrified plane turns smog over a city into artificial clouds, while a second plane clears the air by generating artificial rain (figure 4.4).

Warren claimed that a number of practical applications were just over the horizon: clearing the smoke from cities, removing London fogs, intervening in the course of naval battles, bringing rain to the farmers. As he explained it, the electricity generated by falling sand and rain would cause more rain to generate in the adjacent clouds and set off the entire heavens, much in the manner of a long fuse, thus causing widespread rains. Bancroft, who had been supportive all

4.3 "The Rain Makers": "Go up and bust that there cloud over th' ten-acre field, Noah—before somebody else gets it; an' fer th' love o' peace, keep off th' ol' woman's washin'!" ("THE RAIN MAKERS," *LIFE*, APRIL 5, 1923, 24)

4.4 The way scientists propose to manufacture clouds and rainfall: "The first plane, trailing sparking antennae, condenses the soot and moisture laden air into a cloud by scattering electric charges. The second plane turns this cloud into rain by spraying it with electrically charged sand." (MCFADDEN, "IS RAINMAKING RIDDLE SOLVED?")

along, yet constantly worried about the expenditures, doubted this. Concerned about possible lawsuits downwind of their operations, he recommended that they conduct field trials not over cities but over the Atlantic Ocean, both as a safety precaution and as an opportunity to experiment on marine fogs.

Round 2: Aberdeen and Bolling

In March 1923, seeking better access to government patrons, Warren moved the test flights to the Aberdeen Proving Grounds, on the Chesapeake Bay northeast of Baltimore. It was an adequate but not ideal site. Only about a third of the area was available for tests, since the army's gun-firing range was given first priority. There was no machine shop or other manufacturing facilities, so Warren purchased commercial transformers, which turned out to be too heavy to fly and more costly than originally budgeted. Other delays were caused by problems with workers, a continual lack of funds, and an inordinate amount of red tape.

After more than a year of struggles and setbacks, on July 8, 1924, a plane from Aberdeen carrying Warren's equipment encountered an intense thunderstorm over the mouth of the Susquehanna River. The crew of two aviators reported: "Immediately following the attack [with 10 pounds of negatively charged sand] there was no more lightning or thunder, and a gentle rain of over one hour's duration followed, purely local."[21] Warren claimed that this result was far from coincidental and that the plane's intervention had upset the electrical balance of the storm. More intense lobbying followed. On October 30, 1924, two planes equipped with electrified sand dispensers conducted a demonstration over Washington, D.C. Warren sent messages to President Calvin Coolidge, members of his cabinet, and members of Congress to "watch" as the planes "attacked" the clouds over the city. There is no record from eyewitnesses, but Warren noted, "We scored a most deplorable failure, as not a thing happened" (5). Disappointed but not deterred, Warren moved his operations closer to Washington, to Bolling Field. Less than a month later, *Time* reported a subsequent set of successful sanding demonstrations over Washington. Captain L. I. Eagle and Lieutenant W. E. Melville flew their De Havilland airplanes to an altitude of 13,000 feet, where they "shot down" a series of cumulus clouds with a barrage of electrified sand. As the planes circled overhead, a clearing opened up. "A miracle!" cried some of the watchers.[22] Still, although Warren had busted clouds, he had not yet cleared a fog or made it rain. Positive press clippings notwithstanding, Warren and his A. R. Company had not earned a penny on his invention, and Bancroft was still writing the checks.

Round 3: Hartford, Connecticut

To save money and to be closer to home, Warren moved his operation yet again in 1925, this time to the municipal airfield in Hartford, Connecticut. On the basis of his rather meager successes in busting up fair-weather clouds—forty-three successes and twenty-six failures in six years—he again petitioned for research support from the army, navy, and post office. Warren argued that he needed access to better airplanes that would be capable of reaching the very tops of the clouds, "so that the rays of the sun will freely strike the walls of the wide gashes cut in flight . . . and the electric action of our sand . . . will be reinforced and energized many fold by the radiant energy of the rays of the sun."[23] The army and navy awarded him a grant of $15,000 and the services of some of its pilots, but provided no new airplanes or equipment. Warren thought he deserved more.

Alexander McAdie witnessed the trials in Hartford and noted that the planes had cleared a "figure 8" in the sky with their device—this, two decades before General Electric announced a similar accomplishment using dry ice.[24] Warren was quick to claim success. He telegraphed Bancroft that he had "knocked the stuffings out of two small clouds,"[25] but this was not news; he had claimed this five years earlier. With time running out on the grant, Warren told the press in October, "Our work here is finished. . . . We have clearly proved our theories concerning the art of making it rain through the use of airplanes and are now in position to perfect our apparatus and equipment."[26] But he added, "There are many things yet to be done." Warren admitted that the atmosphere following his cloud-busting test flights had "an uncanny and hard to describe look, effervescent, like dissolved gas escaping under high pressure, or the sudden escape of steam, rolling and tumbling until it quickly disappears."[27] The press, previously enthusiastic, was now turning skeptical. Warren worried that the news stories coming out were surrounded with a halo of unreality, "as wizard or witch-doctor type of news—this needs to be debunked" (17). He could have well said that his own nebulous ideas needed to be debunked. The world's verdict to date was "not proven" (16).

The Business of "Rainmaking"

Ever the businessman, Warren summarized his accomplishments, frustrations, and fantasies in a pamphlet, *Fact and Plans: Rainmaking—Fogs and Radiant Planes* (1928). Here he opined that "once rainmaking is mastered" through good high-tension engineering, "the wealth and prosperity arising from increased production, and decreased cost of living, will reach figures almost 'beyond the dreams of avarice,' not only for our country but for the entire world" (16). With no further prospects for support from the military, it seemed that he would have to realize his dream by raising private capital. But so far, the only investor was Bancroft. Warren's business plan (or vision) included a fleet of airplanes to clear pollution, relieve drought, and suppress forest fires. Clients could be cities, farmers, government agencies, railways, and steamship lines. The London Chamber of Commerce estimated that dense fog cost the city £1 million a day (and he promised he could clear it out in a day). To undertake contracts like this, he proposed the formation of the Warren Company, incorporated in Delaware with 100,000 shares at $8, with four airplanes, two assistants, a machine shop, and a lab, "Warren and his two assistants to devote their entire time, to the exclusion of all else, to the work in hand for at least one year, without salary or expense charges to the company, until the work is satisfactorily completed" (23).

Looking further to the future, Warren waxed philosophical about the possibility of constructing a radiant (ionized) metal plane charged to a potential of 100,000 volts. He wrote that such a plane would operate on what Sir William Crookes had called a fourth state of "radiant matter": "Such a radiant plane will decompose the aqueous vapor immediately in contact with it, creating ozone . . . and the hydrogen, nitrogen, helium, argon, neon, krypton, xenon, etc. or the rare and inert gases will be repelled and forced away, through electric radiation" (10–11). By creating its own partial vacuum, "the resistance to the flight of the plane [would be] reduced to a minimum" and the plane would set new speed records. Pure fantasy! But wait, there's more: Warren wrote that the electric charge would also de-ice the plane so it could fly in bad weather, and the ozone could be collected and used on board "for the benefit of the engines and passengers." A radiant plane would repel and efface everything in nature, "including the frictional action of high winds, storms, tornadoes, cyclones, etc." (11). It could fly at any height in the coldest, iciest conditions; consume less fuel; and attain great speeds. With no drag from the air, the plane would have increased buoyancy, flying on a cushion of highly electrified air. It would be more easily handled and controlled, "immune from all of nature's attacks." The title page of Warren's *Facts and Plans* is marked "Strictly Confidential, Not for General Circulation." The inside cover of Bancroft's personal copy was inscribed by the author: "Kindly keep in your own possession; Sent with supreme confidence in the unexpected; *Don't Worry*."

But Bancroft did worry; he had lost confidence in Warren. His enthusiastic partner, who likened his situation to the struggles of other famous inventors (Morse, Bell, and Marconi), had a tendency to blame others for lack of progress.

Warren blamed the equipment, the hired help, government red tape, his poor health, even interservice rivalries for his shortcomings; however, he never doubted his theory. Bancroft had invested tens of thousands of his own dollars in the fog-clearing and rainmaking project, but after eight years he had little to show for it—only some minor cloud-busting demonstrations, Warren's promises, and worthless shares of stock. By 1927 Bancroft had decided that it was time to cut his losses—and his losses ran deep. Rumor has it that to cover his investments, he even sold a copy of Lincoln's Gettysburg Address given to the family by his grandfather. Warren reacted to Bancroft's pullout with shock and dismay, and then with recriminations of his own, going all the way back to the Dayton experiments, when, he said, Bancroft had "injured rather than helped the cause" with his aloofness and air of superiority. There is evidence that as late as February 1929, Warren was still hanging on, trying to persuade aviation moguls to fund him, trying to issue stock for a new company, and, unbelievably, still trying to solicit money from Bancroft.[28] There the trail fades away, possibly obliterated by the stock market crash, but there is ample material in the Bancroft Papers on this and other ventures to reward a potential biographer. In 1938, on the occasion of the fiftieth anniversary of his college graduation, Bancroft wrote to his Harvard classmates: "Owing to my lifelong habit of being a minority of one on all occasions, my research work does not look convincing to most people. Since I have become avowedly a specialist in unorthodox ideas in the last decade the situation is getting worse, because now I irritate more people."[29] In addition to the electrified sand episode, he was referring to questions raised by the medical community concerning his excursion into the supposed colloidal chemistry of the human nervous system and his theory of anesthesia. In other episodes, Bancroft's attempts to articulate a general chemical explanation of poisoning, drug addiction, alcoholism, and insanity, and his fumbling, and some say unethical, experiments with human subjects brought him into direct conflict with the larger research community and damaged his scientific reputation. There was more to it than just electrified sand.

Fog Research at MIT

"Fog dissipating has, on the one hand, attracted the attention of crack-pot inventors, and on the other, occupied the minds of sober, able investigators. So it is that there have been visionary grandiose ideas of ridding harbors and airports alike of fog. The scale of operations implied together with the lack of factual data relating to fog as a physical entity have at once fascinated the untrammeled mind of the wild inventor and harassed the mind of the cautious investigator."[30] This was written in 1938 by Edward L. Bowles, director of the Round Hill Research Division at MIT and supervisor of its fog research.

Undoubtedly, MIT meteorologist Henry Garrett Houghton Jr. (1905–1987) considered himself a "cautious investigator" engaged in fog research, and most certainly he regarded Warren as a "wild inventor." The theoretical processes involved in precipitation formation—the Wegener-Bergeron-Findeisen ice crystal process (in cold clouds) or the collision-coalescence process (in warm clouds)—had only recently been defined. In 1935, working on the basis of research done by Alfred Wegener, Norwegian meteorologist Tor Bergeron published his hypothesis that the growth of ice crystals in a cloud containing both ice and water droplets could lead to precipitation; three years later, Walter Findeisen clarified and expanded on Bergeron's ideas.[31] The key to fog removal seemed to lie in the reversing of these processes.

In 1938 Houghton and his colleague W. H. Radford surveyed the various approaches to fog removal and categorized them as physical, thermal, or chemical removal methods. Here is a synopsis of his report:

- *Physical methods*. One imaginative approach to freeing airfields from fog involved the installation of powerful fans and ventilation ducts beneath the runways to provide a fresh-air circulation system. This technique would not work, however, if the airport was covered by a large fog bank and the fans merely circulated moist air. Another plan envisioned forcing a stream of air through a set of baffles to slow it down and to condense some of the moisture on contact, but such an apparatus would likely be huge, inefficient, and impractical.

What about high-intensity sound waves? Experiments had demonstrated that they could clear the air of smoke and dust. The theory was that the energy generated by the sound echoing off the walls of a small, enclosed space triggered the precipitation of suspended matter in the air. But an airport is not a tabletop experiment. It is not an enclosed space. Fog particles in the free air are much larger than smoke or dust particles, and air travelers and airport neighbors could not safely or pleasantly be subjected to high-intensity sound waves every time the fog rolled in.

What about electricity? Warren's technique of sprinkling electrically charged sand above fog or clouds should, in theory, lead to the coalescence of the cloud droplets. In practice, however, it was fraught with practical problems and had met with only limited success. Alternatively, spraying charged water drops might also be effective, but could result in the formation of additional fog. An electrical precipitator—long used for removing smoke, dust, and fumes from industrial gases—could be adapted to fog removal, but a medium-size airport installation might require a huge elevated plate suspended some 32 feet above the ground, with a potential difference between plate and ground of 6 million volts. Woe to anyone or anything that short-circuited this apparatus!

- *Thermal methods*. It was well known at the time that supplying heat directly to the atmosphere by burning fuel (discussed in detail later) was a simple, brute-force method of dissipating fog. This technique, however, required an immense amount of energy, since water has such a large latent heat of evaporation. The apparatus (open fires, electric grids, blasts of air or steam) would be large and cumbersome and would probably constitute a dangerous obstruction at an airport. Another approach, using selective absorption of infrared radiation to heat the water vapor and carbon dioxide in the air, lay beyond the capability of current (1938) technology. It was of theoretical interest, however, since it required no cumbersome airport installations, just a properly designed invisible heat ray to zap the fog at a distance.
- *Chemical methods*. Houghton's own research program focused on the physical and radiative properties of condensation, fog, and clouds. His experiments involved the use of calcium chloride as a chemical drying agent, which he sprayed from an array of pipes installed over an airfield. Other possible substances, most with undesirable side effects, included silica gel, sulfuric acid, and certain strong alkalis. For example, calcium oxide (quicklime) releases heat when it reacts with atmospheric carbon dioxide and water vapor, but it is a caustic substance that causes eye and skin irritation and requires proper storage and handling to avoid spontaneous combustion. Thus it was deemed not suitable for field operations involving aircraft.[32]

Houghton was born in New York City and attended high school in Newton, Massachusetts. He was educated at Drexel (B.S. 1926) and MIT (S. M. 1927), receiving his degrees in electrical engineering. From 1928 to 1938, he served on the staff of MIT's Round Hill Research Division, where he and Bowles investigated the behavior of small water droplets as they formed and evaporated, measured the transmission of visible light through fog, and developed chemical techniques for fog dissipation. Houghton became an assistant professor of meteorology at MIT in 1939 and directed the department as associate professor and executive officer (1942–1945) and professor and head (1946–1970). During World War II, Houghton trained weather officers and served on a number of national boards and military committees. After 1945, he chaired the meteorology panel of the Pentagon's Joint Research and Development Board, served on the science advisory board of the Commanding General of the Air Force, and was the first board chairman of the University Corporation for Atmospheric Research (UCAR), in 1959. He also sustained a lifelong interest in weather control. In 1951, in conjunction with the American Meteorological Society, he prepared an appraisal of cloud seeding as a means of increasing precipitation, contributed to discussions about weather warfare (chapter 6), and in 1968 published a review of precipitation mechanisms and their artificial modification in the *Journal of Applied Meteorology*.[33]

A story in *Time* in 1934 described a test of Houghton's chemical "fog broom," conducted at the private airfield of eccentric millionaire Colonel Edward Howland Robinson Green on his Round Hill estate overlooking Buzzards Bay. Houghton had erected a large scaffold across the runway to support a maze of piping and nozzles, "patterned after the business end of a skunk," that he claimed offered "the first practically-tested way of artificially dissipating fog over local areas" (figure 4.5). As a bank of thick fog rolled in from the ocean, Houghton powered up his "secret" apparatus. As the *Time* reporter described it, "Centrifugal pumps sent a

4.5 Henry G. Houghton standing on the chemical fog dissipation apparatus at the MIT research station near South Dartmouth, Massachusetts. Colonel Edward Howland Robinson Green's mansion and the mast of a whaling ship can be seen in the background. (NATIONAL ARCHIVES PHOTO 27-G-1A-8–48)

high-pressure stream of liquid through the overhead pipe. Its nozzles hissed, and jets of Mr. Houghton's chemical cut into the fog like rapiers. The white sea seemed to divide, roll back like the Red Sea before Moses. Soon the watchers were looking through a half-mile tunnel of clear air, 30 feet high, 100 feet wide."[34]

Houghton's research supported the goals of military chemists who were seeking effective smoke screen agents and possible chemical neutralizing agents for use during poison gas attacks. He learned that titanium or zinc chloride could be used in generating smoke screens, but calcium chloride ($CaCl_2$) acted as a hygroscopic drying agent, or desiccant. Calcium chloride is a non-toxic exothermic compound that lowers the freezing point of water. It is used as a water softener, to suppress dust on dirt roads, to melt ice, and as a drying agent in concrete. It can be used as a food preservative, but can also be a powerful abrasive irritant on moist skin tissue and in the eyes, nose, mouth, throat, and lungs. When burned, it produces toxic and corrosive fumes. It attacks zinc in the presence of water to form highly flammable hydrogen gas, and it corrodes steel rebar. Experiments conducted in a laboratory cloud chamber indicated that 1 gram of calcium chloride could clear 3 cubic meters of foggy air, possibly by lowering the vapor pressure of water. Houghton hoped his chemical sprays might be used to clear fog at airports and to add a margin of safety for ocean liners using it to sweep their paths clear. He had basically designed a huge chemical dehumidifier.[35]

Working against the practical adoption of this technique was the enormous amount of chemical needed to keep open even a moderate-sized hole in the fog. Since the ocean fog kept rolling in and re-forming, a constant chemical spray of about 400 pounds a second (!) would have been needed to maintain a half mile opening during the Round Hill experiment. Moreover, as the researchers admitted, "Apparatus of the type described cannot readily be made portable and its size makes it a rather serious obstruction for some applications, notably at airports." As mentioned, the electrified, high-pressure calcium chloride spray was corrosive to metals. It tended to clog the spray nozzles and had to be washed off any metal objects it contacted, especially electrical systems, but also the piping and even the airplanes and ships it was designed to serve. It was also dangerous to personnel, producing skin rashes if not rinsed off, and it killed vegetation (it was sold as a weed killer). The navy had some unfortunate experiences with the chemical and ultimately decided not to use it on its airplane carriers.[36]

Reminiscent of the later distinction between cloud physics and weather modification—perhaps also between basic research and practical applications—MIT researchers were quick to point out that fog research, not fog control, was their ultimate aim: "The end result, whatever the practical application of local fog dissipation, has been a substantial increase in knowledge of the physical properties of fog and of the means for conveniently determining these properties, as well as a more thorough quantitative knowledge of the transmission of electromagnetic waves through fog, whether they be radio, light, or long infrared."[37]

Still, the list of institutions acknowledged for their support in the MIT report reads like a who's who of the military–industrial complex in 1934: Colonel Green for the use of his estate; the American Philosophical Society for a research grant; and the U.S. Navy Bureau of Aeronautics, the U.S. Army Air Corps, and the Bureau of Air Commerce of the Department of Commerce for their support. Huge amounts of chemicals were provided free of charge by the Michigan and Columbia alkali companies and the Dow Chemical Company. Edison Electric of Boston lent the experimenters a large power transformer. This combination of government, commercial, and private philanthropic support was part of a persistent pattern of patronage (6–7).

It is undoubtedly true that a 30-foot-high barrier made of metal pipes and stretched across a runway is dangerous to airplanes landing and taking off, especially in conditions of low visibility. Houghton's chemical mix, although promising, was also impractical, being dangerous and corrosive. More substantial was his basic research on the formation and evaporation of small droplets, on the optical properties of fog, and on the search for possible hygroscopic chemicals to disperse it. Houghton's greatest contribution, however, involved the idea that cloud physics research, as distinct from but related to operational weather modification, had a place in the modern university.

FIDO: A Brute-Force Method of Fog Dispersal

Foggy weather kept aviators grounded in World War I, but by 1921 British meteorologist Sir Napier Shaw discussed the possibility of clearing fog at an airfield by heating it, concluding, "I would not like to say it is impossible with unlimited funds and coal." He noted, however, that "air in the open is very slippery stuff and it has all sorts of ways of evading control that are very disappointing."[38] Professor Frederick A. Lindemann (later Lord Cherwell) agreed with Shaw and chose to emphasize blind landing techniques. Other possibilities, although none of them were proved, included sprays of electrified water, air, or sand (Warren), chemical treatments (Houghton), vigorous fanning, and coating rivers with oil. Yet the brute-force technology of heating the runway was the only one certain to work—although it appeared at the time to be prohibitively expensive.

In 1926 Humphreys estimated that it would require the combustion of 6,600 gallons of oil (or 35 tons of coal) an hour to clear a layer of fog about 150 feet thick from a typical airfield—a cost that he deemed far too large. David Brunt, Shaw's successor at the Imperial College of Science and Technology, revisited the issue in 1939. He estimated that clearing a layer of fog about 300 feet thick would require an average temperature increase of 3.5°C (6°F) (twice this at the ground) and suggested that smokeless burners supplied by an oil pipeline along an airfield could be designed to do the job. Brunt's ideas were field-tested in the winter of 1938/1939, but the results were not promising.[39]

As World War II escalated, fog became an obstacle to successful bombing raids. With more raids scheduled, a surging accident rate, and the large number of flying hours lost to fog, the problem became one of "extreme urgency." In 1942 Prime Minister Winston Churchill directed his scientific adviser, Lord Cherwell, to address the matter and issued the following statement: "It is of great importance to find means to disperse fog at aerodromes so that aircraft can land safely. Let full experiments to this end be put in hand by the Petroleum Warfare Department with all expedition. They should be given every support."[40]

Under the leadership of Britain's minister of fuel and power, Sir Geoffrey Lloyd, and Major-General Sir Donald Banks, the scientific research establishment and industry joined forces to tackle the problem. The Petroleum Warfare Department (PWD), an agency created in 1940 to consider "the possibilities inherent in the use of burning oil as an offensive and defensive weapon in warfare," was charged with developing a reliable, if expensive, brute-force method of clearing fogs over airfields, a system it called Fog Investigation and Dispersal Operation. FIDO was one of the most spectacular but least publicized secret weapons of the war. According to Banks, "We had been making vast preparations to cook the Germans. We would see whether we could cook the atmosphere!"[41]

It was a massive undertaking. The FIDO project brought together pilots, engineers, fuel scientists, industrialists, government bureaucrats, and meteorologists. Given the urgency of the situation, normal research and development plans were shelved in favor of an all-out attack by research teams from the National Physical Laboratory, Imperial College of Science and Technology, Royal Aircraft Establishment, Armament Research Department, and such industries as the Anglo-Iranian Oil Company, Gas Light and Coke Company, General Electric, Imperial Chemical Industries, London Midland and Scottish Railway, and the Metropolitan Water Board. According to Lloyd, the project director, "each was told to get on with the job with the fullest support and freedom of action."[42]

First Successful Tests

FIDO consisted of a system of tanks, pipes, and burners surrounding British airfields and designed to deliver petroleum that, when ignited, raised the ambient temperature by several degrees—enough to disperse fog and light the way for aircraft operations. The first large-scale test of a FIDO system was conducted in a field and did not involve aircraft takeoffs or landings. With strong radiation fog predicted for the morning of November 4, 1942, the FIDO team assembled at Moody Down, Hampshire. An 80-foot fire escape ladder was positioned between two FIDO burners 200 yards in length and 100 yards apart. As a local fireman climbed to the top of the ladder, he disappeared into the fog. When the burners were lit, the fog began to clear and the fireman came into view. To verify the result, the burners were turned down and the fog reappeared. The burners were again ignited, and the fog dissipated. With typical British reserve, it was reported that Lloyd "*almost* whooped for joy" (emphasis added).[43]

On the same day, experiments were also conducted at the airfield in Staines, Surrey, using coke-burning braziers shuttled by miniature rail cars along tracks paralleling the runways. While an even denser fog was cleared with less smoke, the coke took longer to light and required more effort to replenish. Gasoline was much easier to pipe to airfields and ultimately became the fuel of choice for FIDO. The urgency of the situation did not allow much time for further experimentation and research. As a result, the petroleum burner setup at Graveley airfield, Hertfordshire, served as the prototype for other FIDO systems ultimately installed at fourteen Royal Air Force (RAF) fields.

On February 5, 1943, Air Vice Marshal Donald C. Bennett landed a Gypsy Major at Graveley in a midday FIDO light-up. Thirteen days later, in the first night test, he again landed, in a Lancaster. Although it was not foggy, visibility was poor. Bennett recounted seeing the blazing runway when he was still 60 miles out. As he made his approach, he recalled, "I had vague thoughts of seeing lions jumping through a hoop of flames at the circus. The glare was certainly considerable and there was some turbulence, but it was nothing to worry about."[44] Except wildfires. A demonstration test for aircrews on February 23 resulted in grass, hedges, trees, and telegraph poles near the burners going up in flame. All hands, in addition to local bomb spotters and fire companies, were called in to fight the blaze. The first opportunity to land an aircraft in actually foggy conditions occurred in July 1943. A thick fog, approximately 300 to 400 feet deep, blanketed the runway, with visibility less than 200 yards. The FIDO burners were lit at five o'clock in the morning, and within seven minutes, an area 1,500 yards long and 200 yards wide was cleared of fog. Aircraft were then able to land successfully at fifteen-minute intervals.[45]

The futurist Arthur C. Clarke once witnessed a FIDO test in Cornwall:

> The runway was lined on either side with a double row of pipes—four or five miles in all—which conveyed gasoline to long rows of burners. When they were in action, they consumed fuel at the awesome rate of 100,000 gallons an hour and formed multiple walls of flame the full length of the runway.
>
> At night, with the fog rolling in from the Atlantic, a FIDO operation was like a scene from Dante's *Inferno*. The roar of the flames made speech difficult; such an updraft was created that small stones on the edge of the runway were picked up and tossed around by the air currents. The yellow walls of fire, taller than a man, stretched away into the foggy night as far as the eye could see. The miles of burners pumped heat into the air at the rate of 10 million horsepower, cutting a long, narrow trench through the fog down which the retuning bombers found their way to the ground.
>
> I have known nights when the fog was so thick that visibility was less than ten feet; but standing in the middle of the runway, with the flames roaring on either side, you could see the stars shining overhead. FIDO worked by brute force, and the development of radar made it obsolete, but it did show what could be done if the incentive was sufficiently great.[46]

The view from the cockpit was especially exhilarating. Although airmen were thankful for the safety that FIDO provided, they described their first experiences of landing between FIDO burners as frightening. One veteran pilot, echoing Clarke's description, likened it to a descent into hell, remarking that it seemed as if he "was over [the enemy] target once more . . . [and] that the whole place must have caught on fire."[47]

FIDO Becomes Operational

FIDO actually worked. It allowed British and Allied aircraft to take off and land in conditions of poor visibility when the Germans were grounded (figure 4.6). The urgency that Churchill demanded had been met, and FIDO was quickly serving the duty of guiding RAF and Allied airmen home safely. Pilots returning to foggy England after a mission could see the airfield glowing in the distance, beckoning them home to a lighted, fog-free airport. They could also save valuable time getting their shot-up planes and exhausted (and possibly wounded) crews on the ground. Because of FIDO, the Allies could launch patrols and air raids and return their planes safely when enemy

4.6 Boeing B-17 Flying Fortress, 493rd Bomb Group, landing in England with the aid of FIDO, November 16, 1944. Note the giant flames behind the airplane. (NATIONAL ARCHIVES PHOTO A9004, DETAIL)

aircraft were grounded due to poor visibility. RAF Coastal Command aircraft on anti–U-boat patrol used FIDO frequently. On one occasion, a lost Lysander aircraft landed on a runway that had been cleared of fog. When FIDO was turned off, fog once again enveloped the aircraft. Reportedly, the pilot wandered across the tarmac for quite some time before finding the control tower.

The success of FIDO was presented to a war-weary public as almost a miracle. Newspapers proclaimed it as a lifesaver and a triumph for British aviation. Those involved in administering the project credited FIDO with shortening the war and saving the lives of up to 10,000 airmen. Military historians are fond of invoking "the fog of war" as they struggle to reconstruct events. In the case of England, the fog was literal. An ice fog persisted during the opening days of the Battle of the Bulge, when FIDO supported Allied aviation. But during the long campaign, the weather cleared and much of the tactical air support came from the Continent, not England. Thus contemporary evaluations of the overall success of FIDO in "shortening the war" may have been somewhat optimistic and self-serving.[48]

The Aftermath of FIDO

FIDO proved to be one of the innovation success stories of World War II. It was a crash research program that became operational; it saved lives and equipment; and it definitely gave the edge to Allied aviation during the last two years of the war. But FIDO was feasible only under the desperate conditions of wartime. Bomber Command, its chief beneficiary, credited it with introducing a "revolutionary change in the air war," but its success was never replicated. When the FIDO system was ignited at an airfield, up to *6,000 gallons* of gasoline were burned during the time required to land one aircraft. By comparison, a Mosquito bomber might burn between 10 and 20 gallons of fuel during its landing approach. It is estimated that during the two and a half years that FIDO was in operation, airfields that used it consumed a total of 30 million gallons of gasoline. Such expenditures were justifiable only when national survival was at stake. Ironically, FIDO's success was due in large part to the brilliant but modest British defense engineer Guy Stewart Callendar, who was the first scientist to attribute the enhanced greenhouse effect to the burning of fossil fuels, who designed key components of the system (including the trench burners), and who was one of the patent holders on the massive FIDO fuel burner.[49]

After the war, a FIDO system was planned for London's Heathrow Airport, but it was never installed. For a time, FIDO systems were maintained at the Blackbushe and Manston RAF bases, but according to one 1957 estimate, the cost of running a FIDO installation was prohibitively expensive—£44,500 an hour. Experiments using jet engines installed along runways to heat and disperse fog at Orly Airport near Paris and in Nanyuan, China, met with mixed results. The main technique for dealing with fog, developed after the war, was not weather or cloud modification but the widespread use of instrumented landing techniques.[50]

The Airs of the Future

On July 11, 1934, Willis R. Gregg, chief of the U.S. Weather Bureau, presided over the dedication ceremony at the air-conditioned house at the Century of Progress World's Fair in Chicago. It was the Midwest's hottest summer to date, with temperatures that day in St. Louis reaching 100°F (38°C), but Chicago, cooled by a breeze from Lake Michigan, reached only a moderate 82°F (28°C). It was a dust bowl year, with little rain and the average regional temperatures soaring 5 to 10 degrees above normal. Gregg's theme, broadcast over NBC Radio, was weather control, and he began by discounting the "fantastic methods" of the professional rainmakers "who have boasted of their abilities to end drouths by the simple expediency of setting off a few explosives," or of those charlatans who "would mount receptacles containing small quantities of chemicals on poles or platforms in the vicinity of the drouth stricken areas, and then trust to the law of averages and Old Mother Nature to come through with rain at the psychological moment so they may collect rain-making fees."[51] He deemed the prospects for controlling outdoor weather "rather slim" for a great many centuries to come.

Gregg's focus was on the control of indoor weather, on display that day in the air-conditioned house, where there was "no necessity for suffering from weather discomforts." Of course, indoor air-conditioning really began before recorded history, when people sought shelter from the storm to keep them dry and warm. Roofs, doors, windows, screens, fireplaces, stoves, and furnaces function either to keep out undesirable elements like rain, wind, and pests or to allow in or provide desirable elements such as shade, light, and heat. In hot climates, traditional practices of ventilation and evaporative cooling have long served to moderate heat, if not moisture. The inner atmosphere of the show house of 1934, however, had been refrigerated and dehumidified by mechanical means, the science of thermodynamics, the engineering that has come to be known as HVAC, and the power supplied by electricity. According to Gregg, conditioning this indoor air was solving "the one thing that actually has the most lasting effect upon the human body and human activity—weather, if only in a small way."[52] Gregg, speaking from the front porch of the house, speculated about the possible, if impracticable, project of refrigerating an entire city mechanically, but he did point out, prophetically, that air-conditioning would allow cities to expand in areas formerly considered too hot for comfort. He might be amazed today to see air-conditioned mega-malls and domed stadiums, but not really, since even then air-conditioning was becoming more and more popular. On his inspection tour out west, through the dust bowl region, Gregg, at least on occasion, traveled on air-conditioned trains, slept in air-conditioned hotels, and ate in air-conditioned restaurants. He spoke of air-conditioning in relief of hay fever and of living in it from cradle to grave, citing the hospital incubators supporting the Dionne quintuplets, born in May of that year in Canada, and the growing trend for air-conditioned funeral parlors. His weather bureau office in Washington, D.C., however, was not air-conditioned; it had high ceilings and fans that helped alleviate the oppressive heat somewhat. The federal government followed liberal leave policies during heat waves.

But what about the outside air? In the summer of 1938, Gregg sent letters to his colleagues asking them to speculate on what the meteorological profession might look like in fifty years. Most of the responses focused on scientific and technological advances in forecasting. Some emphasized the growing importance of upper-air measurements using radiosondes and broadcasts that would allow "records to be flashed to all parts of the world." Charles Franklin Brooks foresaw remote sensing of the atmosphere using ultra-high-frequency radio transmissions. J. Cecil Alter suggested that "sky-sweeping robots of electric eyes will explore the upper atmosphere for air mass demarcations, depths, direction and velocity movement, moisture content, and other factors. Zigzag tracings or photographic replicas, automatically registered, will be made of the shape of the course of the refracted ray from the electric eye, as it passes through different air masses."[53] Humphreys wrote of "robot reporters—instruments that not only keep a continuous record of the weather elements, but which, at the touch of a button, or automatically at regular intervals, also tell all about the weather there at the time" (215). These predictions were largely realized through the development of weather radar and other forms of remote sensing. Also, in 1939, George W. Mindling foretold, in doggerel, of the "coming perpetual visiontone show" of perfect surveillance and perfect prediction using television and infrared sensors, a technology instituted in the TIROS (Television Infrared Observation Satellite) meteorological satellite program in 1960:

> In the coming perpetual visiontone show
> We shall see the full action of storms as they go.
> We shall watch them develop on far away seas,
> And we'll plot out their courses with much greater ease.
> Then a new day will come in electrical lore
> When the pictures will register very much more. . . .
> Then a day there will be when predictions won't fail,
> Though describing the weather in every detail,
> Just what minute 'twill rain, even when it will hail.[54]

These lines in Mindling's poem are preceded by seven stanzas praising the radiosonde and followed by two stanzas anticipating that weather forecasting might someday attain the accuracy of astronomical predictions.

Two of Gregg's respondents spun wild fantasies involving geoengineering. Major E. H. Bowie of the San Francisco weather bureau office facetiously suggested that the only way to end the dust bowl was through a Works Progress Administration project to lower the height of the Sierra Nevada and the Rocky Mountains. T. A. Blair of Lincoln, Nebraska, issued this ominous forecast for a weather control agency and a "war for the control the air masses," a full century into the future:

> In the year 2038 an American meteorologist discovers how to control the weather. . . . But difficulties arise. This control involves a shifting of the air masses

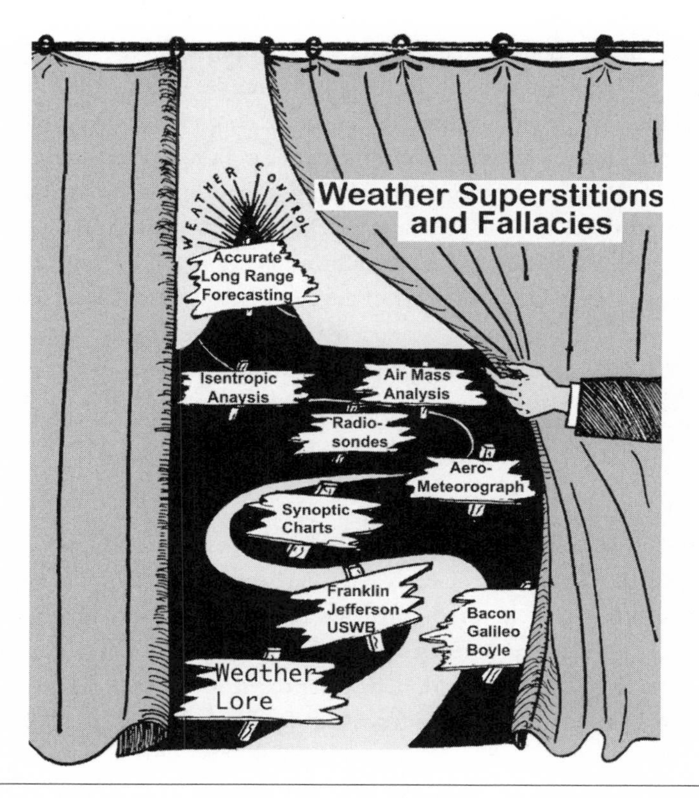

4.7 "Weather Superstitions and Fallacies" (ADAPTED FROM BARBER, *AN ILLUSTRATED OUTLINE OF WEATHER SCIENCE*)

> and means that while the one area is getting the kind of weather it wants, another region is subject to unfavorable weather. . . . Political parties develop on the basis of these differences, and "pressure groups" attempt to control the WDA (Weather Distributing Administration). Soon other nations attempt to manipulate the weather. International complications begin and the human masses of the world are plunged into a war for the control of the air masses.[55]

Representative of the attitudes of the era was Charles William Barber's 1943 illustration of the curtain of "weather superstitions and fallacies" being drawn aside to reveal the progressive path of weather science leading to its ultimate goals (figure 4.7). Perhaps, however, accurate long-range forecasting and weather control are the *real* weather superstitions and fallacies.

I am writing this book in Maine, in the summer, under a tree, without air-conditioning. I do not have it in my office, and I do not need it in my home. With a basement dehumidifier, window screens, fans, and a lake conveniently nearby, I have no need at all to be sequestered from the open air. In fact, I found creative writing to be nearly impossible while on sabbatical, cooped up as I was in the elegant air-conditioned buildings of Washington, D.C. There I focused on doing library and archival research, giving and attending seminars, and otherwise broadening my horizons while avoiding the heat of the day. It seems to me that climate deliberations in the U.S. capital will be conducted indoors, in air-conditioned buildings sequestered from the summer heat of Washington. Some of the people making the decisions might even be advocates for a Weather Distributing Administration.

* * * * *

The rise of civilian and military aviation in the early decades of the twentieth century placed fog clearing at the center of the research-and-development agenda. The airplane provided a new tool and a new research platform, and its vulnerability to fog provided a new urgency. New theories of electrical influence, chemical affinity, and large-scale combustion were put to the test. Of the three case histories presented here, L. Francis Warren was the most speculative, and Wilder Bancroft ended up the biggest loser, in both credibility and financial terms; Henry Houghton's reputation as a careful researcher grew, even if his applications failed; and FIDO actually worked and may even have helped the British war effort, but at an immense cost that rendered it impractical after the war when the question of national survival was no longer at issue.

Cloud physics and chemistry got its start in this era, as did serious attempts to make smoke screens and dissipate clouds. So too did air-conditioning, which grew by leaps and bounds from a novelty to a seeming necessity for larger and larger spaces. In common with later eras, weather control research before 1944 benefited from military patronage and the passing interest, if not support, of large corporations like General Electric. T. A. Blair's 1938 vision of dystopian climate control in the distant future now seems a spooky possibility in the not-so-distant future. These themes, mutatis mutandis, would reemerge in the work of an articulate, highly credentialed spokesperson: Irving Langmuir.

5

PATHOLOGICAL SCIENCE

Pathological Science—the science of things that aren't so.

—IRVING LANGMUIR, "PATHOLOGICAL SCIENCE"

IRVING Langmuir (1881–1957), Nobel laureate in chemistry, quintessential industrial scientist, and associate director of research at the General Electric Corporation in Schenectady, New York, was both a rain king and a friend of weather warriors. He was also the leader of a research team that included Vincent Schaefer (1906–1993), "the snowflake scientist," who developed dry ice seeding, and Bernard Vonnegut (1914–1997), who identified the chemical silver iodide as a cloud-seeding agent. Langmuir's work in surface chemistry was solid, even brilliant, and his scientific intuition was usually quite sound. By some measures he was considered to be a genius and was by no means a charlatan.[1] Yet his work in weather control exemplified his own warnings about pathological possibilities of science gone awry.

In 1953 at GE's Knolls Research Laboratory, Langmuir presented a seminar titled "Pathological Science," on "the science of things that aren't so." He cited a number of examples of this phenomenon, some drawn from the history of laboratory science and some from popular culture. Among them were Prosper-René Blondlot's nonexistent N-rays (1903), so subtle that only a Frenchman could see them; the "mitogenic rays" (1920s) of the Russian biologist Alexandr

Gurwitsch, who claimed to have revealed the secret lives of plants; the extrasensory perception (ESP) of the American parapsychologist Joseph Banks Rhine, whose work convinced many people that they had this sixth sense; and, beginning in the 1940s, worldwide reports of flying saucers. Focusing his argument on basic research rather than on popularizations, Langmuir argued that in many pathological cases there was no dishonesty involved, but researchers were tricked into false results by a lack of understanding about what human beings can do to themselves in the way of being led astray by subjective effects, wishful thinking, or threshold interactions. "Research" is defined as seeking to discover what you do not know. According to Langmuir, science conducted at the limits of observation or measurement—precisely where cutting-edge research is done—may become pathological if the participants make excessive claims for their results. Overly hopeful researchers studying phenomena close to the threshold of delectability may interpret minor variations or even random noise as meaningful patterns. By attributing causation to events that are barely detectable or poorly understood, they may convince themselves and co-workers of the reality of their "discovery." If they persist, weaving theoretical justifications with claims of great accuracy and responding to criticisms with ad hoc excuses, they may cross the boundary into pathological science. If other researchers cannot reproduce any part of the alleged effect, or of the experiment fails repeatedly in the presence of an objective observer, the rules of good scientific practice are supposed to kick in, with support dropping off rapidly until nothing is left to salvage—according to Langmuir.[2]

Many scientists would say that they are working in exciting and rapidly changing fields, in which a breakthrough or named discovery could establish their careers or secure them adequate levels of funding. Otherwise, why bother? Under such conditions, external or social pressures may distort the scientific process and lead into the realm of pathology. Such pressures may include the rush to publish questionable or speculative results, to claim priority, or to avoid priority disputes; intervention of the press, the courts, or government regulators in the process; or competitions for prizes. The patentability and potential profitability of proprietary discoveries may also short-circuit the scientific process and result in the violation or circumvention of established standards of evidence. When things begin to go awry, investigators may suspect a conspiracy to discredit their results, which, depending on the personality of the leading figure, may be convincing to others.

Pathological science is by no means limited to esoteric physics experiments done in darkened rooms or at high temperatures and pressures where the subjectivity of the experimenter or malfunctioning equipment may be the source of the deception. In fact, at the very same time Langmuir presented his seminar on pathological science, he was deeply involved in making highly dubious and unsupportable claims for the efficacy of cloud seeding in creating rain, otherwise modifying the weather, and perhaps even altering the climate. We can thus add one final criterion supporting pathological outcomes that Langmuir did not mention in his lecture—over-reliance on the credentials of a scientist, for example a Nobel laureate, instead of proof.[3] When Robert N. Hall transcribed Langmuir's talk, he added, editorially, "Pathological science is by no means a thing of the past. In fact, a number of examples can be found among current literature, and it is reasonable to suppose that the incidence of this kind of 'science' will increase at least linearly with the increase in scientific activity."[4] If Langmuir's lecture were to be given today, one might include such pathologies as polywater, an illusory form of water promoted by Soviet physicists Nikolai Fedyakin and Boris Derjaguin in the 1960s, and cold fusion, purportedly discovered by Martin Fleischmann and Stanley Pons in 1989. A 2008 Purdue University report on "bubble" fusion contained the following line about the misconduct and unsustainable claims of one of the school's physicists, who publicly purported to have produced nuclear energy in a tabletop experiment by making tiny bubbles collapse: "From small beginnings there developed a tangled web of wishful thinking, scientific misjudgment, institutional lapses and human failings."[5] This is pathological science. Langmuir's obsessive and unbridled enthusiasm for weather control and his unsubstantiated claims for it represented a serious lapse in judgment. Thus his final, major undertaking—his foray into weather control—deserves to be scrutinized in light of the criteria developed in his own lectures on pathological science.

Blowing Smoke

During World War II, General Electric held contracts with the National Defense Research Council, the Office of Scientific Research and Development (OSRD), the Chemical Warfare Service, and the U.S. Army Air Force for research on gas mask filters, screening smokes, aircraft icing studies, precipitation static, and other aspects of what came to be known as aerosol or "cloud" physics. In 1941, following German successes in using a smoke generator to hide the battleship *Bismarck* in the fjords of Norway, Langmuir asked his associate Schaefer to enlarge a small smoke generator he had built under military contract for testing air filters for gas masks. Using a mercury diffusion pump originally designed by Langmuir attached to a pot of boiling oil, Schaefer proceeded to "smoke up the whole room," getting him into trouble with his laboratory neighbors and with the local fire department when he tested it, without advance notice, on the laboratory roof.[6]

A full demonstration for the military was held at dawn on June 24, 1942, at Vrooman's Nose, a 600-foot cliff that provided a panoramic view of an agricultural region in upstate New York known as the Schoharie Valley. Notables in attendance included Vannevar Bush and Alan Waterman of the OSRD, Vladimir K. Zworykin of RCA Research Labs (prominent in chapter 7), and top military officers. On cue, at sunrise, with drainage winds flowing down the valley, a tiny puff of smoke from a single Langmuir–Schaefer generator rose in the distance and quickly spread to fill the valley floor. The device worked by forcing 100 gallons of lubricating oil at a temperature of about 450°C (842°F) at supersonic speeds through a hot manifold. As the oil vapor hit the cold air, it formed a dense white cloud of tiny particles. Within minutes, the generator had belched out a persistent, thick smoke screen 1 mile wide, 10 miles long, and 1,000 feet deep, totally obscuring the valley. The army had its smoke screen, GE its contract, and Langmuir and Schaefer had taken their first steps in the new field of cloud physics. Since they had made an artificial cloud successfully, why not modify an existing one?[7]

In 1943, again under military patronage, Langmuir's research team shifted its attention to studies of electrical effects such as precipitation static, which interfered with radio communications during snowstorms. The Mount Washington Observatory in New Hampshire provided ideal conditions for their experiments and, serendipitously, led them into the study of the behavior of clouds containing water droplets as cold as −40°C (−40°F) ("quite a bit below freezing"). Such conditions represented the typical environment for clouds in the free atmosphere and provided insights into the nature of ice nuclei, ice deposition, and other aspects of cloud physics. "In the process," according to an interview with Schaefer, "Langmuir and I became very much interested in the whole business of supercooled clouds, and whether you could modify them." The military roots of weather control research should not be surprising, given the earlier history of army aviators using electrified sand and chemicals for cloud busting and the contemporaneous effort to clear fog in England using the FIDO system.

Langmuir and his team read the latest articles by meteorologists Alfred Wegener, Tor Bergeron, Walter Findeisen, and other European researchers on the initiation of ice-phase precipitation. Schaefer again took the lead, seeding supercooled clouds of water droplets with "dozens of different materials": talc, carbon, graphite, volcanic dust, various smokes, and quartz crystals—following an idea attributed to Findeisen that the crystals might provide suitable cloud condensation nuclei (this "didn't work at all").[8] Yet Schaefer persisted in his "cut and try" methods, emulating Thomas Edison's search for a suitable lamp filament. Rather than theory, it was Schaefer's use of dry ice to cool his cloud chamber in the summer of 1947 that opened up a new chapter in the history of weather control.[9] Others had tried this before.

Liquid and Solid Carbon Dioxide

In 1891, Louis Gathman of Chicago obtained a patent to encourage and enhance rainfall by chilling the atmosphere through the release of "liquefied carbonic acid gas" shot from a projectile or released from a balloon. In Gathman's plan, the liquid carbonic acid sent into the clouds would vaporize and expand, chilling the surrounding air.[10] Even though one method involved delivery of the agent by artillery burst, Gathman's idea was quite distinct from those of the concussionists. Senator Charles Farwell (R-Illinois), who had supported Robert Dyrenforth, was reportedly interested in the patent, but did not pursue the idea. Fernando Sanford, a physics professor at Stanford University, praised Gathman's theory, since he thought that cooling the air was a physically sound technique to enhance rainfall artificially. Sanford categorized Dyrenforth's recent Texas rainmaking expedition as a "national fiasco," since the explosions of the concussionists actually heated the air and encouraged the proliferation of charlatans. Real scientists conducted carefully controlled tests and were published in the technical literature; they did not petition Congress for money on the basis of their brainstorms. Sanford wrote: "Unquestionably we have [in Gathman's proposal] the proper kind of an agent for producing rain. The only question to be considered is one of finance."[11] Unfortunately, the scale of the atmosphere worked against the idea. Sanford calculated that it would take an astronomical amount of carbonic acid, 406 million pounds of it, to cool a cubic mile of air sufficiently to generate a quarter of an inch of rainfall over 640 acres. With carbonic acid selling for $1 a pound, Sanford estimated that the cost of the rainfall per acre was a prohibitively expensive $600,000.

If Gathman had taken the next step, proposing the use of solid carbonic acid (dry ice); if Sanford had seen a triggering effect to the cloud seeding rather than a brute-force approach to chilling the entire atmosphere; and if someone had actually tried the experiment, perhaps by shelling a growing cumulus congestus cloud . . . but those are a lot of "ifs." In 1948 the *Stanford Law Review*, in an examination of the science behind the current cloud-seeding rage, briefly mentioned Gathman's patent, pointing out that minute particles of dry ice and even artificial clouds must have been formed in the rapid cooling process. They speculated that if Gathman were alive, and if his patent had not long since expired, "he might have an action for patent infringement against those who are using dry ice to cause rainfall."[12] Two more big "ifs."

In the late nineteenth century, supercooled cloud conditions were known, and meteorologists were hinting at the possibility that ice-phase processes could initiate precipitation. In 1895 Alexander McAdie wrote that, by analogy, "a snowflake or ice crystal falling into [a supercooled cloud] may suffice to start a sudden congelation, just as we see ice needles dart in all directions when a chilled surface of a still pond is disturbed." Speaking of towering convective clouds—which are certainly large but not quiescent like a pond—McAdie noted, "We liken this monstrous cloud to a huge gun, loaded and quiet, but with a trigger so delicately set that a falling snowflake would discharge it."[13] He predicted that "successful rain engineers will come in time . . . from the ranks of those who study and clearly understand the physical processes of cloud formation" (77). The key word here is "trigger," which is just what the General Electric scientists were attempting to do in 1946.

Readers of the September 1930 issue of *Popular Mechanics* learned that a Dutch scientist, August Veraart, had recently "succeeded" in producing rain by throwing dry ice power (solid CO_2) on clouds.[14] Veraart also claimed to be able to produce more sunshine by conducting his seeding in the early morning, which cleared the sky of fog, mist, and clouds for the rest of the day. From a small airplane flying above the Zuider Zee, Veraart scattered some 3,300 pounds of crushed ice particles cooled to a temperature of −78°C (−108°F) into growing cumulus clouds. Observers testified that the intervention was followed by falling streaks of rain, although there is no evidence that the rain actually reached the ground.[15] In 1931 Veraart published a small popular book in Dutch, now quite rare, titled *More Sunshine in the Cloudy North, More Rain in the Tropics*. Here he presented a history of his involvement in rainmaking, an overview of his experiments and theories, and a summary of his wide, sweeping claims.[16]

Over the years, Veraart said, he had tried an assortment of seeding techniques involving dry ice, supercooled water-ice, and ammonium salts. He theorized that seeding particles could upset the stability or release instability in clouds, release latent heat of condensation, and perhaps influence their electrical charges to either dissolve them or condense their moisture into rain. As a kind of budding climate engineer, he speculated that the widespread application of such techniques could produce both more rain (at night) and more sunshine, while serving to purify the air and reduce the frequency and severity of storms. Veraart thought that this would make the world better by rearranging climate zones that were either too hot, too cold, too wet, or too dry.

Veraart died in 1932, before Bergeron and Findeisen published their work on cloud physics. Meteorologists have minimized Veraart's contribution, even though he was using the "right" substance, by claiming that he probably did not understand the mechanism involved in the precipitation process he triggered, he did not realize that the dry ice was effective in development of ice crystals by cooling supercooled clouds, and his success was likely only a coincidence. Veraart's lack of an academic affiliation and his excessive enthusiasm led Dr. E. van Everdingen, head of the Royal Dutch Meteorological Service, to brand him a "non-meteorologist and charlatan."[17] Meteorologist Horace Byers wrote in 1974 that Veraart's vague concepts on changing the thermal structure of clouds, modifying temperature inversions, and creating electrical effects were not accepted by the scientific community.[18] Thus, instead of Veraart, it is the scientists at GE who are remembered as the pioneers in weather control.

Schaefer was trained as a machinist and toolmaker at GE and joined Langmuir's research team in 1932, specializing in building models, devices, and prototypes. He was involved in outdoor activities, including nature study, preservation, and hiking in the Adirondack Mountains. In 1940 he became widely known for his method of replicating individual snowflakes using a thin plastic coating. On July 12, 1946, Schaefer attempted to cool off a home freezer that he was using as a cloud chamber by dropping a chunk of dry ice into it. To his surprise, he saw the cold cloud instantly transform into millions of tiny ice crystals (figure 5.1). Later measurements indicated that he had reduced the temperature of the chamber from −12 to −35°C (10 to −31°F) and had generated an ice cloud from "supercooled" water droplets. His GE laboratory notebook for the day reads: "I have just finished a set of experiments in the laboratory which I believe points out the mechanism for the production of myriads of ice crystals."[19] Schaefer later recalled

> It was a serendipitous event, and I was smart enough to figure out just what happened . . . so I took the big chunk out of the chamber and used the smaller one and a still smaller one until I finally found that by producing the supercooled cloud, and then scratching a piece of dry ice held above the chamber, a tiny grain would just flood the chamber with ice crystals. So I knew I had something pretty important.[20]

The following week, on July 17, when Langmuir returned from a trip and witnessed the effect, he scribbled in his laboratory notebook "Control of Weather" above his analysis of Schaefer's discovery.[21] Schaefer recalled that Langmuir "was just ecstatic and he was very excited and said, 'Well, we've got to get into the atmosphere and see if we can do things with natural clouds.' So I immediately began to plan . . . to seed a natural cloud."[22]

Speculation was rampant that summer about the possibilities of weather control. On July 31, Schaefer made some rough calculations that indicated that if a 50-pound block of dry ice, costing $2.50, could be ground up and dispersed

5.1 Vincent Schaefer reenacting his discovery on July 12, 1946, that sparked fresh weather-control experiments, as Irving Langmuir (*left*) and Bernard Vonnegut watch. Colleagues have said that he did this on innumerable occasions for anyone who would watch. (SCHAEFER PAPERS)

into a cloud from an airplane, hundreds of thousands of pounds of snow could be generated.[23] Like the electrified sand researchers of the 1920s, Schaefer supposed that "precipitation" not reaching the ground would serve to dry out the clouds and dissipate them. "Thus," he speculated, "it would seem possible with the right arrangements—barrage or captive balloons, rockets, etc., etc., to clear areas around airports, on flight paths, or possibly to precipitate snow in mountainous regions where it could be used for water storage and sport and prevent it from being deposited in cities!"[24] Langmuir too was engaged in calculations of his own about the vast economic and practical consequences of seeding natural clouds with dry ice.

The Rainmaker of Yore

The public had not yet heard about cloud seeding in September 1946 when the midwestern novelist and screenwriter Homer Croy reminisced in *Harper's Magazine* about the rainmakers of his youth. The article was an instant anachronism: "One day when I was just a boy, my father said, 'Get ready and we'll go to town and see the rainmaker.' No work! Maybe a candy mouse. Maybe some 'lick-orish.' . . . There were always wonderful things to be had in town. It was not long before we were in the hack and jogging along the dusty road. There, on each side of us, was the suffering, gasping, dying corn."[25] Croy recalled that in the 1890s, especially during times of drought, many people sought the services of rainmakers. He and his family gathered that day with other citizens at the railroad depot where the Rock Island Railroad had sent its rainmaker to work his magic from a specially equipped boxcar. There was

> a great stirring inside the mysterious car and in a few minutes a grayish gas (that was going to save our corn) began coming out the stove-pipe hole in the roof. In no time the gas hit our noses—the most evil-smelling stuff we had ever encountered. But if it took that to make it rain, why, all well and good, we could stand it. The theory, as most of us knew by this time, was that this gas went up and drops of moisture coagulated around the particles and down came the rain. . . . It seemed simple and logical to us. Up went the gas and up went our eyes and up went our hopes . . . sometimes it took only two or three hours, sometimes it took two or three days. (215)

But by the end of the afternoon, only a little cloud, "about as big as a horseblanket," appeared and suddenly disappeared in the otherwise cloudless sky. Croy and his family returned home that evening disappointed but not disillusioned. As they prepared for bed, they heard, on the tin roof of the shed, a hopeful pitter-pat that soon became a downpour—the soaking rains had started. The next morning "everything in all the world was all right. The drought was broken. And we knew why it had been broken. . . . And we were thankful to God for the wonderful man who had come among us" (217).

In his essay, Croy relegated these events to the gullibility of a bygone era, concluding, "There is now not a farmer in all the corn belt who believes in rainmakers. . . . It hardly seems possible today that I once went to town to see a rainmaker save our crops, but I believed in it then and so did most people" (220). The timing could not have been more ironic. Croy's article was published in *Harper's* just *after* Schaefer's discovery of dry ice seeding and just *before* General Electric announced it to the public, initiating a new wave of faith and hope in weather control—and a resurgence of commercial rainmakers.

GE Tells the World

On November 13, 1946, the General Electric News Bureau announced that laboratory cold box experiments had succeeded in making snowflakes and that scientists would soon conduct an outdoor experiment to see if they could exercise "some human control over snow clouds."[26] The *New York Times* headline read, "Scientist Creates Real Snowflakes."[27] November 13 was also the day that Schaefer conducted an airborne test by dropping 6 pounds of dry ice pellets into a cold cloud over Mount Greylock in the nearby Berkshires, creating ice crystals and streaks of snow along a 3-mile path. This marked the beginning of a new era of cloud seeding.[28] Here is Schaefer's account of the test flight:

> At 9:30 am Curtis G. Talbot of the GE Flight Test Division at the Schenectady airport piloted a Fairchild cabin plane taking off from the east west runway. I was in the plane with Curt with a camera, 6 pounds of dry ice, and plans for attempting the first large scale test of converting a supercooled cloud to ice crystals. As we took off of the ground, temperature was 6°C [43°F]. In the sky were long stratus clouds isolated from each other and at an altitude of what appeared to be about 10,000 feet.
>
> We started climbing immediately and continued for more than an hour . . . [reaching a cloud at 14,000 feet that appeared to be supercooled, with temperature estimated to be −18.5°C [1.3°F]. Some brilliant iridescent colors on the edges, and the thermometer bulb beginning to show a light deposit of ice]. At 10:37 am Curt flew into the cloud and I started the dispenser in operation. We dropped about three pounds [of dry ice] and then swung around and headed south.
>
> About this time I looked toward the rear and was thrilled to see long streams of snow falling from the base of the cloud which we had just passed. I shouted to Curt to swing around and as we did so we passed through a mass of glistening snow crystals! We then saw a brilliant 22° halo and adjacent parhelia. . . . We made another run through a dense portion of the unseeded cloud during which time I dispensed about three more pounds of crushed dry ice (pellets from 5/16" down to sugar size). This was done by opening the window and letting the suction of the passing air remove it. We then swung west of the cloud and observed draperies of snow which seemed to hang for 2–3,000 feet below us and noted the cloud drying up rapidly. . . . While still in the cloud we saw the glinting crystals all over.[29]

The next lines in Schaefer's notebook reveal the true excitement of the moment: "I turned to Curt and we shook hands as I said 'We did it!' Needless to say we were quite excited. The rapidity with which the CO_2 dispensed from the window seemed to affect the cloud was amazing. It seemed as though it almost exploded the effect was so widespread and rapid." Later, back at the airport, Langmuir rushed out enthusiastically to congratulate the experimenters, praising the remarkable view from the airport control tower and exclaiming that only minutes after the cloud-seeding run had begun, he had seen long streamers of falling snow pouring out of the base of the cloud more than fifty miles away.

C. Guy Suits, GE vice president and director of research, immediately wrote a memo recommending access to a better airplane, either commercial or military, since the one operated by GE could not fly over 14,000 feet. Demonstrating his easy access to the military, he wrote, "We might want the Army Air Force to give us some help. I think a call to [Major General Curtis E.] LeMay would be helpful in this connection, particularly if [he knows] about the preliminary result of the experiment."[30]

The following day, GE told the story in detail, framing it as a triumph of scientific prediction with seemingly limitless practical possibilities: "Schenectady, NY, Nov. 14, 1946—Scientists of the General Electric Company, flying in an airplane over Greylock Mountain in western Massachusetts yesterday, conducted experiments with a cloud three miles long, and were successful in transforming the cloud into snow."[31] Langmuir claimed that this result "completely fulfilled" predictions based on laboratory experiments and calculations. If one pellet of dry ice, "about the size of a pea," could precipitate several tons of snow, he predicted that "a single plane could generate hundreds of millions of tons of snow" over mountain ski resorts, possibly diverting the snowfall from major cities. Or, depending on conditions, perhaps the seeding technique could be used to clear fogs over airports and harbors or prevent aircraft icing problems. A flurry of news reports followed leaving the lab "snowed under" by hundreds of clippings (figure 5.2). The *New York Times* read, "Three-mile cloud made into snow by dry ice dropped from plane . . . opening vista of moisture control by man." A banner headline in the *Boston Globe* announced, "Snowstorm Manufactured."[32] Louis Gathman and August Veraart rolled over in their graves.

Letters, postcards, and telegrams flooded in, too. One of them asked for indoor snow for a Christmas pageant to replace the white corn flakes used the previous year; another asked for artificial snow for a college winter carnival; and a ski operator seeking market advantage asked for advice. A search-and-rescue operation on Mount Rainier urgently asked GE to clear out the clouds so the team might be able to spot a downed aircraft. Movie producers requested tailor-made blizzards.

5.2 Avalanche of news articles received by General Electric after press releases of November 13 and 14, 1946. (SCHAEFER PAPERS)

A Los Angeles air pollution officer wrote to Schaefer, asking him for advice on how to clear the air over the city. The chairman of the Kansas State Chamber of Commerce sent a telegram to President Harry Truman, asking for relief of the drought conditions using GE technology. This stimulated a reply from Francis W. Reichelderfer, chief of the U.S. Weather Bureau, to the effect that dry ice seeding worked only in special circumstances, and even then the results were controversial, since no one had established a method to determine how much was caused by human intervention and how much by natural processes. A cane sugar producer in Hawaii wrote that he, too, had tried, in 1941, to make it rain, cooling the clouds by launching slabs of dry ice into the valley fog from a huge slingshot on the mountain summit. Since he was working with warm clouds, he would have needed an enormous amount of dry ice. A newspaper editorial wondered if GE would be forming a "snow cartel" to sell us a white Christmas.[33]

Threat of Litigation

An extremely optimistic announcement of progress in weather modification appeared in the *General Electric Annual Report* for 1947: "Further experiments in weather control led to a new knowledge which, it is believed now, will result in *inestimable* benefits for mankind."[34] When one of Schaefer's cloud-seeding attempts coincided with an 8-inch snowfall in upstate New York—earlier the weather bureau had forecast "fair and warmer"—Langmuir was quick to claim that cloud seeding had "triggered" the storm. Cloud seeding was becoming a controversial issue, and Langmuir's exaggerated claims threatened to take the company into litigious territory, far beyond the limits of normal corporate support for research.[35]

On November 18, 1946, just three days after the public learned about cloud seeding, Simeon H. F. Goldstein, an insurance broker in New York City, wrote to General Electric warning of the need for liability coverage and offering insurance services "to protect your Company against lawsuits for bodily injury and property damage resulting from artificial snowstorms produced at your direction":

> The newspapers report that your Company has developed a method of manufacturing snow, and will soon use it in the field. This is likely to produce lawsuits against your Company. Traffic accidents, as well as injurious falls by individuals, frequently result from natural snow, and are similarly likely to be caused by artificial snow. Government units, as well as large property-holders, will be put to extra expense in removing snow from roads and thoroughfares. When it melts, snow causes floods. It may also cause direct damage to property which happens to be in the open, as well as to structures which are not fully enclosed. . . . In addition to the foreseeable results, the complete novelty of the operation means that other sources of liability—unforeseeable both in their nature and extent—may exist. It would therefore seem dangerous to leave yourselves unprotected in these circumstances. May I hear from you?[36]

GE lawyers, fearing a deluge of property damage and personal inconvenience suits, immediately tried to silence Langmuir and his team. Langmuir and Schaefer, however, were riding high on a wave of publicity. They were both outspoken, enthusiastic promoters and popularizers of large-scale weather control. But Langmuir had extra clout and flaunted his Nobel laureate status. In the press and before the meteorological community, Langmuir repeatedly expounded his sensational vision of large-scale weather control and even climate control, with possible military implications. It was beginning to get pathological.[37]

Project Cirrus

In February 1947, General Electric research director Suits hurriedly called a halt to outdoor experimentation on cloud seeding and instructed Langmuir's team to serve only as advisers on Project Cirrus, a new classified cloud-seeding effort to be conducted by the U.S. Army Signal Corps, the Office of Naval Research, and the U.S. Air Force. As stated in the GE contract, the general purposes of the project were "research study of cloud particles and cloud modifications" by seeding, including investigations of liquid water content, particle sizes and distribution, and vertical cloud development.[38] They were searching for fundamental knowledge of cloud physics and chemistry to improve operational forecasting as well as practical techniques of cloud modification for military purposes or possible economic development.[39] An important clause in the contract further stipulated that "the entire flight program shall be conducted by the government, using exclusively government personnel and equipment, and shall be under the exclusive control of such government personnel." Suits notified his staff that "it is essential that all of the GE employees who are working on the project refrain from asserting any control *or* direction over the flight program. The GE research laboratory responsibility is confined *strictly* to laboratory work and reports."[40]

GE argued that the whole matter properly belonged to the government, and that the government, by suitable legislation, should both regulate the inducing of rainfall and indemnify for loss any contractor acting on the government's behalf—especially themselves. Secretary of Defense James Forrestal asked Congress for a law "to protect contractors engaged in cloud modification experiments against claims for damages by third parties,"[41] but no such legislation was forthcoming. The *Harvard Law School Record* reported:

> Today "Project Cirrus" has an annual budget of $750,000 from military and naval funds because of its war implications—bogging down enemy troops in snow and rain, clearing airfields of fog at lowest cost, and infecting induced storms with bacteriological and radiological materials. The Battle of the Bulge, in which the Nazis mobilized and attacked under supercooled fog, could have been much altered by a few pounds of dry ice.[42]

Between 1947 and 1952, Project Cirrus conducted about 250 experiments involving modification of cold cirrus and stratus clouds, warm and cold cumulus clouds, periodic seeding, forest fire suppression, and a notable attempt to modify a hurricane. Researchers in the project developed a suite of modern techniques applicable to cloud physics, including instruments for measuring temperatures and cloud properties in flight, collecting cloud droplets and ice crystals, and generating artificial nuclei.[43] Military aircraft (a B-17, later a B-29, and eventually as many as six planes) equipped with seeding devices, new instrumentation, and camera equipment operated over a 1,000-square-mile restricted flight area just north of the Schenectady airport, where the team was based. Under the auspices of Project Cirrus, Langmuir consulted with cloud seeders in Central America and corresponded with cloud seeders in Hawaii who were seeking to generate rainfall from warm convective clouds. This stimulated Langmuir's thinking about possible chain reactions in cumulus clouds seeded by as little as a single drop of water. Although the Project Cirrus staff collected and analyzed mountains of photographic and other data, the response of the atmosphere to seeding was erratic and the researchers could not obtain any definitive measures of the efficacy of artificial nucleating agents. The results from several experimental runs were spectacular, however, and the Department of Defense decided to expand the work of Project Cirrus to include rain enhancement experiments in New Mexico, forest fire suppression trials in New England, liquid water seeding of warm clouds in Puerto Rico, and hurricane modification in the Atlantic Ocean.[44]

One parallel study, the joint Air Force–Weather Bureau Cloud Physics Research Project, found that seeding did indeed produce striking visual changes in clouds, including dissipation of cold stratus decks. However, experiments with clouds over Ohio in 1948 and over California and the Gulf states in 1949 led the researchers to conclude that cloud seeding could not initiate self-propagating storms or relieve drought. The weather bureau spent $85,000 on the project in 1948 and $100,000 in 1949, with the air force supplying aircraft, personnel, and ground radar facilities.[45]

Hurricane King

In October 1947, GE announced that Project Cirrus would be intercepting a hurricane, not to "bust" it but to experiment on the effects of seeding with dry ice on a portion of a storm. Atlantic tropical storm number eight, unofficially dubbed Hurricane King, had just made a devastating pass over southern Florida and was churning in the Atlantic Ocean about 400 miles northeast of Orlando. It was expected to head farther out to sea. On October 13, the Project Cirrus team, led by navy lieutenant commander Daniel Rex and accompanied by Schaefer, bombed the heart of the storm with 80 pounds of dry ice and dropped 100 pounds more into two embedded convective towers.

The newspapers initially reported that the task force had "attacked" the storm in a "hurricane-busting" effort to reduce its winds or redirect it. It was reported in the press as "history's first assault by man on a tropical storm," an experiment with energies of nature far greater than those unleashed by the atomic bomb.[46] The official results were classified as military secrets, and Schaefer told the press

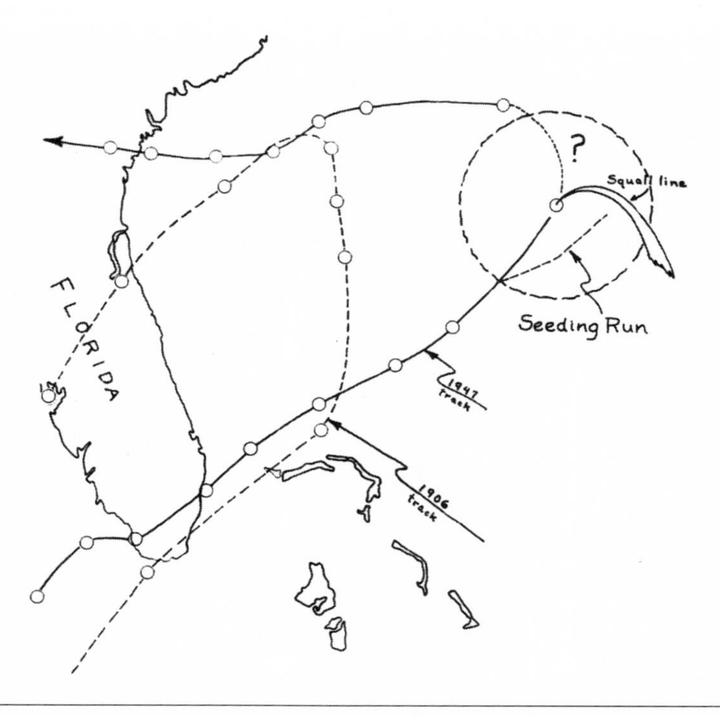

5.3 Project Cirrus hurricane-seeding experiment and the subsequent path of Hurricane King in 1947 (*solid line*), compared with the path of a 1906 hurricane (*dashed line*) that also had turned suddenly. A retrospective study by the weather bureau showed that upper-level steering currents, not seeding with dry ice, had likely caused the storm to veer suddenly. (SCHAEFER PAPERS)

that he was "not allowed to say" whether the seeding had had any visual effects.[47] Commander Rex's official report, not yet released, claimed a pronounced modification of the cloud deck that had been seeded. What happened after that, according to Langmuir, "nobody knows," since Hurricane King made a "hairpin" turn and headed west, smashing into the coast along the Georgia–South Carolina border near Savannah (figure 5.3). In Charleston, a tree fell, killing one person, and the storm caused more than $23 million in damage during its second landfall.[48] A letter in a St. Petersburg newspaper from J. M. Enders and addressed to GE research director Suits placed the blame for the devastation on "the weather tinkers of your lab" and pointed out that the people of Savannah were not so sure it was a coincidence. In fact, they were "pretty sore at the army and navy for fooling around with the hurricane."[49]

No one held the "hurricane busters" officially liable, but that would certainly not be the case today. The storm's unexpected turnaround following—if not necessarily because of—seeding dampened GE's hopes of making grandiose claims about storm control. Schaefer participated in a press conference at which evading questions was the order of the day, and he wrote in his official report: "Change in plans of the publicity angles to the project caused considerable delay and should be completely eliminated. This should be done by the assignment of a [public relations officer] to the project if it's again tried."[50] An unrepentant Langmuir admitted, "The main thing we learned from this flight is that we need to know enormously more than we do at present about hurricanes."[51] Langmuir was already looking ahead to future hurricane seasons—he hoped that the Project Cirrus team could intercept hurricanes far out at sea, fly multiple tracks through them, "and see if we cannot, by seeding them, in some way modify or shift their positions. . . . The stakes are large and, with increased knowledge, I think we should be able to abolish the evil effects of these hurricanes" (185).

Six decades later, the case of Hurricane King might serve as a warning to the Department of Homeland Security, which, as of 2008, wants to fund a new wave of research aimed at weakening the strength of tropical storms and steering them "off course." But, of course, hurricanes do not run on tracks or on a schedule, so *everyone* damaged by a modified hurricane could sue for damages—unless the government tried to place an embargo on such lawsuits.[52]

Silver Iodide

The exciting news from GE about weather control took another step in January 1947 when physical chemist Bernard Vonnegut discovered that molecules of silver iodide act as artificial nuclei and can "fool" cloud water droplets into crystallizing.[53] During World War II, Vonnegut worked in the Department of Chemical Engineering at MIT on projects related to gas warfare and with the Department of Meteorology on problems of aircraft icing. He moved to GE in 1945 and worked closely with Langmuir and Schaefer. His brother, the famous writer Kurt Vonnegut, also worked at GE as a publicist.

Following Schaefer's cold box discovery, Langmuir asked Vonnegut to do quantitative work "on the number of ice crystals produced by dry ice." This led Vonnegut to search for other agents that might initiate ice phase processes in a cloud. As he told the story five years later, "It occurred to me that if I could get something that was awfully close to ice in its crystal structure that might do the job, and I looked up in the handbook to find out what substances were close. I came across . . . lead iodide, antimony, and silver iodide."[54] Powdered lead iodide produced "a reasonable number of crystals" in Schaefer's cold box, a phenomenon that Schaefer attributed to the hexagonal shape of the molecules, but still they could not get a good result: "Well, I couldn't figure this out; so I was just puttering around and I decided to see just what happened when I put metal smoke in there [from silver]. I was amazed—the ice box was just swimming with ice crystals—colossal numbers. . . . Then I remembered silver iodide and made smoke. . . . Vaporizing silver iodide worked like a charm." Vonnegut's laboratory notebook indicates that he had identified lead iodide as an artificial nucleating agent by November 6, 1946. After numerous trials, he finally got iodine vapor and metallic silver to work on November 14: "Hallelujah! the nucleation was even more wonderful." By November 18, just four days after Schaefer's outdoor experiment, Vonnegut had found out that it was the silver iodide that did the job. Because of its hexagonal structure, silver iodide imitates ice condensation nuclei, causing "explosive ice growth" in supercooled clouds.[55]

Vonnegut soon started seeding experiments with ground-based silver iodide generators, but found it hard to tell where the smoke was going and what effect it was having. He did some inconclusive aircraft tests in December 1947, but "felt" that the experiments he conducted with GE meteorologist Ray Falconer with a silver iodide generator on the summit of Mount Washington were the most satisfying: "I feel darn sure we got some nice results . . . we caused quite a nice snow squall downwind from the generator on air as it goes up over the mountain."[56] He had yet to develop reliable techniques for following the particles and measuring their concentrations. The General Electric News Bureau, however, was quick to claim credit:

> If generators can be used on the ground to introduce silver iodide or other foreign particle nuclei into huge masses of air, it might be possible to alter the nature of the general cloud formation over the northern part of the United States during winter. . . . It would prevent all ice storms, all storms of freezing rain, and icing conditions in clouds. The amount of heat absorbed by sunlight would be changed. It should be possible to change the average temperature of some regions during winter months. [57]

Vonnegut, who was less sanguine, pointed out that silver iodide has its own problems. It is persistent in the environment and can activate long after its release in clouds of proper temperature; dry ice, though, works immediately and then sublimates. He recalled in an interview, "This is bad, I think, for commercial cloud-seeding operations, because I think they're playing with fire releasing this stuff all over the place and I think it's a shame they haven't shown any sense of public responsibility particularly when they deny it has any large scale effect . . . to stink up the atmosphere for hundreds or thousands of miles down wind producing God knows what effect is a dangerous thing."[58]

Nevertheless, Langmuir touted chemical seeding agents as being superior to natural ice nuclei because they act at higher temperatures, they do not melt or evaporate, and they can be spewed into the atmosphere over widespread areas to remain active until it snows. He echoed the GE News Bureau in making unsubstantiated claims that the chemical might eliminate severe aircraft icing, suppress hailstorms, and perhaps, since by his estimates only 200 pounds of silver iodide would be needed to seed the entire atmosphere of the United States, could result in large-scale weather or even climatic changes.[59]

The New Mexico Seedings

Project Cirrus was operating in Socorro, New Mexico, in July 1949, with Vonnegut running a test burner on the ground while the military air crews, with Langmuir and Schaefer as observers, seeded clouds aloft with dry ice. One day, one of Vonnegut's cumulus clouds "really whooped it up, and the first thing I knew there was lightning, and it was all very exciting, and I thought, 'Gee, I wonder if I'm responsible for it.'"[60] When his colleagues returned that evening with no results to report, Vonnegut ventured the suggestion that perhaps his ground generators had glaciated all the clouds, so the dry ice seeding planes could not find any for their experiments.

Noting that widespread rains were reported downwind on this day, Langmuir ordered seeding to be done periodically, once a week for eighty-two weeks, from December 1949 to July 1951. Then, even before the data were collected and other possibilities explored, he proceeded to make the outrageous claim that large-scale seven-day periodicities in the nation's weather were being caused by Vonnegut's single ground-based silver iodide generator located in New Mexico. Langmuir supported his claim by noting that the Midwest and East were moister than normal during this period, while the Southwest was drier than normal—conditions that occur naturally on a regular basis, although he did have to admit, sheepishly, that in the weeks when the generators were not operating, "rains were about forty percent greater than the other weeks."[61] Nevertheless, Langmuir went so far as to claim that severe flooding in the Midwest and the Ohio Valley, accompanied by widespread property damage and loss of life, was the result of these experiments. He apparently "proved" his result using unconventional statistical methods of his own devising.

In his insightful autobiographical memoir, Sverre Petterssen, a leader in the field of weather analysis and forecasting, reviewed his involvement with Langmuir in the 1950s. Petterssen was trained in the Bergen School of meteorology, chaired the Department of Meteorology at MIT, and served in Norwegian uniform in World War II, preparing forecasts for bombing raids by the British Royal Air Force, the Anzio landing, and, notably, the D-day invasion of Normandy. After World War II, Petterssen served as head of the Norwegian Forecasting Service, scientific director for the U.S. Air Force Weather Service, and professor and chair of the Department of Meteorology (later Geophysics) at the University of Chicago. He explained the situation:

> About 1947 Irving Langmuir, a Nobel Prize laureate in chemistry, and his group working at the General Electric Laboratories had discovered that silver iodide had a structure similar to small ice crystals. Since natural clouds, even at very low temperatures, are generally deficient in ice crystals while silver iodide can readily be produced, it seemed possible to supply silver iodide dust to cold clouds, hoping that the clouds might be "fooled into believing" that natural ice crystals were present. Thus, if the clouds could be so misled (and few doubted it) weather modification (or control) would not be much of a problem.
>
> Langmuir was unlucky and became a victim of one of the many pitfalls that nature so generously provides for scientists who venture too far outside their own field of specialization. Though a leading authority on the chemistry of crystal points and surfaces, a philosopher, and a polyhistor in general science, Langmuir did not appreciate the complexity of meteorology as a science. In the atmosphere, processes of vastly different spatial scales and life spans exist together and interact; impulses and energy are shuttled through the whole spectrum of phenomena—all the way from molecular processes to global circulations and the changes in the atmosphere as a whole. No chemist, physicist, or mathematician who has not lived with and learned to understand this peculiar nature of meteorology can pass valid judgment on how the atmosphere will react if one interferes with the details of the natural processes. Moreover, to determine whether or not the atmosphere has responded to outside interference, it is necessary to predict what would have happened had it been left alone.
>
> As I have just said, Langmuir was unlucky. For no profound reason he had left a silver-iodide generator somewhere in New Mexico and made arrangements with a local person to "burn" the generator on a weekly schedule. Using a set of readily available weather reports, Langmuir found that the rainfall had begun to vary in a weekly rhythm. The amazing thing was that the response was not just local; it was nationwide and might well be of hemispheric proportions. Langmuir, and

> many with him, concluded that the weekly injection of silver iodide from a single generator in New Mexico had excited a hitherto undiscovered natural rhythm of the atmosphere, with the result that the rainfall had yielded to the will of man. . . . In his mind, and in the minds of many others, there was but little doubt that the weather processes could be intensified or repressed to suit human needs.[62]

Langmuir was unable to accept the criticism of Petterssen or the analysis of weather bureau statistician Glenn Brier that the atmosphere frequently exhibits a *natural* seven-day periodicity.[63]

After the New Mexico incident, Suits again warned Langmuir that his field experiments and unsupported claims might put Project Cirrus at risk and expose the lab to litigation. He pointed out that Schaefer and Vonnegut were "a great deal less certain" about the interpretation of the New Mexico results than he was, and that ground-based seeding would again raise legal questions for General Electric: "If the [cloud seeding] program develops in such a direction as to subject the Company to serious hazards from a liability standpoint, it may very well become impossible for us to continue with this work." In a long letter to Langmuir, with carbon copies to Schaefer and Vonnegut, Suits reminded the team that "there has been no recent change in the law which makes it less necessary at present for General Electric personnel to be cognizant of the hazards from the standpoint of legal liability than when the agreement referred to above was reached." Suits again reminded the team that GE employees were to serve only as advisers to the government: "GE personnel *must not* engage directly or indirectly in seeding experiments which might lead to harmful weather phenomena. They *may* engage in laboratory experiments which they consider advisable and in very small scale weather experiments for confirming laboratory tests with actual meteorological conditions." Suits could not approve their publications that reported the results of large-scale modification. He issued a similar embargo on technical talks, claiming that GE was doing the experiments for the sake of humankind and was earning no profits from the activity: "I do not believe that our obligation extends to the taking of exceptional risks of damage suits as a result of any work which we may do in this field."[64]

Langmuir retired from GE on January 2, 1950, after a forty-year career with the company. The press release referred to him as a "world-famous scientist who is regarded as the greatest of modern times," a man "who continually embarks upon mental voyages in regions so nearly airless that only the mind can breathe in comfort." He invented the gas-filled incandescent lamp, the high-vacuum power tube, atomic hydrogen welding, a highly efficient smoke screen generator, and methods for artificial production of snow and rain from clouds, and received the Nobel Prize in Chemistry in 1932. GE also announced that he would continue working in a consulting capacity, primarily with Project Cirrus.[65]

A Pathological Passion

Even in retirement, Langmuir continued to make increasingly outrageous claims. He was a featured speaker at the National Academy of Sciences annual meeting held in Schenectady in October 1950. There, on the occasion of receiving the John J. Carty Award for the Advancement of Science, presented by NAS president D. W. Bronk, he reiterated his claim that silver iodide seeding in New Mexico on July 21, 1949, had produced 0.1 inch of rain over an area of 33,000 square miles and could have led to unusually heavy rains and flooding in Kansas a few days later, 700 to 900 miles downwind. Suits must have been in the audience, and he must have been seething.[66]

Distinguished meteorologist Charles Hosler tells of an encounter with Langmuir at a symposium at MIT in 1951 where the seventy-two-year-old scientist was again describing how cloud seeding had apparently changed the course of a hurricane off the coast of Florida, causing it to veer westward and hit Savannah, Georgia. When the twenty-seven-year-old Hosler, with a newly minted doctorate in meteorology, pointed out that forecasters had attributed the change in the hurricane's direction to steering currents in the larger-scale circulation, and that the small amount of ice generated by cloud seeding would have been overwhelmed by naturally occurring ice in the storm, Langmuir, in essence, replied that Hosler "was so stupid that [he] didn't deserve an explanation and that [he] should figure it out."[67] During a meeting break, Henry G. Houghton, the chair of the Department of Meteorology at MIT, took Hosler aside and explained to him that Langmuir's attitude stemmed from his belief that cloud seeding was his greatest scientific discovery and he had no time or patience to listen to objections—yet another characteristic of pathological science.

Langmuir had made such claims early and often. He spoke about how Project Cirrus had redirected Hurricane King at a meeting of the National Academy of Sciences in November 1947, only a month after the event. He made similar claims on national television on the *Today Show* when it broadcast from Schenectady, the hometown of host Dave Garroway.[68] Throughout his life, Langmuir made claims for weather control that could not be substantiated by meteorologists. Storms of controversy raged for years between Langmuir and the U.S. Weather

Bureau, although the bureau, too, had a vested interest in the techniques (chapter 8). When the weather bureau's cloud physics experimenters failed to produce significant precipitation from either summer cumulus or winter stratus, they concluded that the findings of Project Cirrus were largely unsubstantiated and the redirection of Hurricane King was a "colossal meteorological hoax."[69] They found no evidence to show spectacular precipitation effects and filed a conservative assessment of the economic importance of cloud seeding.[70]

Although Langmuir remained enthusiastic about the potential benefits of large-scale weather and even climate control, he changed his tactics in 1955 by warning of possible dangers of experiments gone awry and suggested that the cloud-seeding trials be moved to the wide-open spaces of the South Pacific. In a speech presented in Albuquerque, New Mexico, he expressed his personal belief that widespread and devastating droughts, such as the Southwest had experienced in 1949, could be triggered by trying to make it rain elsewhere, and that the floods in Kansas in 1951, in which forty-one people had died, were caused by a military cloud-seeding experiment. "We need research, much more research," he said, and it would be best to move the experiments to the South Pacific, "where there is less population" (and less likelihood of litigation).[71]

In his 1955 report to the congressionally mandated Advisory Committee on Weather Control, Langmuir said that experimenters look for "big effects," extending over continental distances, and interactions between seeding and planetary circulation patterns, including hurricanes and especially typhoons in the South Pacific: "There are obvious reasons for not experimenting with hurricanes near the coast of North America, but it would seem very important to learn how such storms can be controlled. This would require experimentation with typhoons far from any inhabited lands."[72]

Langmuir recommended three types of Pacific experiments: (1) intervention in mature storms, as Project Cirrus had done with Hurricane King, but with no one living in the way; (2) large-scale experiments across the entire region to see if regular seeding with silver iodide could trigger typhoons to start prematurely, perhaps producing more-frequent storms of lower intensity; and (3) intervention in nascent storms, not necessarily to stop the storm or prevent it from forming, but to "*control its path*" (emphasis added).[73] Langmuir wanted to go to Bikini Atoll to redirect typhoons or possibly slosh the entire Pacific basin circulation, as El Niño is now known to do. In doing so, he was expanding the nuclear analogy from "chain reactions in cumulus clouds" (with energy similar to the detonation of an A-bomb) to control of typhoons on and even beyond the energy released by H-bomb tests.[74]

Commercial Cloud Seeding

While researchers were struggling for verifiable results, an uncritical, determined, and enthusiastic band of private meteorological entrepreneurs, operating primarily in the West and Midwest, had appropriated the new technology and succeeded in placing nearly 10 percent of the land area of the country under commercial cloud seeding. The annual cost of this plan to farmers and municipal water districts was $3 million to $5 million. The spread of this practice generated numerous public controversies that pitted weather control entrepreneurs and their clients against weather bureau officials. Third parties often claimed damages purportedly caused by cloud seeding. In 1951, for example, New York City was facing 169 claims totaling more than $2 million from Catskill communities and citizens for flooding and other damages attributed to the activities of a private rainmaker, Dr. Wallace E. Howell. The city had hired Howell to fill its reservoirs and, at least initially, claimed that Howell had succeeded. When faced with the lawsuits, however, city officials reversed their position and commissioned a survey to show that the seeding was ineffective. Although the plaintiffs were not awarded damages, they did win a permanent injunction against New York City, which terminated further cloud-seeding activities; further litigation stopped just short of the Supreme Court. As discussed earlier, this prompted Colonel John Stingo to comment on the incivility of it all.[75]

"State Farmers Wage Fight For, Against Rain," reported the *Seattle Times* on June 14, 1952: "Cloud formations moving toward the Yakima and Wenatchee Valleys are being bombarded daily in secret, opposing experiments financed by wheat-growers who want rain and fruit growers who don't. One set of attacks is designed to punch holes in the clouds to bring rain. The other seeks to disperse the clouds without rainfall." Both "wet" and "dry" campaigns were being waged with competing ground-based silver iodide generators. One array was deployed by the Water Resources Corporation of Denver, which was attempting to make rain for the wheat growers; the other array, deployed by Olympia meteorologist Jack M. Hubbard, was run continuously to "overseed" the clouds and ward off rain for the fruit growers—a domestic version of cloud wars (figure 5.4).

Disasters

Although cloud seeding has never been proved to cause or augment precipitation directly, it has been implicated in weather-related disasters. On the night

5.4 Cartoon emphasizing commercial applications of weather control, accompanying Vincent Schaefer's lecture for the meeting of the Institute of Aeronautical Sciences at Rensselaer Polytechnic Institute. (SCHAEFER PAPERS)

of August 15, 1952, a sudden and appalling tragedy struck the little seaside resort of Lynmouth in Devonshire, England, when 6 to 9 inches of torrential rain drenched the area and a flash flood ripped through the town's main street, killing thirty-five people outright and injuring many others. A contemporary newsreel called it the "most destructive storm in British history," but was it a natural one? Within days of the catastrophe, there were rumors of government-sponsored experiments being conducted nearby, which the Meteorological Office and the Ministry of Defence flatly denied. Decades later, requests for weather control documents and research in the archives revealed only one thing: a gap in the records for that year. In preparation for the fiftieth anniversary of the tragedy, the British Broadcasting Corporation (BBC) obtained Royal Air Force flight records and interviewed one of the participants in the experiment, glider pilot Allan Yates, now deceased, who described the secret cloud-seeding trials going on at the time, called Operation Cumulus, alternatively known as Operation Witch Doctor.

Yates recalled, "We'd assembled in Cranfield in Bedfordshire in mid-August 1952 studying clouds. On the day I'm recalling, the weather was superb, but the cotton ball cumulus clouds were going everywhere, and it was decided to make them rain." After he and his team had sprayed "salt" into the clouds, he was told that it had rained heavily some 50 miles away: "I circled down between the clouds, still doggedly noting temperature, height, time and the rest. Eventually far below I saw a sodden-looking countryside. Toasts were drunk to meteorology. [We] certainly had made it rain." But when he and his colleagues heard the news of the flood, he recalled, "a stony silence fell on the company." More than fifty years after the event, it is impossible to say if cloud seeding really did trigger the flooding, or if it was just an unfortunate coincidence. What is clear is that the British government, anxious not to be blamed, closed down the project and denied that it had ever taken place.[76]

Other cases where cause and effect cannot be proved include Langmuir's claims that seeding redirected Hurricane King in 1947 and could have caused the Midwest floods of 1949 and 1951. So, too, was a famous flood disaster in 1972 in Rapid City, South Dakota, where, at the time, large-scale weather modification trials were under way. In that case, there was widespread official state and public support for the experimentation. A much more sinister case, however, occurred in 1986 in the Soviet Union.

For decades, the Soviet Union had seeded clouds before they reached Moscow, hoping to prevent rain from falling on big military parades. This seemed both harmless and foolish, like hail shooting or the weather control promised by the Chinese for the 2008 Summer Olympics. Evidence has recently come to light that the Soviet authorities also used cloud-seeding technology to clear the air after the Chernobyl nuclear power plant in the Ukraine exploded in 1986, melted down, and caught on fire, spewing hundreds of tons of radioactive material into the atmosphere. It was the most horrific nuclear power plant accident in history. Again, the BBC took the lead in uncovering the links to cloud seeding.[77]

Following the incident, there was a buildup of heavily radioactive rain clouds above Chernobyl. The prevailing winds were blowing toward Russia and its major cities, like Moscow and St. Petersburg, but the rain never reached that far. Instead, very heavy rains fell in rural Belarus, in a region located between Chernobyl and Moscow. In 1992 Dr. Allan Flowers, a British ecologist studying radioactivity patterns downwind of the reactor, discovered that extremely high levels of fallout had been deposited in the Gomel area of Belarus, some 60 miles north of the power plant. Many children were showing the effects of internal radiation poisoning, but how was that possible more than 100 miles from the reactor? One possibility was Soviet cloud seeding.

Eyewitnesses told Flowers of experiencing very heavy rain after the incident and noticing airplanes in the sky trailing colored smoke. After the planes had passed, the black rainfall started. Dr. Zianon Pazniak, a Belarussian scientist and politician, is convinced that to save Moscow, the Soviet authorities deliberately rained down radioactive material on Belarus without notifying local inhabitants. After issuing this accusation, he feared for his life and decided to live in exile. Authorities in Moscow denied such allegations, but in 2006, on the twentieth anniversary of the disaster, Major Alexsei Grushin, a former military pilot who received an award for cloud-seeding operations during the Chernobyl cleanup, shared his testimony:

> I am proud to say that I took part in the operation back in 1986; my comrades are proud as well. The area where my crew was actively influencing the clouds was near Chernobyl, not only in the 30-kilometer zone, but out a distance of 50, 70, even 100 kilometers. The plane was equipped with artillery shells which were filled with a seeding material, silver iodide, and we were following orders from [Moscow] regarding which zones we were required to seed. The wind direction was moving from west to east and the radioactive clouds were threatening to reach highly populated areas of Moscow, Voronezh, and Yaroslavl. If the rain had fallen on these cities it would have meant a catastrophe for millions.[78]

During this operation, Grushin, his crew, and his plane were heavily exposed to radiation. He recalls that after he landed at the airport, he was ordered to park far from the hangar and his plane was greeted by technicians wearing anti-radiation suits and carrying sensors. They approached from an upwind direction, but soon turned around and ran away from the highly contaminated plane. Grushin and others said they flew seeding missions as early as two days after the explosion and continued their operations for months.

The decision makers applied a dose of utilitarian ethics in their attempt to use the technology at their disposal to spare millions from the radioactive cloud, yet inbred secrecy, ethnic prejudice, and a horrendous and criminal lapse of judgment prevented them from trying to mitigate the effects of their actions by warning the population in advance to stay indoors and by distributing potassium iodide tablets. According to Flowers, "It is quite clear that these actions did not take place." The high levels of radiation found in Belarus have led to the frequent occurrence of leukemia, thyroid cancers, and birth defects. His informant Pazniak said the area had been devastated, and he blamed Moscow:

> I had lots of friends in the area where the cloud seeding took place. They had to move away from their homes and their children became ill. Thyroid cancer has increased among children by fifty times in the area where the cloud seeding took

> place. If there had been no cloud seeding, there would be no radiation, even if the radiation had reached that far, it would never have been on such a huge scale. They decided to keep the cloud seeding quiet. They thought that the public would never find out about it. Everything would stay a secret and nobody would need to take responsibility.[79]

This was truly a pathological situation.

* * * * *

Intervention in a weather system, any weather system, carries immense ethical considerations. One of the moral pitfalls could be that trying to modify the weather in one place could actually cause a disaster elsewhere. This is true in local- and regional-scale situations (Lynmouth and Chernobyl) as well as in large-scale instances (intervening in hurricanes and synoptic weather). It may also be true of the Earth's climate in general. Weather and climate are essentially very chaotic systems, and although they may be somewhat predictable on short timescales, in many cases surprises will arise from non-linear interactions. There are all sorts of unknowns. As Irving Langmuir told his audience at the 1953 lecture on pathological science, when you are examining new or threshold phenomena in science, it means that you do not know—*you really do not know*—whether you are seeing something important or not. In such cases, it is much better to err on the side of caution than to try to operationalize what is unknown. Ironically, by his own criteria, Langmuir's final undertaking—his involvement in weather and climate control—must be judged a pathological obsession and somewhat of a scientific dead end. Pathological science may be bad enough, but pathological engineering can actually create disasters. Remember this as you analyze the odd mixtures of wild speculation, faulty logic, poor experimental design, passionate certainty, and appeals to authority that so often arise in the fields of weather and climate control.

6

WEATHER WARRIORS

Conflict over weather control [is] the likely cause of "the last war on earth."

—EDWARD TELLER, QUOTED IN CHRISTOPHER STONE, "THE ENVIRONMENT IN MORAL THOUGHT"

IN an interview conducted in 2008, Colonel Don Berchoff, chief of U.S. Air Force Weather Resources and Programs, denied knowledge, interest, or involvement in techniques for controlling the weather: "I personally don't believe weather modification is a good thing, and I don't think the military believes in it. . . . The military does not conduct any kind of experimentation, that I understand, to control the environment to become more advantageous on the battlefield against our enemies. . . . We don't do that. . . . As far as I know."[1] We might take this at face value, or we might assume that particular individuals, of whatever rank, however highly placed or seemingly well informed, simply have no knowledge of ongoing top-secret research projects.

Just what do we know? We know that throughout history, weather has been a crucial factor in the outcome of wars and battles, and we know that the military has been a major patron in the development of weather science and services, providing logistical support and leadership for scientific field campaigns and running large-scale, even national, weather services. We know that the military has supplied important equipment for meteorological research, in some cases through new research-and-development projects in aviation, electronics, digital computing, and space and in other cases through dual-use or surplus hardware. We know that the emergence of modern meteorology is, in many ways, a product of two world wars and the cold war. We also know that in the Vietnam era, only a very few people knew about secret cloud seeding over the Ho Chi Minh Trail, originating as it did directly from the White House. This dynamic continues today. Geoscientists with high-level security clearances share associations, values, and interests with national security elites. Both groups agree on the necessity of preserving deniability in top-secret programs. We know with certainty that historically, weather and climate control have been portrayed as weapons that might be used against enemies without their knowledge—or the knowledge of lower-level operatives and the wider public.

The military roots of meteorology can be traced from the deep past through the history of the cold war and Vietnam eras. In addition to traditional goals of being able to function and prevail under all environmental conditions, weather warriors have attempted to weaponize weather control. In the early cold war era, they were particularly active in experimentation on cloud seeding, in hurricane modification efforts such as Project Stormfury, and in rainmaking efforts in Vietnam. The United Nations Convention on the Prohibition of Military or Any Other Hostile Use of Environmental Modification Techniques (ENMOD), which entered into force in 1978, marks the end of this era and serves as a landmark treaty that may have to be revisited soon to avoid or at least try to mitigate possible military or hostile use of climate control or geoengineering. If, as has been recently asserted but not yet demonstrated, "[c]limate change has the power to unsettle boundaries and shake up geopolitics, usually for the worse,"[2] it is certain that the governments of the world will have their strategic military planners working in secret on both worst-case scenarios and technological responses.

Weather in Wars and Battles

The weather has often been called the most violent variable in human affairs; that characterization could also apply to military affairs. Generals "mud" and "winter" and admiral "storm" have always had a big influence on the outcome of battles. Historians attribute the devastating defeat of the Roman general Varus and three of his legions in Germany in 9 C.E. to a combination of treachery, poor strategy, rough terrain, and bad weather; the kamikaze (divine winds), legendary protectors of Japan, destroyed Mongol emperor Kublai Khan's invading force not once but twice, in 1274 and 1281; and British history teaches that favorable winds and gales contributed mightily to the defeat of the Spanish Armada in 1588.

Military history is filled with weather lore. The Revolutionary Army's successful retreat from Long Island in August 1776 was said to have been enabled by night fog and favorable tides; five months later, General George Washington crossed the Delaware River by boat in a driving snowstorm and surprised the Hessian troops in New Jersey; and the ambush of General Nicholas Herkimer's volunteers in upstate New York in August 1777 was interrupted by the onset of a violent thunderstorm. Napoleon's attack on Russia, like those of generals before and after him, was thwarted by winter weather, while his battle plan at Waterloo was interrupted by heavy rains. In World War I, it was all quiet on the western front during mud season. In World War II, the miracle at Dunkirk took place under the cover of heavy fog, the Japanese carrier fleet skirted Pacific storms to launch its attack on Pearl Harbor, and the outcome of the Battle of Midway hinged on the ability of American dive-bombers to shield their approach behind the clouds.[3] The D-day invasion of Normandy in the unusually stormy month of June 1944 proceeded on the basis of the most critical set of forecasts in history.[4] Of course, there are many more examples, with the winning side often considering a favorable outcome as an act of Providence.

Science and the Military

A mutually supportive relationship has long existed among science, engineering, and the military. According to engineering legend, long before the birth of modern science, Archimedes designed and built a series of machines to keep the Romans at bay during the siege of Syracuse. Leonardo da Vinci's Renaissance drawings of war engines are also legendary. And in 1638, Galileo's *two* new sciences were astronomy and the strength of materials as applied to military engineering.[5]

In later centuries, scientists often "hitched a ride" with army and navy exploring expeditions; scientists utilized military scope, organization, and discipline to collect data during field campaigns; and military institutions forged their identities in part around scientific and engineering agendas, leadership, and training. As the prestige of scientists grew, linked as it was to their power over nature, whether actual or perceived, military planners took note and became major patrons. Scientists gained state funding, approbation, and political power through governmental channels with direct links to the military. The French Academy had long supported scientific research for national interests, and the Russian Academy, founded in 1724, served the interests of the tsars and, after 1917, the technical needs of the Communist Party. The National

Academy of Sciences was established by President Abraham Lincoln in 1863 to "investigate, examine, experiment, and report upon any subject of science or art" whenever called on to do so by any department of the government. In the twentieth century, the task-oriented National Research Council coordinated scientific research during America's brief involvement in World War I; and during World War II, the National Defense Research Committee, the Office of Scientific Research and Development, and the Manhattan Project demonstrated convincingly the absolute importance of promoting and supporting scientific research, development, testing, and evaluation of new weapons systems. In the cold war era, science became a prominent and permanent component of all modern militaries.[6]

Links between science and the military—in perspectives, personnel, values, budgets, scale—have grown inexorably over the years. This "pursuit of power" in modern states, however, has come at a steep price. As was the case in Irving Langmuir's pathological enthusiasm for weather control, and in many other instances of what has come to be called "big science," the military can act as a distorting force in the ongoing development of natural and engineering knowledge, specifically by imposing secrecy on new discoveries in the name of national security and seeking to weaponize every new technique, no matter how new or speculative, such as James Van Allen's discovery of the magnetosphere (chapter 7).[7]

In the relationship between scientists and the military, it is safe to say that scientists seek support from the state and access to political power, while the state (especially the military) seeks power over nature as promised (and sometimes delivered) by scientists. Of course, transcending this dichotomy is the coproduction of the military-scientific-industrial state in which the various components are by no measure independent of one another. As geoscientists pursue knowledge of the Earth, they tend to focus their investigations on those areas in which technology and military interests have made resources readily available. In doing so, they go far beyond availing themselves of commercial, state, or military patronage; they actually contribute to the commodification, nationalization, and militarization of the natural world.[8]

Meteorology and the Military

During the War of 1812, U.S. Army Surgeon General James Tilton, motivated by prevailing environmental theories of disease that linked illness and epidemics to weather and climate, issued a general order directing all the medical personnel under his command to prepare quarterly reports as part of their official duties and to "keep a diary of the weather."[9] For the next six decades, the Army Medical Department continued its support for meteorology by observing, recording, and analyzing airs, waters, and places for the protection of the health of the troops. In the 1830s, the U.S. Navy also initiated a program to collect meteorological data at navy yards and aboard its ships. As discussed earlier, the army and navy supported James Espy's storm studies in the 1840s and 1850s while simultaneously downplaying his weather control eccentricities. Not long after, Charles Le Maout and Generals Edward Powers and Daniel Ruggles developed their notions about cannonading leading to disturbed weather and enhanced rainfall.

Between 1870 and 1891, the U.S. Army Signal Office administered the national weather service, providing daily weather reports and forecasts for the benefit of commerce and agriculture. Linked to Washington by military and commercial telegraph networks, the weather service served as a national surveillance force reporting to the government on a variety of threats to the domestic order, such as striking railroad workers, Indian uprisings on the frontier, locust outbreaks, and natural hazards to transportation, commerce, and agriculture.[10] In World War I, meteorology took on new roles in warfare. Knowledge of lift, lob, and loft was needed for planes, shells, and poison gas, all of which rode the air currents. Meteorologists developed principles of battlefield climatology as they advised on how to launch and possibly survive poison gas attacks. In the newly minted field of aeronomy, or the study of conditions in the upper atmosphere, data collection from balloons, airships, and airplanes supported reconnaissance flights, the siting of aerodromes, and computations of the ballistic wind needed for long-distance artillery shelling. One proposal suggested using wind currents to carry balloons over enemy territory so that they might drop propaganda leaflets. As discussed earlier, with the rise of aviation, a desire to alter the weather, especially fogs, for the benefit of pilots got under way under military patronage.[11]

During World War II, the U.S. Army Air Forces and the U.S. Navy trained approximately 8,000 weather officers, who were needed for bombing raids, naval task forces, and other special and routine operations worldwide. Personnel of the army's Air Weather Service (AWS), an agency that was nonexistent in 1937, numbered 19,000 in 1945. Even after demobilization, the AWS averaged approximately 11,000 soldiers during the cold war and Vietnam eras. In 1954 a National Science Foundation (NSF) survey of 5,273 professional meteorologists in America revealed that 43 percent of them were still in uniform on active duty, 25 percent held Air Force Reserve commissions, and 12 percent were in the Navy Reserve. Thus almost a decade after World War II, 80 percent of American meteorologists still had military ties. Postwar meteorology also benefited from new tools such as radar, electronic computers, and satellites provided by or pioneered by the military.[12]

The importance of weather to war and weather science to the military is reflected in the history of military interest in weather and climate control, a long-term relationship that deepened and intensified after World War II.

Cold War Cloud Seeding

Early in 1947, the new cloud-seeding techniques developed at the General Electric Corporation led to crash military programs in weather control research. Could there be a weather weapon that would release the violence of the atmosphere against an enemy, tame the winds in the service of an all-weather air force, or, on a larger scale, perhaps disrupt (or improve) the agricultural economy of nations and alter the global climate for strategic purposes? At the time, Langmuir was very interested in the idea of starting a "chain reaction" in clouds—using a tiny amount of a "nucleating" agent such as dry ice, silver iodide, or even water—that could release as much energy as an atomic bomb. If this technique could be weaponized and controlled, it could be used surreptitiously and without radioactive fallout; moreover, it would be unidirectional, in that clouds seeded upwind (for example, west of the Soviet Union) would be carried to their targets by the prevailing winds. This was an attractive idea for cold warriors, since the use of weather modification as a weapon could easily be denied and any damage could be blamed on natural causes. Given the military and economic implications of the technique and the powers it promised its masters, meteorologists advised the military to launch an "intensive research and development effort."[13]

Edward Teller—cold warrior extraordinaire, father of the H-bomb, and possibly the "real Dr. Strangelove"—recalled in his memoirs that Langmuir visited him at Los Alamos in the summer of 1947 and that he was "mostly interested in talking about cloud seeding; he talked so much about the amount of damage done by a storm his seeding had caused that I began to wonder whether he saw the technique as competition to the atomic bomb."[14] Although the timescales are different by many orders of magnitude, the total amount of energy released by a single thunderstorm is equal to that of a 20-kiloton atomic bomb. Moreover, a mature hurricane of moderate strength and size releases as much energy in a day as that of about four hundred 20-megaton hydrogen bombs. Such impressive numbers—despite the technical uncertainties involved in attempting to control storms—made Langmuir's comparisons between weather modification and nuclear weapons very popular in military circles. Langmuir and his GE team had the security clearances needed to work on the Manhattan Project—but they had not participated. Metaphorically, a seeded thunderstorm became Langmuir's

A-bomb, and like his nuclear peers, he tested his techniques in the desert of New Mexico and bombed clouds with a B-29 aircraft, the sister of *Enola Gay* and *Bockscar*, the planes that had delivered atomic bombs on Hiroshima and Nagasaki. A seeded hurricane was, by analogy, Langmuir's "Super" or H-bomb, and he yearned to take his techniques to the South Pacific for basin-scale tests near Bikini Atoll.[15]

Langmuir talked openly to the press about the analogy. From a military perspective, he pointed out, cloud seeding could produce widespread drought and thus play havoc with an enemy's food supply and hydropower plants, or trigger torrential downpours sufficient to cause flooding, immobilize troop movements, and put airfields out of commission. In 1950 he claimed that weather control "can be as powerful a war weapon as the atom bomb."[16] Invoking the famous letter written by Albert Einstein to President Franklin Roosevelt in 1939 describing the potential power of an atom-splitting weapon, Langmuir recommended that the government seize on the phenomenon of weather control as it did on atomic energy. GE research director C. Guy Suits reinforced the nuclear analogy in his Senate testimony of March 1951, pointing out the "many points of similarity between the release of atomic energy and the release of weather energy,"[17] including the immense energies involved, the chain reaction mechanisms common to both, the trans-boundary problems and the need for international agreements, and similar national defense and economic implications. Suits also highlighted key differences—such as the early stage of weather modification research, its small-scale experimental needs, and its lack of top-secret processes—but he ended up "placing his bets" on Langmuir's scientific judgment and argued that a central authority was needed, modeled after the Atomic Energy Commission.

Weather control had tactical dimensions as well. In Washington, D.C., on August 28, 1947, Langmuir and Vincent Schaefer demonstrated cloud-seeding techniques for the military's top brass. Invited to the show were the chief of naval operations, Admiral Chester W. Nimitz; the commander in chief of the Army Air Forces, General Carl A. Spaatz; and the U.S. Army chief of staff, General Dwight D. Eisenhower. A total of sixteen generals, seven colonels, and two GE vice presidents attended the demonstration. Among them were the deputy chief of staff of the War Department, the chief of the Army Engineer Corps, the chief of the U.S. Army Signal Office, the head of War Department Intelligence, and the chief of research and development for the Air Corps. The Pentagon's Joint Research and Development Board had the task of examining all the implications.[18]

By October, GE was discussing plans to develop "bullets of compressed carbon dioxide or silver iodide. Shot from the nose of a plane, these fifty-caliber tracer

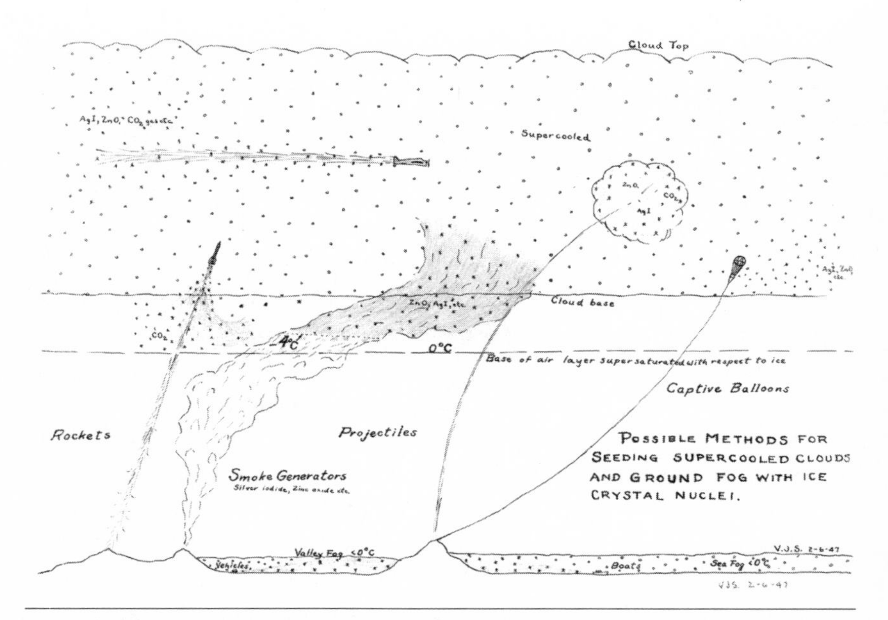

6.1 "Possible Methods for Seeding Supercooled Clouds and Ground Fog with Ice Crystal Nuclei," February 6, 1947: Vincent Schaefer's military-inspired diagram for barraging the clouds. Techniques include delivery of seeding agents silver iodide (AgI), zinc oxide (ZnO), carbon dioxide (CO_2), and gas using aircraft, rockets, smoke generators, projectiles, and captive balloons. (SCHAEFER PAPERS)

bullets might cover a range of two miles in 30 seconds, change supercooled moisture in the path of the plane to ice crystals, and thus continuously dissipate an icing condition as the plane travels"—sort of shooting your way through the clouds.[19] Remington Arms was the initial contractor, but the Army Ordnance Department soon took over the task.[20] Schaefer sketched pseudo-military schemes to barrage the clouds with seeding agents, an array of armaments for battling the clouds that would have made General Robert Dyrenforth jealous (figure 6.1).

The commercial cloud seeder Irving Krick was a weather warrior, too. He was so sure that weather modification worked that he once testified before a Senate subcommittee, "The nation that first acquires control of the weather shall be the leading power in the world." He envisioned wild strategic applications, such as enhancing artificial precipitation, increasing radioactive fallout over nuclear targets, and dispersing bomb plumes in case of an attack on our own cities. In his view, waging weather warfare by producing both droughts and floods was well within the realm of possibility.[21]

Because of the terrifying implications of the new technology, Senator Clinton P. Anderson (D-New Mexico) proposed federal regulation of rainmaking and related weather activities and introduced a bill in 1951 to provide for studies of the possible use of weather control in military operations. The Department of Defense viewed this idea as a threat to its autonomy and categorically opposed any new laws or agencies. In this, the department found strong support from the American Meteorological Society and the U.S. Weather Bureau. Meteorologist Horace Byers pointed out how unfortunate the analogy between weather modification and atomic weapons was at the time, since the weather bureau "was in the midst of a difficult task of assuring the public that atomic explosions were not changing and could not change large-scale weather."[22] As the agency responsible for guiding public policy in such matters, "it was forced into the unpleasant position of trying to restrain Langmuir who, because of his high standing in the scientific community, had strong support from scientists and the general public alike" (13).

Military Research

Cloud-seeding technology seemed to have such great military potential that, at the urging of Langmuir and Teller, Vannevar Bush, an MIT-trained engineer and a Washington insider, brought the issue to the attention of Secretary of Defense George C. Marshall and General Omar Bradley, chairman of the Joint Chiefs of Staff. This was in 1951. Bradley immediately convened a "cushion committee" consisting of an admiral, a general, and weather bureau chief Francis Reichelderfer, which in turn appointed the special scientific Ad Hoc Committee on Artificial Cloud Nucleation (ACN), chaired by meteorologist Sverre Petterssen, director of scientific weather services for the U.S. Air Force. In his memoirs, Petterssen referred to the ACN as an innocuous-sounding name "that did not suggest interest in secret weapons."[23] To add camouflage (Petterssen's words), Dr. Alan T. Waterman, director of the NSF, was appointed a member.

At the direction of General Bradley, and with the hope that a secret weapon might emerge from this technology, Petterssen's ACN conducted a brief survey of the state of the field and recommended a program of technology development and statistically controlled experiments "to clarify major uncertainties."[24] The U.S. Air Force, working with the University of Chicago, tried to modify thunderstorms by bombing cold clouds in the Midwest with dry ice and bombing warm clouds in the Caribbean with liquid water to test Langmuir's chain reaction theory. For what purpose? The navy seeded mid-latitude storm systems in an attempt to evaluate Langmuir's claim that large-scale weather systems could be controlled. Again, for what purpose? The army tried to cut holes in cold stratus clouds using dry ice seeding. These tests were done in strategically sensitive areas of West Germany, Greenland, and Ellesmere Island. An air force project, however, determined that the most likely cause of ice fogs at air bases in the Arctic regions was pollution from the military's own motor vehicles and aircraft. The army contracted with the consulting firm Arthur D. Little to try to modify warm stratus and fog using electrical and chemical agents, but with little success. Other ACN-sponsored experiments involved documenting the meteorological effects of nuclear explosions, trying to suppress jet contrails, and, in accordance with Schaefer's vision, developing small tactical seeding rockets.[25]

This series of experiments, run by the air force, the navy, and the army, was every bit as military-oriented as Project Cirrus, with better scientific advice and much better statistical controls (Petterssen claimed the "meteorological lambs" could no longer be thrown to the "statistical wolves"). The experiments ended in 1954, but because of security requirements, the final report was not published until 1957, when it appeared in a limited-circulation monograph published by the American Meteorological Society. The report claimed that the experiments were inconclusive and did not produce any significant results and that more basic research in cloud physics was needed before attempts at weather modification could be justified.[26] But did the military reveal all that it had learned? Could it?

Public Perceptions

The ongoing debate over private, public, and military cloud seeding prompted Congress to pass a law establishing the Advisory Committee on Weather Control (ACWC) in 1953. Chaired by a presidential appointee, retired navy captain Howard T. Orville, the committee conducted no new experiments of its own but made site visits and collected testimony.[27] Orville placed the military theme squarely before the public in a 1954 cover article in *Collier's* that included scenarios for using weather as a weapon of war (figure 6.2). In one scenario, airplanes would drop hundreds of balloons containing seeding crystals into the jet stream and then return to their air bases. Downstream, when fuses on the balloons exploded, the seeding agents would fall into the clouds, initiating

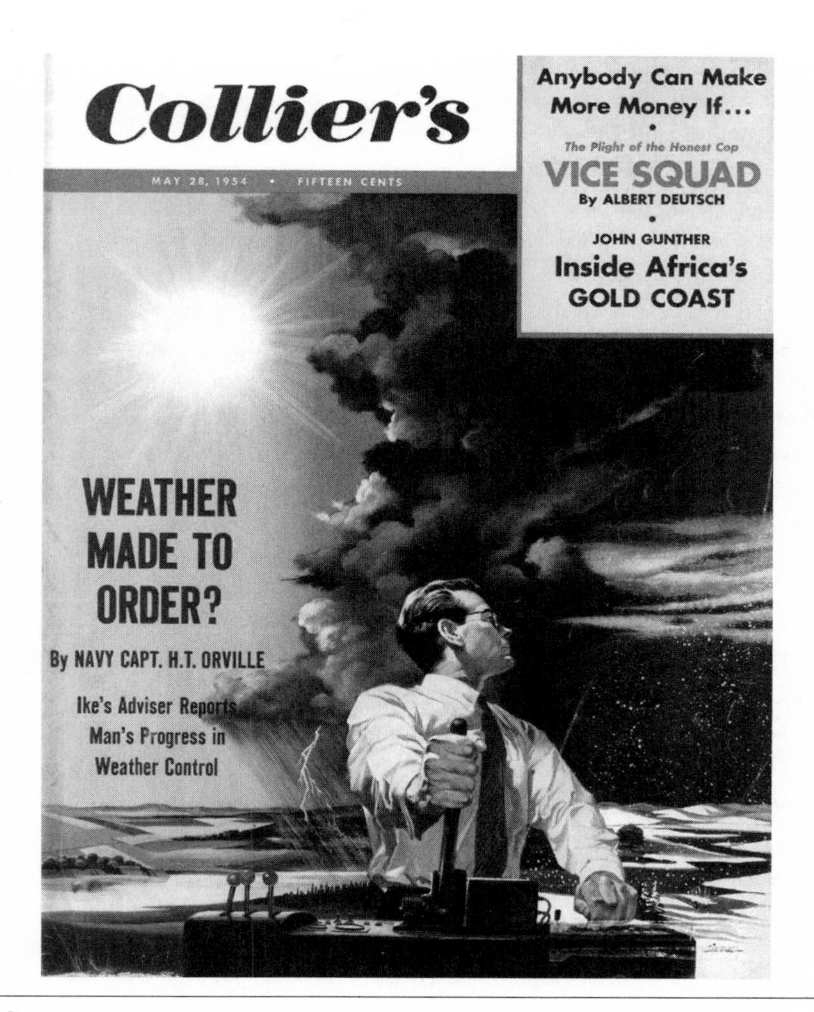

6.2 Technocrat pulling the levers of weather control. Howard T. Orville's article "Weather Made to Order?" focuses on weather warfare. (*COLLIER'S*, MAY 28, 1954)

thunderstorms and disrupting enemy operations. The illustrations indicated that because of prevailing westerly winds, this technique might work as a unidirectional weapon for NATO nations confronting a tank column from the eastern bloc.

Although in Orville's assessment total weather mastery would be possible only after four decades of intensive research, the spin-offs from this work, when combined with the maturation of electronic computers, would provide a completely accurate system of weather forecasting, perhaps within a decade: "I think it entirely probable that, in 10 years, your daily weather forecast will read something like this: 'Freezing rain, starting at 10:46 A.M., ending at 2:32 P.M.' or 'Heavy snowfall, seven inches, starting today at 1:43 A.M., continuing throughout the day until 7:37 P.M.'"[28] This sort of accuracy of prediction, even without weather control, would have major consequences for military operations. Orville, like Krick, was echoing the surety of earlier determinists. Orville thought it was "conceivable that we could use weather as a weapon of warfare, creating storms or dissipating them as the tactical situation demands" (25). A more insidious technique would strike at the enemy's food supply by seeding clouds to rob them of moisture before they reached enemy agricultural areas. Orville wrote: "We might deluge an enemy with rain to hamper a military movement or strike at his food supplies by withholding needed rain from his crops. . . . But before we can hope to achieve all the *benefits* I have outlined, hundreds of meteorological unknowns must be solved at a cost of possibly billions of dollars" (25–26; emphasis added). Although speculative and wildly optimistic, such ruminations from an official source helped fuel a weather race with the Russians and the rapid expansion of meteorological research in all areas, but especially in weather modification.

In December 1957, while Americans were still reeling from the psychological impact of the launch of the Soviet Union's first Earth-orbiting satellite, the *Washington Post and Times Herald* informed its readers that there was a "new race with the reds" in the form of weather warfare.[29] *Newsweek* picked up the story in its next issue. Again, Orville, whose final ACWC report was about to be released, was quoted indicating that the need to keep ahead of the Russians was more clear than ever: "If an unfriendly nation gets into a position to control the large-scale weather patterns before we can, the result could even be more disastrous than nuclear warfare."[30] The article also quoted Teller, an expert on hydrogen bombs but not on weather control, who told the U.S. Senate Preparedness Subcommittee: "Please imagine a world . . . where [the Soviets] can change the rainfall over Russia . . . and influence the rainfall in our country in an adverse manner. They will say, 'we don't care. We are sorry if we hurt you. We are merely trying to do what we need to do in order to let our people live'" (54). Henry Houghton at MIT expressed the same concerns: "I shudder to think of the consequences of a prior Russian discovery of a feasible method of weather control. . . . An unfavorable modification of our climate in the guise of a peaceful effort to improve Russia's climate could seriously weaken our economy and our ability to resist" (54). At the time, by some estimates, the Soviet Union employed some 70,000 hydrometeorologists, more than three times as many as the United States. Harry Wexler reported that the Russians seemingly had unlimited access to funding. One of their leading academicians, K. N. Fedorov, had wondered if the Soviet Union was engaged in an international "struggle for meteorological mastery," but paranoid cold warriors thought perhaps that meant also meteorological mastery over the free world.[31] The distinguished aviator-engineer Rear Admiral Luis de Florez urged the U.S. government to "start now to make control of weather equal in scope to the Manhattan District Project which produced the first A-bomb." He added the by-now-obvious militant twist: "With control of the weather the operations and economy of an enemy could be disrupted. . . . [Such control] in a cold war would provide a powerful and subtle weapon to injure agricultural production, hinder commerce and slow down industry."[32]

Project Stormfury

In 1954 three damaging landfalling hurricanes, Carol, Edna, and Hazel, convinced members of Congress that funding was needed for a National Hurricane Research Project (NHRP), to be directed by the weather bureau using equipment on loan from the air force. Official histories claim that the NHRP was established to measure and model the storms, but in 1958 the research group employed silver iodide in an unreported and unpublicized attempt to modify Hurricane Daisy off the coast of Florida—in spite of the public relations disaster that had followed the Project Cirrus seeding of Hurricane King in 1947. Again, in 1961 Hurricane Esther reportedly displayed some apparent weakening after seeding. This encouraged meteorologists to develop a more aggressive hurricane modification project called Project Stormfury, a collaboration, initially between the weather bureau and the navy but later involving the air force, that operated from 1962 to 1983.[33]

Both Robert H. (Bob) Simpson and Joanne (Malkus) Simpson were early directors of the project, which involved a team of scientists and technicians flying into mature hurricanes and seeding them using military equipment. According to an oral interview with Bob Simpson, Project Stormfury was conceived after a high-altitude visual reconnaissance flight into Hurricane Donna made by meteorologist Herb Riehl in 1960 that indicated a concentrated, perhaps supercooled, outflow region above the storm. Riehl called this feature the "chimney cloud"; Simpson thought it was worth trying to seed it to attempt to cause the eye of the hurricane to expand and perhaps weaken the storm. In 1967 one of the directors of Stormfury compared hurricane hunting to big-game hunting: "For scientists concerned with weather modification, hurricanes are the largest and wildest game in the atmospheric preserve. Moreover, there are urgent reasons for 'hunting' and taming them."[34]

The NSF provided some initial funding for Stormfury, but it was the U.S. Navy that was most interested in modifying and hopefully controlling the air–ocean environment. The navy's vision of weather control involved using fog and low clouds as screens against enemy surveillance, calming heavy seas, and redirecting violent storms both to enhance its own operations and to interfere with enemy plans and operations. The wish list included the capability to change the intensity and direction of hurricanes and typhoons; produce rain, snow, or drought as desired; and "modify the climate of a specific area"—all for the sake of military operations. As the navy saw it, the military problem in the field of weather modification and control was "to alter, insofar as possible, the environment surrounding the task force or target area so that the success of the naval operation is enhanced."[35] The Navy Weather Research Facility in Norfolk, Virginia, was designated a center for weather control experiments aimed at better understanding and controlling a vast array of atmospheric phenomena. The Naval Research Laboratory was involved in developing the equipment and instrumentation, while the Naval Ordinance Test Station, in China Lake, California, led by atmospheric scientist Pierre St. Amand, specialized in pyrotechnic units for seeding clouds with silver iodide. The navy's vast array of instrumentation for basic cloud physics and atmospheric research and the availability of aircraft and crews made it a logical partner for scientists seeking support for field studies.

Frustration mounted as Stormfury scientists began to realize that their hurricane-seeding hypotheses were flawed. First of all, hurricanes contain very little of the supercooled water that is necessary for effective silver iodide seeding. Also, the effects of seeding were so small that they were impossible to measure. Morale plummeted when Stormfury scientists learned that the navy intended to weaponize their research. St. Amand, in particular, wanted to learn how to intensify and steer hurricanes, certainly for tactical advantage but also perhaps as weapons of war. Bob Simpson recalled, without specifying the details, that St. Amand did not share his scientific values and "succeeded in throwing monkey wrenches into the works."[36] In 2007 Joanne Simpson, then a retired NASA employee, recalled in an on-camera interview, "I thought it was terrible—I mean all my life I've tried to work for the betterment of the planet and the people in a small way—and to use what I have done as some kind of a military thing. I obviously am very concerned and not happy about it."[37]

In October 1962, just as Stormfury was getting under way, the Cuban missile crisis brought the world to the brink of nuclear war. A year later, Fidel Castro accused the United States of having waged strategic weather warfare by changing the course of Hurricane Flora. Although Flora was not seeded, its behavior was indeed suspicious. It hit Guantánamo Bay as a Category 4 storm and made a 270-degree turn, lingering over Cuba for four full days, with intense driving rains that caused catastrophic flooding, resulting in thousands of deaths and extensive crop damage. Nor was Cuba alone. Mexico denounced the United States for having caused a protracted drought "resulting from cloud seeding." The response to these complaints, according to Bob Simpson, involved "restrictions of area and of conditions in which seeding would be allowed, restrictions to such degree that little hope remained to demonstrate statistically that hurricanes could be usefully seeded."[38] Meanwhile, plans were afoot to use operational cloud seeding in a real war—over the jungles of Vietnam.

Cloud Seeding in Indochina

Weather warfare took a macro-pathological turn between 1966 and 1972 in the jungles over North and South Vietnam, Laos, and Cambodia when the U.S. military conducted secret operations intended to generate rain and reduce "trafficability" along portions of the Ho Chi Minh Trail, which Hanoi used to move men and matériel to South Vietnam. In March 1971, nationally syndicated columnist Jack Anderson broke the story about U.S. Air Force rainmakers in Southeast Asia in the *Washington Post*, a story confirmed several months later with the publication of the *Pentagon Papers* and splashed on the front page of the *New York Times* in 1972 by Seymour Hersh.[39] These reports confirmed that the U.S. government had tested its techniques in Laos in 1966 and had begun a top-secret program of operational cloud seeding in and around Vietnam in 1967. The code name of the field trial was Project Popeye, and the much larger operational program was known to the air force fliers as Operation Motorpool, sometimes referred to in news reports as Intermediary-Compatriot.

In October 1966, Project Popeye, a clandestine, all-service military/civilian experimental program, seeded the skies over southern Laos to evaluate the concept of impeding traffic on Viet Cong infiltration routes by increasing the amount of rainfall and the length of the rainy season. It was hypothesized that excess moisture would soften road surfaces, trigger landslides, wash out river crossings, and in general maintain saturated soil conditions longer than would normally be expected. After seeding about sixty-eight cloud targets, the Popeye experimenters concluded, using their own techniques of analysis, that their interventions had caused "significant" increases in both cloud growth and precipitation and the operational feasibility of the technique had been clearly established.[40] St. Amand, who had designed the seeding flares and was leading the project, claimed that "the first [cloud] we seeded grew like an atomic bomb explosion and it rained very heavily out of it and everybody was convinced with that one experiment that we'd done enough."[41] General Dyrenforth would have concurred.

Operation Motorpool, which began on March 20, 1967, was conducted by air force fliers each year during the rainy monsoon season until July 5, 1972. This was done with the full and enthusiastic support of President Lyndon B. Johnson, Secretary of Defense Robert S. McNamara, and the U.S. State Department. General William Westmoreland was one of the few individuals privy to the details. The governments of Thailand, Laos, and South Vietnam were not informed, nor were the American ambassadors to those countries. After 1969, the administration of President Richard M. Nixon continued the program—and the secrecy.[42]

Operating out of Udorn Air Base, Thailand, the Fifty-fourth Weather Reconnaissance Squadron flew three WC-130 and one or two RF-4 aircraft in more than 2,600 seeding sorties, expending almost 50,000 flares over a period of approximately five years at an annual cost of approximately $3.6 million.[43] Air force pilot Howard Kidwell told how, out of curiosity, he volunteered for a secret mission, code-named Motorpool, and once he was approved for a higher-level security clearance, was involved in trying to make rain over the Ho Chi Minh Trail:

> During the rainy season each crew flew once a day, on the average, in addition to regular missions. A "scout" plane (WC-130) would call back and "scramble" us—giving us a flight level, which was usually 19,000 [feet]. We would go into the roll cloud (or whatever you WX guys call it) by the side of each thunderstorm. When it got to raining like crazy we would pickle off a cart [fire a rack of silver iodide flares], count to 5, pickle off another one, and then you were out in the blue, made a 360 degree turn and, like magic, another thunderstorm had usually formed and you did the same thing again.[44]

Although some claimed that Operation Motorpool induced from 1 to 7 inches of additional rainfall annually along the Ho Chi Minh Trail, no scientific data were collected to verify the claim. General Westmoreland thought there was "no appreciable increase" in rain from the project. Even if the cloud seeding had produced a tactical victory or two in Vietnam (it did not), the extreme secrecy surrounding the operation and the subsequent denials and stonewalling of Congress by the military resulted in a major strategic defeat for military weather modification.[45]

Typical of the cover-up during this period was the Air Weather Service annual survey report on weather modification for 1971, which contained brief accounts of cold and warm fog dissipation experiments, one precipitation augmentation trial, and illustrations of its equipment; of course, there was no mention of the (still) top-secret Operation Motorpool.[46] Even after the scandal broke, the official history of AWS weather modification in the period 1965 to 1973 contained no mention of military cloud seeding in Vietnam, admitting only, in vague and bland terminology, that "AWS's *current* operational weather modification capabilities include airborne and ground based cold fog dissipation and precipitation augmentation" (emphasis added).[47] Under the heading "Precipitation Augmentation," the report claimed that AWS efforts "have been few indeed" but did admit to having seeded over the entire Philippine archipelago in 1969 for drought relief for the benefit of agriculture. A short section titled "Other Activities" mentioned hurricane seeding in Project Stormfury and "participation" in several non-AWS weather modification projects, both as observers and as project workers, to keep abreast of the field and to find new techniques applicable to air force and army operations. Read between the lines.

In 1973 the National Academy of Sciences issued a report, *Weather and Climate Modification:* Problems *and Progress* (emphasis added). The panel, chaired by Thomas Malone, a cold war–era meteorologist with high-level security clearances, prefaced its report with this bald statement: "During the course of this study, no attempt was made by the Panel to examine . . . or to ascertain the existence of classified experimental programs in weather modification."[48] Yet the field's largest *problem* at the time was the recently revealed militarization of cloud seeding in Vietnam. The prime example of stonewalling, however, came from President Richard Nixon's secretary of defense, Melvin Laird, who told the Senate Foreign Relations Committee in 1972 that there was no cloud seeding going on over North Vietnam but never mentioned that Operation Motorpool was still functioning over Laos, Cambodia, and South Vietnam.[49]

Project Popeye and Operation Motorpool were neither the first use of weather modification as a weapon of war nor the first use in Asia. Cloud seeding was apparently used in Korea in 1950 to clear out cold fogs. In 1954 the French High Command announced in connection with the besieged French forces at Dien Bien Phu that "it will try to wash out Vietminh communication routes from Red China with man-made rainstorms as soon as cloud conditions permit."[50] Confirming this, a Vietnamese account of the battle reported that the French had shipped 150 baskets of activated charcoal and 150 bags of ballast from Paris "for the making of artificial rain aimed at impeding our movement and supply."[51] Moreover, the Central Intelligence Agency seeded clouds in South Vietnam as early as 1963 in an attempt to disperse demonstrating Buddhist monks after it was noticed that the monks resisted tear gas but disbanded when it rained. Cloud-seeding technology had also been tried, but proved ineffective, in drought relief efforts in India, Pakistan, the Philippines, Okinawa, and elsewhere. All the programs were conducted under military sponsorship and had the direct involvement of the White House. In 1967 St. Amand participated in Project Gromet, a secret effort to employ weather modification in India to mitigate the Bihar drought and famine and to achieve U.S. policy goals in this strategically important region.[52]

Operation Motorpool, made public as it was at the end of the Nixon era, was called the Watergate of weather warfare. Some argued that environmental weapons were more "humane" than nuclear weapons. Others suggested that inducing rainfall to make travel more difficult was preferable to dropping napalm; and the Fifty-fourth Weather Reconnaissance Squadron was directed, in the jargon of the era, to "make mud, not war." St. Amand tried to put a benign spin on the project when he claimed that "by making the trail more muddy and trafficability difficult, we were hoping to keep people out of the fight."[53] Philip Handler, president of the National Academy of Sciences, represented the mainstream of scientific opinion, however, when he wrote to Senator Claiborne Pell (D-Rhode Island): "It is grotesquely immoral that scientific understanding and technological capabilities developed for human welfare to protect the public health, enhance agricultural productivity, and minimize the natural violence of large storms should be so distorted as to become weapons of war."[54] Prominent geoscientist Gordon J. F. MacDonald observed that the key lesson of the Vietnam experience was not that rainmaking is an inefficient means for slowing logistical movement on jungle trails but "that one can conduct covert operations using a new technology in a democracy without the knowledge of the people."[55] The dominant opinion was that seeding clouds—like using Agent Orange or the Rome Plow, setting fire to the jungles or bombing the irrigation dikes over North Vietnam—was but one of many sordid techniques involving war on the environment that the military used in Vietnam.

ENMOD: Prohibiting Environmental Modification as a Weapon of War

In 1972 Senator Pell, following the hearings, introduced a resolution calling on the U.S. government to negotiate a convention prohibiting the use of environmental or geophysical modification activities as weapons of war. Testifying to the Senate, Richard J. Reed, president of the American Meteorological Society, cited earlier bans on chemical and biological warfare and atmospheric nuclear testing and urged the government to present a resolution to the United Nations General Assembly that pledged all nations to refrain from engaging in weather modification for hostile purposes. Citing a 1972 public policy statement of the society, he referred to the primitive state of knowledge in the field and the difficulties of controlled experimentation during military operations. The testimony of other prominent atmospheric scientists stressed the need to protect open and peaceful international scientific cooperation.[56] Despite the opposition of the Nixon administration, the Senate adopted the resolution in 1973 by a vote of 82 to 10. Representative Donald M. Fraser (D-Minnesota) led a parallel effort in the House.

In May 1974, Senator Pell placed the formerly top-secret Department of Defense briefing on cloud seeding in Vietnam into the public record. Less than two months later, at the Moscow summit, President Nixon and Soviet General Secretary Leonid Brezhnev signed the "Joint Statement Concerning Future Discussion on the Dangers of Environmental Warfare," expressing their desire to limit the potential danger to humankind from the use of environmental modification techniques for military purposes whose effects would be "widespread, long-lasting and severe." This wording of the communiqué, favored by the National Security Council, presented the fewest constraints on the military, since it seemed to indicate that only conjectural and highly impractical techniques of climatic and large-scale environmental modification, such as climate engineering, would be covered, while more or less operational techniques of weather modification, including rainmaking and fog dispersal, whose effects were considered limited in time and place, were to be excluded from the discussion.[57]

Within a month, the Soviet Union, realizing the weakness of the U.S. position on cloud seeding in Vietnam and taking full advantage of the Watergate crisis, seized the diplomatic initiative by unilaterally bringing the issue of weather modification as a weapon of war to the attention of the United Nations. The Soviet proposal did not limit the treaty to a bilateral agreement, nor did it limit it to effects that were "widespread, long-lasting and severe." According to Soviet ambassador Andrei Gromyko, "It is urgently necessary to draw up and conclude an international convention to outlaw action to influence the environment for military purposes."[58] The draft convention unveiled by the Soviet Union in September 1974 sought to forbid contracting parties from using "meteorological, geophysical or any other scientific or technological means of influencing the environment, including weather and climate, for military and other purposes incompatible with the maintenance of international security, human well-being and health, and, furthermore, never under any circumstances to resort to such means of influencing the environment and climate or to carry out preparation for their use."[59]

The UN General Assembly, taking note of the Soviet draft convention, decided that the subject deserved further attention and, with the United States abstaining, voted to turn it over to the Conference of the Committee on Disarmament. To avoid further embarrassment, the administration of President Gerald R. Ford (Nixon had resigned) insisted that the qualifiers "widespread, long-lasting and severe" be put back into the convention. The final treaty, Convention on the Prohibition of Military or Any Other Hostile Use of Environmental Modification Techniques, was a watered-down instrument that applied only to environmental effects that encompass an area on the scale of several hundred square miles, last for a period of months (or approximately a season), and involve serious or significant disruption or harm to human life, natural and economic resources, or other assets. Such language implicitly legitimized the use of cloud seeding in warfare, the diversion of a hurricane, and other, smaller-scale techniques. The convention, however, does not prohibit "the use of environmental modification techniques for peaceful purposes."[60] It was designed to be of unlimited duration and contains provisions for periodic meetings of the parties to assess its effectiveness and for emergency meetings to respond to perceived violations.

ENMOD was opened for signature in Geneva on May 18, 1977. It was signed initially by thirty-four states, including the United States and the Soviet Union, but did not enter into force until October 5, 1978—ironically, when the Lao People's Democratic Republic, where the American military had tested Project Popeye and had used weather modification technology in war only six years earlier, became the twentieth nation to ratify it. After a delay of more than a year, the convention entered into force for the United States on January 17, 1980, when the U.S. instrument of ratification was deposited with the United Nations Secretariat.[61]

When the wording of ENMOD was being negotiated, environmentalists were disappointed with the process and urged the United States not to ratify the treaty. They saw many flaws in the document, including its vague wording, its unenforceable nature, its overly high threshold for violations, and the fact that it dealt only with intentionally hostile environmental modification. Moreover, it did not prohibit research and development in the field and applied only to parties that had ratified or acceded to the convention. Jozef Goldblat, vice president of the Geneva International Peace Research Institute, commented: "Evidently, certain powers preferred not to foreswear altogether the possibility of using environmental methods of warfare and sought to keep future options open."[62] This was precisely what the U.S. military wanted. The Air Weather Service was of the opinion that the treaty's language was so vague that it did not affect its program in weather modification at all, and the Military Airlift Command was instructed to retain its capabilities in this area. For the military, the deciding factor was not the ENMOD convention but the fact that weather modification technology had "little utility" or "military payoff" as a weapon of war. Federal funding for all weather modification programs was collapsing by this time, and by 1978 the Department of Defense claimed that its operational programs were directed solely at fog and cloud dispersal, while military research funding continued in cloud physics, computer modeling, and new observational systems. Dan Golden, a senior scientist with the National Oceanic and Atmospheric Administration (NOAA) and an alumnus of Project Stormfury, observed in a 2008 interview that ever since ENMOD, "our defense department has, at least to my knowledge, not engaged in weather modification activities, and if you ever ask them if they are supporting weather modification activities, they strongly deny it. However, there have been recent workshops sponsored by the defense department on various types of weather modification."[63]

Two conferences reviewing ENMOD have been held since its ratification, one in 1984 and one in 1992. The 1984 review conference pushed, without success, to expand the scope of the treaty and to reduce the threshold for violations. The 1992 meeting was influenced by the first Persian Gulf War, which included belligerent environmental acts such as torching oil wells. This conference expanded the convention to cover herbicides and various "low-tech" interventions such as using fire for military purposes. As of this writing, however, the ENMOD treaty has not been used formally to accuse a country of a violation.[64]

Of relevance to climate engineering, ENMOD prohibits environmental modification techniques that change "through the deliberate manipulation of natural processes—the dynamics, composition or structure of the Earth, including its biota, lithosphere, hydrosphere and atmosphere, or of outer space."[65] This restriction would be relevant on the scale of climate or ocean engineering. Today's climate engineers emphasize the altruistic aspects of their work—saving the world from global warming, while also declaring "war" on global warming. They are fully aware, however, of the military implications of the techniques they are developing. In this case, a revised version of ENMOD may be the best hope of providing the international community with sufficient diplomatic leverage to stop any unacceptable collateral damage from geoengineering, or even to intervene if rogue states or terrorist groups were to employ these techniques.

While purposeful military or hostile intent would be required to trigger the existing convention, all climate-engineering schemes involve deliberate manipulation of the dynamics, composition, or structure of the Earth, and all such schemes carry the potential for "widespread, long-lasting and severe" harm on national, regional, and possibly global scales. At a minimum, ENMOD will have to be revisited under its provisions for consultation of the parties before any large-scale climate engineering projects are field-tested or deployed and before any human or environmental damage is either threatened or done. The present war on global warming must be viewed as the outgrowth of a long historical process in which military metaphors are much more than metaphors. They are hard-nosed realities influencing the course of scientific research, military policy, and perhaps most tellingly, our attitudes toward nature.

* * * * *

The history of meteorology and military history have many points of significant overlap and mutual influence. Weather warriors have long sought to take advantage of natural phenomena and, in the twentieth century, to manipulate them for military advantage. The interaction of science and the military seems to be well on its way to fulfilling a Faustian bargain struck in the early modern era if not before. Weapons systems of the past and current centuries have increasingly been based on science; they have also been increasingly lethal (especially to civilians), increasingly toxic, and increasingly pathological. Physics, chemistry, and biology have weaponized the atom, molecule, virus, and bacterium, while the geosciences have militarized the global environment in the air, under the seas, and in outer space. In the cold war era, it was presumed that clouds, storms, and even the climate, like any other natural phenomenon, could be controlled and weaponized. Nano-scale warfare meets geo-scale warfare. It was further presumed that a weather warfare race, analogous to the space race, was under way and that the other side was probably ahead. All was fair in war, especially surreptitious programs.

The cases presented here go beyond simple military support or patronage for science. They clearly document the interpenetration of values and perspectives among meteorologists and military officers. Project Cirrus, Project Stormfury, and their kin were all too common during the cold war. When military cloud seeding in Vietnam was revealed in the press, it caused an immediate firestorm of controversy. People were concerned at the time that we had opened a Pandora's box of evil and we really did not know where it might lead. Ultimately, it led to international embarrassment for the United States and the ratification of the rather toothless ENMOD convention. But if ENMOD was born from abuse, can it be revised and reinvigorated to prevent larger abuses?

In the decades following the ratification of ENMOD, the rhetoric of the meteorological community emphasized scientific internationalism and the free exchange of data and information, even as much of its funding continued to flow from cold war military sources. The showcase international research collaboration of the 1980s, the Global Atmospheric Research Programme, served the dual needs of the U.S. world-spanning military.[66] After the collapse of communism, the U.S. national laboratories, showcases for the talents of weapons scientists, suddenly became "greener," providing a boost to Edward Teller's ongoing program of training atmospheric scientists, many of them armed with basic physics and access to military funding and hardware, hell-bent on "fixing the sky."[67] After 2001, classified meteorological research, funded by the deep pockets of the military and the Department of Homeland Security, was dedicated, for example, to detecting and predicting the spread of plumes of heavier-than-air gases in urban settings, especially Washington, D.C., or to seeking effective chemical sniffers for toxic, explosive, or radiological sources in and around government buildings, airports, train stations, and harbors.

With the reputation of the field of weather control severely tarnished as it is, the military's semi-official public line is "you can't control how the world is changing around you, so you have to be able to control how you react to that change."[68] Spokesmen for this view emphasize training, discipline, and vigilance. The air force will say it is in the business of improving the accuracy and usefulness of its forecasts and its capabilities in general by applying operational risk management techniques to both routine and exceptional weather services. This is true so far as it goes, but there is probably much more that the military simply does not know or cannot say—most likely the latter.

On the other side of the coin are conspiracy theorists who see a toxic cloud on every horizon. Their fears are fueled by statements such as those made in 1997 by Secretary of Defense William S. Cohen, who warned of "an eco-type of terrorism whereby [adversaries] can alter the climate, set off earthquakes [and] volcanoes remotely through the use of electromagnetic waves. . . . It's real, and that's the reason why we have to intensify our efforts, and that's why this is so important."[69] Cohen, known to levitate on occasion, at least rhetorically, was responding, off the cuff, to questions about the possibility of all sorts of futuristic weapons falling into the hands of terrorists, and his remarks should not be misconstrued. Nevertheless, conspiracy theorists have focused on his words in support of their suspicions that the military is supporting secret geoengineering projects involving directed energy beams, chem trails, or other technologies. The historical record, rather than such speculation, is actually much more revealing—and chilling.

7

FEARS, FANTASIES, AND POSSIBILITIES OF CONTROL

Present awful possibilities of nuclear warfare may give way to others even more awful. After global climate control becomes possible, perhaps all our present involvements will seem simple. We should not deceive ourselves: once such possibilities become actual, they will be exploited.

—JOHN VON NEUMANN, "CAN WE SURVIVE TECHNOLOGY?"

CLIMATE fears, fantasies, and the possibility of global climate control were widely discussed by scientists and in the popular press in the third quarter of the twentieth century. Some chemists, physicists, mathematicians, and, yes, meteorologists tried to "interfere" with natural processes, not with dry ice or silver iodide but with new Promethean possibilities of climate tinkering opened up by the technologies of digital computing, satellite remote sensing, nuclear power, and atmospheric nuclear testing. Aspects of this story involve engineers' pipe dreams of mega-construction projects that would result in an ice-free Arctic Ocean, a well-regulated Mediterranean Sea, or an electrified and well-watered Africa. Pundits also fantasized about engineering the climate and possibly weaponizing it, using, for example, nuclear weapons as triggers. Far from being a heroic story of invention and innovation, global climate control has had, from its first mention in the literature of science fiction, a dark side, hinting at the possibility of global accidents or hostile acts. The warnings of two close scientific associates, John von Neumann (1903–1957) and Harry Wexler (1911–1962), one famous and one as yet relatively unknown, provide a framework for examining such issues. Von Neumann was a mathematician extraordinaire at the Institute for Advanced Study (IAS) in Princeton, New Jersey, and a pioneer in the application of digital computing techniques to the problem of numerical weather prediction and climate modeling. It was the dark side of climate control that led von Neumann to ponder the brave new world of such techniques. Wexler was chief of scientific services at the U.S. Weather Bureau. He was instrumental in advancing the agenda for climate modeling and promoted many other new technologies, especially meteorological satellites. It was Wexler who conducted the first serious technical analysis of climate engineering and issued an early warning about the possibilities of climate control. It was the very real possibility of purposeful destruction of the stratospheric ozone layer that led Wexler to spell out, in great technical detail, the dangers of both inadvertent and intentional climate tinkering. The interplay of such technical, scientific, and social issues moves beyond the timeworn origin stories of the modern atmospheric sciences into another dimension, a marketplace of wild ideas, a "Hall of Fantasy" or "Twilight Zone" whose boundaries are that of imagination.

Fears

We are apprehensive about climate change, we seek to understand it, and some may seek to stop it. The word "apprehension" signifies several distinct meanings: (1) fear, (2) awareness, and (3) intervention. In *Historical Perspectives on Climate Change* (1998), I examined how people became aware of and sought to understand phenomena in which they were immersed, that covered the entire globe, that had both natural and anthropogenic components, and that changed constantly on a multiplicity of temporal and spatial scales. What do climate scientists know about climate change and how do they know it? By what authority and by what historical pathways have they arrived at this knowledge? How have they established privileged positions? I offered some reflections on the ways such perspectives emerged historically: from appeals to authority, data collection, fundamental physical theory, critical experiments, models (including computer models), new technologies (including space-based observations), to consensus building and the beginnings of coordinated action.

I also examined climate-related fears, including drought, crop failures, volcano weather, apocalyptic visions of the return of the deadly glaciers, and global warming. The cultural geographer Yi-fu Tuan once observed, "To apprehend is to risk apprehensiveness."[1] For much of history, people feared that the powers of evil were active during inclement weather or that when the rains failed to arrive or it rained too much something was terribly wrong with either nature or, more likely, the social order. Many also saw themselves as agents of climate change. Even in fictional accounts of weather and climate control, much of the dramatic tension is derived from fundamental fears. An incomplete understanding, fueled by fear, may result in ineffective or even dangerous interventions. In the field of climate change, the two main approaches seem to be big technical fixes and social engineering.

In 1955 in a prominent article titled "Can We Survive Technology?" von Neumann referred to climate control as a thoroughly "abnormal" industry. He thought that weather control using chemical agents and climate control through modifying surface albedo or otherwise managing solar radiation were distinct possibilities for the near future. He argued that such intervention could have "rather fantastic effects" on a scale difficult to imagine. He pointed out that it was not necessarily rational to alter the climate of specific regions or purposely trigger a new ice age. Tinkering with the Earth's heat budget or the atmosphere's general circulation "will merge each nation's affairs with those of every other more thoroughly than the threat of a nuclear or any other war may already have done." In his opinion, climate control, like other "intrinsically useful" modern technologies, could lend itself to unprecedented destruction and to forms of warfare as yet unimagined. Climate manipulation could alter the entire globe and shatter the existing political order. He made the Janus-faced nature of weather and climate control clear. The central question was not "What can we do?" but "What should we do?" This was the "maturing crisis of technology," a crisis made more urgent by the rapidity of progress.

Banning particular technologies was not the answer for von Neumann. Perhaps, he thought, war could be eliminated as a means of national policy. Yet he ultimately deemed survival only a "possibility," since elements of future conflict existed then, as today, while the means of destruction grew ever more powerful and was reaching the global level.

In Baconian terms, do we consider climate to be based on the unconstrained operations of nature, now modified inadvertently by human activities, or do we seek to engineer climate, constrain it, and mold it to our will? Certainly, the ubiquity and scale of indoor air-conditioning could not have been imagined less than a century ago, but what about fixing the sky itself? In attempting to do so, we run the risk of violently rending the bonds of nature and unleashing unintended side effects or purposely calculated destruction. After all, von Neumann identified frenetic "progress" as a key contributor to the maturing crisis of technology. Fumbling for an ultimate solution, but falling well short, he suggested that the brightest prospects for survival lay in patience, flexibility, intelligence, humility, dedication, oversight, sacrifice—and a healthy dose of good luck.[2]

Fantasies

Global climate control has a history rooted in the quest for perfectly accurate machine forecasts and supported by the dream of perfectly accurate data acquisition. Calculating the weather has long been a goal of meteorologists. By the turn of the twentieth century, Felix Exner and Vilhelm Bjerknes had identified the basic equations of atmospheric dynamics. In 1922 Louis Fry Richardson had actually tried to solve the equations numerically with rather poor data and without the use of a computer. Their dreams—to solve the equations of motion for the atmosphere faster than the daily weather develops—were fulfilled with the advent of numerical weather prediction in the 1950s. This story has been told often, always as a heroic saga—a quest to do what no one has ever done before. Kristine Harper is the latest in a long line of historians and meteorologists to illuminate the "genesis" of modern meteorology, its "exodus" from weather bureau captivity, and its arrival at the edge of a "promised land" of digital computer modeling.[3] As complex (and familiar, at least in outline) as this story might be, there is nevertheless a story as yet untold, a darker tale of digital climate modeling, prediction, and control.

Just after World War II, in October 1945, Vladimir K. Zworykin, associate research director at the Radio Corporation of America (RCA) Laboratory in Princeton, New Jersey, wrote his influential but now all but forgotten mimeographed "Outline of Weather Proposal" (figure 7.1). He began by discussing the importance to meteorology of accurate prediction, which he thought was entering a new era. Modern communication systems were beginning to allow the systematic compilation of scattered and remote observations, and, he hoped, new computing equipment would be developed that could solve the equations of atmospheric motion, or at least search quickly for statistical regularities and past analog weather conditions. He imagined "an automatic plotting board" that would instantly digest and display all this information.

Zworykin suggested that "exact scientific weather knowledge" might allow for effective weather control. If a perfectly accurate machine could be developed that could predict the immediate future state of the atmosphere and identify the precise time and location of leverage points or locations sensitive to rapid storm development, effective intervention might be possible. A paramilitary rapid-deployment force might then be sent to intervene in the weather as it happened—literally to pour oil on troubled ocean waters or use physical barriers, giant flame throwers, or even atomic bombs to disrupt storms before they formed, deflect them from populated areas, and otherwise control the weather. Zworykin suggested a study of the origins and tracks of hurricanes, with a view to their

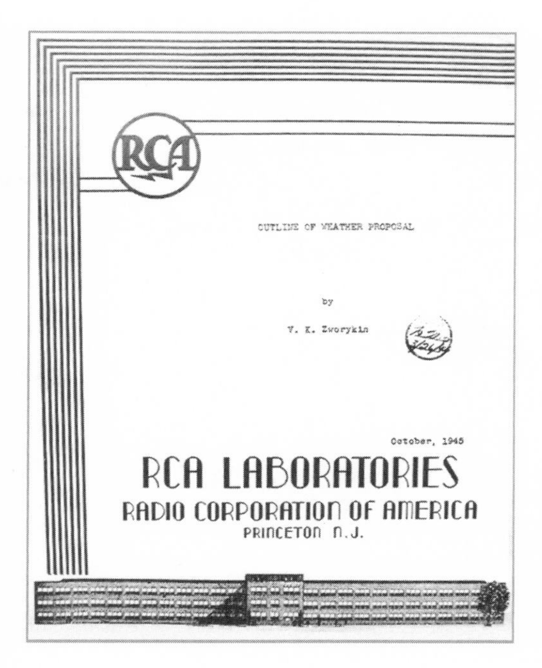

7.1 Cover of Vladimir K. Zworykin's "Outline of Weather Proposal," October 1945. (WEXLER PAPERS)

prediction, prevention, and even diversion. Long-term climatic changes might be engineered by large-scale geographical modification projects involving such climatically sensitive areas as deserts, glaciers, and mountainous regions. In effect, numerical experimentation using computer models would guide field experiments and interventions in both weather and climate. According to Zworykin,

> [t]he eventual goal to be attained is the international organization of means to study weather phenomena as global phenomena and to channel the world's weather, as far as possible, in such a way as to minimize the damage from catastrophic disturbances, and otherwise to benefit the world to the greatest extent by improved climatic conditions where possible. *Such an international organization may contribute to world peace by integrating the world interest in a common problem and turning scientific energy to peaceful pursuits. It is conceivable that eventual far-reaching beneficial effects on the world economy may contribute to the cause of peace.*[4]

Zworykin's proposal gained a powerful formal endorsement when von Neumann attached a letter to it dated October 24, 1945, stating: "I agree with you completely. . . . This would provide a basis for scientific approach[es] to influencing the weather." Using computer-generated predictions, von Neumann envisioned that weather and climate systems "could be controlled, or at least directed, by the release of perfectly practical amounts of energy" or by "altering the absorption and reflection properties of the ground or the sea or the atmosphere." It was a project that neatly fit von Neumann's overall agenda and philosophy: "All stable processes we shall predict. All unstable processes we shall control."[5] Zworykin's proposal also contained a long endorsement by the oceanographer Athelstan Spilhaus, then a U.S. Army major, who ended his letter of November 6, 1945, with these words: "In weather control, meteorology has a new goal worthy of its greatest efforts."[6]

Popularizations

Complicating the picture at the time were suggestions about the use of atomic weapons for climate control and announcements of new discoveries in cloud seeding. In 1945 the prominent scientist-humanist Julian Huxley, then head of UNESCO, had spoken to an audience of 20,000 at an arms control conference at Madison Square Garden about the possibilities of using nuclear weapons as "atomic dynamite" for "landscaping the Earth" or perhaps using them to change the climate by dissolving the polar ice cap. Captain Eddie Rickenbacker was on record as advocating the use of atomic bombs for "cracking the Antarctic icebox" to gain access to its known mineral deposits.[7] "Sarnoff Predicts Weather Control" read the headline on the front page of the *New York Times* on October 1, 1946. The previous evening, at his testimonial dinner at the Waldorf Astoria, RCA president Brigadier General David Sarnoff had speculated on worthy peaceful projects for the postwar era. Among them were "transformations of deserts into gardens through diversion of ocean currents," a technique that could also be reversed in time of war to turn fertile lands into deserts, and ordering "rain or sunshine by pressing radio buttons,"[8] an accomplishment that, Sarnoff declared, would require a World Weather Bureau in charge of global forecasting and control (much like the Weather Distributing Administration proposed in 1938). A commentator in the *New Yorker* intuited the problems with such control. "Who," in this civil service outfit, he asked, "would decide whether a day was to be sunny, rainy, overcast . . . or enriched by a stimulating blizzard?" It would be "some befuddled functionary," probably bedeviled by special interests such as the raincoat and galoshes manufacturers, the beachwear and sunburn lotion industries, and resort owners and farmers. Or if a storm was to be diverted, "Detour it where? Out to sea, to hit some ship with no influence in Washington?"[9] Recall that all this was just one month before the General Electric Corporation announced the news of Vincent Schaefer's cloud-seeding exploits and Irving

Langmuir began making his fantastic claims about weather control. But such notions are still around. In 2009 a fantastic proposal was floated for just such a bureau or administration to implement regional-scale geoengineering—in the Arctic, in certain ocean regions, or for certain storms—to attempt to moderate specific climate change impacts.[10]

In the era of cloud seeding, von Neumann and Zworykin, especially the latter, continued to feed public speculation about control. In January 1947, both men spoke in New York at a joint session of the American Meteorological Society and the Institute of the Aeronautical Sciences, chaired by incoming AMS president Henry G. Houghton. Von Neumann's talk, "Future Uses of High Speed Computing in Meteorology," was followed by Zworykin's much more controversial "Discussion of the Possibility of Weather Control." According to the *New York Times*: "Hurricanes may be dispersed, Dr. Zworykin said, and rain may be made, first through the speed which an electronic computer now approaching completion can synthesize all elements in weather problems, and second, through application of energy in small doses from spreads of blazing oil to heat critical portions of the atmosphere or blackened-over areas to cool them."[11] Zworykin focused on "trigger" mechanisms such as artificial fogs or even cloud seeding as examples of adding small amounts of energy to cause enormous effects, claiming that the missing ingredient was not the techniques but how to "make the most of our weather information mathematically." A follow-up story the next day ended with the comment "If Dr. Zworykin is right the weather-makers of the future are the inventors of calculating machines."[12]

Most scientists thought this speculation was premature. Wexler and a colleague who dined at Zworykin's home discussed weather control with him, including techniques such as igniting oil on the sea surface to redirect hurricanes, but Wexler indicated that Zworykin's views "were not shared by most tropical meteorologists."[13] The distinguished oceanographer Harald U. Sverdrup at the Scripps Institution of Oceanography was not convinced by Zworykin's claim that "the underlying general physical principles governing weather behavior are mostly well understood." Regarding weather control, he wrote: "It seems that only in rare cases can we expect to know the initial conditions in sufficient detail to predict the consequences of a 'trigger action.'"[14] Yet talk of triggers was something the military understood. In a 1947 fund-raising speech presented at the annual alumni dinner at MIT, General George C. Kenney, commander of the Strategic Air Command, speaking of future weapons systems, asserted: "If rain could be kept from falling where it has been falling for ages," it is conceivable that "the nation which first learns to plot the paths of air masses accurately and learns to control the time and place of precipitation will dominate the globe."[15] He urged MIT to look into this, suggesting a multimillion-dollar research program.

That spring, Martin Mann, an associate editor of *Popular Science Monthly*, sent a draft of his article "War Against Hail," to von Neumann for corrections. The article began with a declaration of war against hail, announcing that the army would soon be using Schaefer's cloud-seeding techniques in its quest for triggers—"With enough triggers, weather could be made to order!"[16] Mann then highlighted the quest at the IAS to develop numerical "model experiments," which he said would reveal additional triggers:

> Once the weather equations are perfected [von Neumann and Zworykin] foresee their use to control weather. . . . Figures corresponding to imaginary weather conditions will be fed into the computer. It will then forecast what final weather would have resulted from the imaginary starting conditions. For example, the machine would show how a higher temperature over the Caribbean Sea would have affected the weather in Miami. (10)

Mann had also interviewed Zworykin for the article, citing his opinion that because of the vast energy generated by weather systems, the use of brute-force methods, even nuclear bombs, to divert a hurricane was futile. Zworykin favored smaller "trigger mechanisms" such as modifying the surface of an island in the path of the storm to make it either darker or more reflective and to upset the storm's already shaky balance of forces. Covering an island with a thin layer of carbon black would absorb heat, while generating a white smoke screen would make it more reflective: "These reflecting-absorbing areas would have to be placed in exactly the right spots and used at exactly the right times" (11)—a job for the decision-making power of the digital computer. Pity the poor Caribbean islanders whose tropical paradises would be invaded and possibly brutalized each hurricane season by paramilitary forces trying to save Miami. How could they clean up all the soot? Mann concluded his article by juxtaposing the more proximate goal of hail suppression and the distant goal of climate control. He also contrasted Vincent Schaefer, the Edison-like everyman "who never even finished high school," and the Princeton eggheads working on a big military–industrial project.

Hurricane Control

Over five decades later, Ross Hoffman, principal scientist with Atmospheric and Environmental Research, used a little butterfly as a logo in his presentations to symbolize his notion that chaos theory, developed by his graduate school adviser, Edward Lorenz, might be used to control weather systems such as hurricanes. Since the atmosphere is chaotic and exhibits extreme sensitivity to small changes, he argued, a series of "just right" perturbations might be used to control the weather. Echoing the perennial hope of William Suddards Franklin, who believed in perturbations, and the pathological hype of Irving Langmuir, who advocated control, Hoffman asked us to imagine a world with no droughts, no tornadoes, no snowstorms during rush hour, and no killer hurricanes.

Hoffman proposed an atmospheric controller similar in its characteristics to feedback systems common in many industrial processes. The components of his integrated system included numerical weather prediction (NWP), data assimilation systems, and satellite remote sensing—all part of today's normal meteorological practices. Where Hoffman's system differed was in adding a fourth component, what he called "perturbations," into natural weather systems to move them off course. These are planned interventions in natural systems that are then to be monitored by remote sensing and modeled via NWP in an endless management loop. He provided examples of global-scale interventions that included changing the altitude and flight paths of airplanes to optimize contrail formation for perturbing solar and infrared radiation, launching reflectors into low Earth orbit to produce bright spots on the night side and shadows on the day side of the Earth, running wind turbines as high-speed fans at a sufficient scale to transfer atmospheric momentum and influence storm tracks, and the pièce de résistance, a space solar power generator that would "downlink microwave energy" to provide a tunable atmospheric heat source—an orbiting death ray to zap hurricanes or anything else in the way.

His goal was not to eliminate hurricanes but just to "control their paths" in order to prevent them from striking population centers. To demonstrate his ideas, Hoffman increased the sea surface temperature by 5°C (9°F) in one quadrant of a computer model of Hurricane Iniki. The model responded by "steering" the storm away from the Hawaiian Islands. Hoffman concluded by gesturing in the direction of legal and ethical questions, asking, "If we can do it, do we want to?"[17] Just imagine the lawsuits from those poor folks over whose property Hoffman steers the storm! And isn't "steering" too strong a term to use to describe the result of a chaotic perturbation? Hoffman's effort was supported by the NASA Institute for Advanced Concepts, which funded wild futuristic ideas such as space elevators, robotic asteroid patrols, antimatter propulsion, and genetically modified organisms for terraforming other planets.[18] Hoffman cited as his inspiration Arthur C. Clarke's vision of a Global Weather Authority:

> It had not been easy to persuade the surviving superpowers to relinquish their orbital fortresses and to hand them over to the Global Weather Authority, in what was—if the metaphor could be stretched that far—the last and most dramatic example of beating swords into plowshares. Now the lasers that had once threatened mankind directed their beams into carefully selected portions of the atmosphere, or onto heat-absorbing target areas in remote regions of the Earth. The energy they contained was trifling compared with that of the smallest storm; but so is the energy of the falling stone that triggers an avalanche, or the single neutron that starts a chain reaction.[19]

But Hoffman also alluded to possible future misuse and militarization. Every prophetic call to "beat their swords into plowshares, and their spears into pruning hooks" (Isaiah 2:4) can be countered by another prophet calling, "Beat your plowshares into swords and your pruning hooks into spears" (Joel 3:10).

Hurricane expert Kerry Emanuel has been generally supportive of Hoffman's hurricane control quest: "Weather modification will occur, almost inevitably . . . the recent results, that suggest you can do it with a little tiny bit of energy placed in the right place, will prove irresistible, and one can only hope that it is done for the right reasons and to good ends."[20] On another occasion, Emanuel opined: "We might be able to prevent or reduce vulnerability to serious hurricanes by controlling the storm, by reducing its intensity, by steering it out to sea. I don't think it would take very many years to come up with a technology."[21] Recall that it was Emanuel who advanced the dubious notion of taking up Phaethon's reins. In June 2009, Microsoft's Bill Gates announced that he intended to fight hurricanes by manipulating the sea, "draining warm water from the surface to the depths, through a long tube."[22] One commentator on the proposal suggested not to "mess with Mother Nature"; another included the hope that this technique might work better than the Windows operating system!

Soviet Fantasies

Vladimir Lenin set the tone for Soviet attitudes toward "the mastery of nature." According to his philosophy, a new era was dawning through "objectively correct reflection" on independently occurring phenomena and processes embedded in the absolute and eternal laws of nature. Mastery would be manifest in praxis.[23] Two decades later, in 1948, Soviet leader Joseph Stalin announced his "Great Plan for the Transformation of Nature," an ultimately futile attempt to expand the Soviet economy by harnessing nature and controlling the weather and climate.[24]

Throughout the cold war era, authors from at least nineteen research institutions in the Soviet Union published numerous books, articles, and reports on weather and climate modification.[25] Several popularizations of this literature are notable for their geoengineering fantasies. In *Soviet Electric Power* (1956), Arkadii Borisovich Markin outlined the progress of electrification in the Soviet Union and provided a forecast to the year 2000, when, he supposed, electrical power output would be one hundred times greater than at present. Markin gave special emphasis to the future role of nuclear power, including using nuclear explosions for geoengineering purposes:

> Gigantic atom explosions in the depths of the earth will give rise to volcanic activity. New islands and colossal dams will be built and new mountain chains will appear. Atom explosions will cut new canyons through mountain ranges and will speedily create canals, reservoirs, and sea, carry[ing] out huge excavation jobs. At the same time we are convinced that science will find a method of protection against the radiation of radioactive substances.[26]

Such ideas were derived from the Soviet program Nuclear Explosions for the National Economy, which, like Edward Teller's Project Plowshare, proposed techniques to employ nuclear explosives for peaceful construction purposes. Surely, Markin concluded, the Soviet power engineers can achieve "magnificent results" when inspired by the "omnipotence of human genius" (135).

In *Man Versus Climate* (1960), Soviet authors Nikolai Petrovich Rusin and Liya Abramovna Flit surveyed a large number of schemes for climatic tinkering. Invoking a Jules Verne–style fantasy, the book's cover is illustrated by the Earth surrounded by a Saturn-like ring of dust particles intended to illuminate the Arctic Circle, increase solar energy absorption, and ultimately melt the polar ice caps. Chapters in the book are dedicated to mega-engineering projects such as damming the Congo River to electrify Africa and irrigate the Sahara, diverting the Gulf Stream with a causeway off Newfoundland or harnessing it with turbines installed between Florida and Cuba, and, of course, Petr Mikhailovich Borisov's proposal to dam the Bering Strait to divert Atlantic waters into the Pacific and melt the Arctic sea ice. The authors' ultimate goal was to convince the reader "that man can really be the master of this planet and that the future is in his hands."[27]

In a much more politically oriented book, *Methods of Climate Control* (1964), Rusin and Flit admitted that "we are merely on the threshold of the conquest of nature," attributing the nascent ability to control nature to the emergence of the new Soviet man: "Before the Revolution, under the autocracy, nine-tenths of the territory of Russia had not been studied at all. The Soviet man, taking ownership of the greatest natural wealth, learned not only how to use it, but how to subordinate nature to his will. And now we are not surprised when we learn that a new sea has been developed or the desert has blossomed."[28]

Referring to the macro-engineering projects discussed in their earlier book, Rusin and Flit argued that deeper scientific insight into the laws of nature would result in ever more "grandiose" plans for developing immense energy reserves, controlling the flow of rivers, and subjugating permafrost, to name but a few of the advances that they expected. Science was not just about observing and understanding nature; it was about exploiting and controlling it as well. They cited the program of the Communist Party of the Soviet Union on this: "The progress of science and technology under the conditions of the Socialist system of economy is making it possible to most effectively utilize the wealth and forces of nature for the interests of the people, make available new forms of energy and create new materials, develop methods for the modification of climatic conditions and master space" (3).

I. Adabashev reviewed many of the same projects in his book *Global Engineering* (1966), with his utopian hopes tinged by strong ideological commitments. Concerning the "second Nile" project in Africa, he wrote: "The great new man-made inland seas would transfigure the Sahara . . . and create a new climate in Northern Africa. . . . Millions and millions of fertile acres would be made to yield two and even three crops a year for the benefit of mankind."[29] This would enhance the "struggle of African peoples for national liberation" against the vested interests of American and European capitalists seeking to control the African economy (161). In essence, Adabashev envisioned in the not-too-distant future a new global hydrologic era of gigantic dams and dikes, pumping stations capable of handling entire seas, and other facilities that would "trigger" various meteorological processes. He called it "a better heating system for our planet, better able to serve all the five continents" (201). But with world population and energy needs increasing, why should a visionary engineer stop with the surface of the Earth? Adabashev concluded his book with a fanciful account of a "Dyson sphere," one astronomical unit in radius, a new home for humanity roughly a trillion times greater than that of the Earth, synthesized from the remains of the outer planets and capturing all the energy of the Sun—solar-powered sustainable development in action—at least for the next 300 million years! For Adabashev, however, implementation of such projects had been delayed by the continued existence of capitalism, which he likened to "a ball and chain hampering man in his progress towards a happier lot" (237).

Warming the Arctic

The idea of melting the Arctic ice cap dates at least to the 1870s, when Harvard geologist Nathaniel Shaler suggested channeling more of the warm Kuroshio Current through the Bering Strait:

> Whenever the Alaskan gates to the pole are unbarred, the whole of the ice-cap of the circumpolar regions must at once melt away; all the plants of the northern continents, now kept in narrow bounds by the arctic cold, would begin their march towards the pole. . . . It is not too much to say that the life-sustaining power of the lands north of forty degrees of latitude would be doubled by the breaking down of the barrier which cuts off the Japanese current from the pole.[30]

In 1912 Carroll Livingston Riker, an engineer, inventor, and industrialist, proposed a scheme to change the climate of polar regions by tinkering with the ocean currents of the Atlantic. This was to be accomplished by preventing the cold Labrador Current from colliding with the Gulf Stream. To do this, he proposed building a 200-mile causeway extending east from Cape Race off the coast of Newfoundland. The theory was that the causeway could be built by suspending a long rope cable, or "Obstructor," in the ocean that would act to slow the southward flow of the Labrador Current, causing it to deposit its sediment load. Potential benefits of diverting the Gulf Stream farther east (shades of Thomas Jefferson) included fewer fogs and a general warming of northern climates. Riker's proposal was inspired by recently completed mega-projects such as Henry Flagler's railroad bridge from Key West, Florida, to the mainland and the ongoing excavation of the Panama Canal. The tragic sinking of the *Titanic* also lent urgency to his proposal, since his causeway might help remove icebergs from shipping lanes. Riker was supported in Congress by Representative William Musgrave Calder (R-New York), who proposed the creation of a Commission on the Labrador Current and Gulf Stream. Secretary of the Navy Josephus Daniels was not at all convinced by the proposal, but thought that a general survey of the currents of the Grand Banks would be useful.[31]

An ice-free Arctic Ocean was one of the largest-scale and most widely discussed climate-engineering projects of the time. Jules Verne's story *The Purchase of the North Pole* (1889) may have been inspired by such ideas. Ironically, an ice-free Arctic Ocean is something we may actually see sooner or later through a combination of natural and anthropogenic influences. In 1957 Soviet academician Borisov, alluding to the centuries-old quest of the

Russian people to overcome the northland cold, proposed building a dam across the Bering Strait to melt the Arctic sea ice.[32] In numerous articles and then again in his book *Can Man Change the Climate?* (1973), Borisov detailed his vision of a dam 50 miles long and almost 200 feet high with shipping locks and pumping stations. He proposed that the dam be built in 820-foot sections made of prefabricated freeze-resistance ferroconcrete that could be floated to the construction site and anchored to the sea bottom with pilings. He further suggested that the top of the dam be shaped so that ice floes would ride up over the dam and break off on the southern side. An alternative design included an intercontinental highway and railroad. According to Borisov, "What mankind needs is war against cold, rather than a 'cold war.'"[33]

To liquidate Arctic sea ice, Borisov wanted to pump cold seawater out of the Arctic Ocean, across the dam, and into the Bering Sea and the North Pacific. This displacement would allow the inflow of warmer water from the North Atlantic, eliminate fresh water in the surface layer in several years, and thus prevent the formation of ice in the Arctic Basin, creating warmer climate conditions:

> In this day and age, with mankind's expanding powers of transforming the natural environment, the project we are advancing does not present any technical difficulties. The pumping of the warm Atlantic water across into the Pacific Ocean will take the Arctic Ocean out of its present state of a dead-end basin for the Atlantic water [and] drive the Arctic surface water out into the Pacific Ocean through the Bering Strait.[34]

His goal was to remove a 200-foot layer of cold surface water, which would be replaced by warmer, saltier water that would not freeze. Inspired by Markin's popular book *Soviet Electric Power*, Borisov also assumed that huge amounts of electricity would soon be available to run the pumps, perhaps from hydroelectric generators or nuclear reactors.

The dam was, of course, never built, but if it had been attempted, would the nations of the world have confronted the Russians? The net climatic effect of the project, if it had been carried out, is still highly uncertain. A good argument can be made that the effect would be less than that of naturally occurring variations in the Atlantic influx, but none of the computer models at the time were sophisticated enough to show any robust results.

Other ocean-engineering schemes included installing giant turbines in the Strait of Florida to generate electricity and adding a thin film of alcohol to the northern branch of the Gulf Stream to decrease surface water evaporation and warm the water by several degrees, although the cod might become rather tipsy.

In Japan, engineers imagined that the icy Sea of Okhotsk could be tamed by deflecting the warm Kuroshio Current with a dam or one-way water valve built at the Tatarsk Strait. And in a 1970 geoengineering experiment thought suitable only for testing on a computer model (aren't they all?), the Japanese geoscientific speculator Keiji Higuichi wondered what would happen to the global atmospheric and oceanic circulation and thus the world's climate if the Drake Passage, between the tip of South America and Antarctica, was blocked by an ice dam. One possibility was the onset of a new ice age.[35]

Russian scientists warned of possible climate disruption from such megaprojects. Borisov admitted that the large-scale climatic and ecological effects of his Bering Strait dam could not be fully predicted, nor could they be confined within the borders of any one national state; rather, they would directly involve the national interests of the Soviet Union, Canada, Denmark, and the United States and indirectly affect many countries in other areas that might experience climate change caused by the project. With such a dam in place, the middle-latitude winters would be milder due to the warming of Arctic and polar air masses. He thought areas such as the Sahara would be much better watered and would perhaps turn into steppe land or savannah. Direct benefits of an ice-free Arctic Ocean would include new, more-direct shipping routes between East Asia and Europe, while, by his overly optimistic calculations, sea-level rise would be modest, even with the melting of the Greenland ice cap. Yet such climatic changes elsewhere were of little concern to the Soviets.

Larisa R. Rakipova noted that a substantial Arctic warming could cool the winters in Africa by 5°C (9°F), "leading to a complete disruption of the living conditions for people, animals, and plants,"[36] and Oleg A. Drozdov warned that the warming of the Arctic would lead to a total breakdown of moisture exchange between the oceans and continents with excess rain in the Far East and great aridity in Europe. The resulting drastic changes in the soils, vegetation, water regime, and other natural conditions would have widespread negative ecological, economic, and social consequences (25). As in the fictional case described earlier in *The Evacuation of England*, Rusin and Flit also wondered what might happen if the Americans implemented one of their projects and turned the Gulf Stream toward the shores of America: "In Europe the temperature would drop sharply and glaciers would begin to advance rapidly" (22). In his book *The Gulf Stream* (1973), T. F. Gaskell pointed out, "This is why such natural phenomena as the Gulf Stream have political implications."[37] Geoengineers should realize that the same is true of a wide range of natural phenomena.

In addition to sea ice, the Soviets were also battling the "curse of the Siberians"—permafrost as thick as 1,600 feet in places. One suggestion to remove it involved applying soot to the snowfields to absorb more sunlight; or perhaps cheaper materials such as ash or peat could do the job. Reminding their readers that "everyone knows what permafrost is," Rusin and Flit recounted its horrors: "A newly constructed house unexpectedly begins to shift, a Russian stove suddenly begins to sink into the ground, deeply driven piles spring from the ground," and when it melts and refreezes, the trees of the mysterious "drunken forests" lean akilter, like a Siberian full of vodka.[38] In the twenty-first century, permafrost has reemerged not as a local curse but as something to be saved, in part to preserve the migration patterns of the reindeer and caribou, and as a global environmental issue because of its high methane gas content. In 1962 Rusin and Flit opined, "Much has been learned, but it has been impossible to completely eliminate permafrost" (27).

Rehydrating and Powering Africa

The completion of the Suez Canal in 1869 under the direction of the French diplomat Ferdinand de Lesseps led to a number of mega-engineering proposals for rehydrating Africa. One was proposed by an eccentric British adventurer and entrepreneur, Donald Mackenzie, who proposed flooding the Sahara Desert in Algeria with water from the Mediterranean Sea to improve transportation, benefit commerce, and spread Christianity. The *Daily Telegraph* reported:

> Instead of a pathless wilderness across which once in the year a line of camels carry merchandise, the envious but admiring ears of M. de Lesseps are destined to hear the fleets of merchantmen sailing over the conquered Sahara. Liverpool will only be fourteen days from the Upper Niger, and while a magnificent new market will be opened for British and other goods, the regeneration of Africa will be advanced as if centuries had suddenly rolled over.[39]

A colleague wrote to Mackenzie that the project "would recommend itself to every Christian mind, spreading a net of Christianity over Africa" (274). The French, not to be outdone, appointed geographer François Elie Roudaire to lead a commission that suggested that the French Academy of Sciences explore the idea.[40] This discussion raised the possibility that an inland sea might enhance rainfall and thus agricultural production in the Sahara, but also might adversely affect the climate of Europe.

Jules Verne's novel *L'Invasion de la mer* (1905) was based on the premise that French engineers returned to Africa to complete Roudaire's project. The book raised a number of environmental, cultural, and political concerns, including the possibility of warfare triggered by macro-engineering projects.[41] Verne's idea was revived in whole cloth in 1911 by a French scientist named Etchegoyen, who again proposed to convert large portions of the Sahara into an inland sea by digging a 50-mile canal on the north coast of Africa. He touted the ease of construction and the massive benefits: more fertile soil and cropland, a cooler local climate, and a great new colony for France along the "Sea of Sahara." Critics warned that the massive redistribution of water, up to half the volume of the Mediterranean Sea, might tip the Earth's axis, adversely affect regional precipitation patterns, or even trigger an ice age in northern Europe.[42]

In the 1930s, the German architect Herman Sörgel's "Atlantropa Project" promoted the idea of lowering the level of the Mediterranean Sea and developing more than 3 million acres of new territory (an area as large as France) for European settlement. According to Sörgel, the construction of gigantic dams at Gibraltar and the Dardanelles to drain much of the Mediterranean and generate massive amounts of power "would assure Europe a utopian future of expanded territory; abundant, clean, and cheap energy; and the revival of its global economic and political might." Sörgel tried to sell his ideas first to the Nazis and then, during the cold war, to Western governments as a hedge against Soviet expansionism in Africa.

But lowering the Mediterranean Sea was only part of Sörgel's vision. He also wanted to irrigate much of Africa by building a massive system of dams and artificial lakes. Damming the Congo River, Africa's mightiest and the second-most-voluminous river in the world, near its outlet at Brazzaville, Congo, would create a huge new lake that Sörgel dubbed the "Congo Sea," basically covering the entire surface area of that nation. A chain of events, including the drowning of natives, wildlife, and ecosystems, would then occur. By his calculations, the Ubangi River would reverse its course, flowing northwest into the Chari River and finally into the greatly enlarged "Chad Sea." These two new seas would cover about 10 percent of the continent, and the northern outlet could be dubbed the "Second Nile," flowing north across the Sahara to create an irrigated settlement corridor in Algeria similar to that in Egypt. Sörgel's plan also included a giant hydroelectric plant at Stanley Falls, with sufficient surplus electric power to illuminate and industrialize much of the continent (figure 7.2).[43]

American and Soviet hydrological engineers, too, dreamed of such macro-scale projects. In the 1950s and 1960s, the North American Water and Power Alliance proposed to channel 100 million acre-feet of water per year from Alaska and Canada for use in the southwestern United States and Mexico. Soviet

7.2 Herman Sörgel's plan for transforming Africa and the Mediterranean. (RUSIN AND FLIT, *MAN VERSUS CLIMATE*)

engineers dreamed of creating a massive new "Siberian Sea" east of the Ural Mountains by damming the Ob, Yenisei, and Angara rivers, for irrigation of crops and climate modulation. As recently as 1997, Robert Johnson, a retired geoscientist at the University of Minnesota, commandeered the front page of *EOS: Transactions of the American Geophysical Union*, to warn that the Mediterranean Sea was being starved of fresh water because human activities have diverted the outflow of rivers, mainly the Nile. He called for a dam across the Strait of Gibraltar to block the outflow of salt water into the Atlantic Ocean, paradoxically making the Mediterranean even saltier than at present. All of this was for a good cause, however, since his computer models indicated that the mega-dam would stave off a little ice age in northern Europe while preserving the holy grail of climate change, preventing the West Antarctic ice sheet from collapsing, and raising the worldwide sea level by 20 feet. It seems that all current geoengineering schemes should be able to do this, at least.[44]

Space Mirrors and Dust

In July 1945, a classified U.S. Army Air Force memorandum on the subject of German liquid rocket development included speculations on "future possibilities," including ideas on intercontinental ballistic missiles, Earth-orbiting satellites, space station platforms, and interplanetary travel. Significantly, a section of the memo titled "Weather Control" cited a 1923 proposal by Herman Oberth to launch large mirrors, a mile or so in diameter, into orbit to be used to concentrate the Sun's energy on the Earth's surface "at will," and in this way influence the weather.[45] *Time* further popularized Oberth's idea in 1954, describing the space mirror as made of "shiny metal foil reinforced with wire" and spinning slowly around a space station as its hub.[46] The space mirror would be positioned in such a way as to illuminate the Earth's nighttime hemisphere. It would bathe cold countries in reflected sunlight, making them productive and habitable. Areas with excess rainfall could be heated and dried with the mirrors. Conversely, rainfall might be generated in an arid region by concentrating the Sun's rays on the nearest lakes to evaporate water and form clouds. Then the rain clouds could be directed toward arid regions by thermal currents and pressure gradients generated by "proper manipulation of the mirrors."

The army report speculated that these mirrors could be used by "the world group of nations" against a country that became aggressive or obnoxious to persuade it "to be more friendly and reasonable by the concentration of intensive heat on their country,"[47] but did not discuss other possible hostile applications of these death rays. *Time* was considerably more blunt in its account: "If war should start on the earth below, the 'aggressor' . . . could be handily incinerated by making the mirror concave to concentrate its beam."[48] *Time* also reported that the Nazis gave serious consideration to a space mirror for military purposes during World War II.

Other radiative effects on climate were also being considered. Beginning in 1913, William Jackson Humphreys explored the idea that volcanic dust might control the climate.[49] Two decades later, astronomer Harlow Shapley and his associates realized that space is filled with interstellar dust that might be influencing their calculations by obscuring distant stars. Astronomers Fred Hoyle and R. A. Lyttleton further speculated that space dust may affect the solar constant and thus cause climatic change.[50]

Early in the space age, Leningrad mathematician Mikhail Aleksandrovich Gorodskiy proposed creating an artificial dust ring passing over both poles.[51] Shaped like a flat washer with its lower boundary at an altitude of 750 miles and its upper boundary at 6,000 miles, the Saturn-like ring would be made of

7.3 Mikhail Aleksandrovich Gorodskiy's plan for launching a Saturn-like ring of reflective particles into Earth orbit to warm the Arctic. (RUSIN AND FLIT, *MAN VERSUS CLIMATE*)

metallic potassium particles that were highly reflective, lightweight, and relatively inexpensive. Gorodskiy wanted the ring turned full face to the Sun in summer and oriented on edge in the winter, but his back-of-the-envelope calculations provided no details on the coherence or lifetime of the ring or how to shift its orientation. He imagined, however, that the ring would increase shortwave radiation between 55° and 90°N to values up to 50 percent greater than those at the equator! Permafrost would disappear, polar ice would melt, the cities of Siberia would flourish, and the entire planet would warm considerably (figure 7.3).

Another Soviet engineer, Valentin Cherenkov, proposed a much smaller orbiting cloud, formed from only 1 ton of opaque particles, that would direct the Sun's rays earthward (63–65). He estimated that the cloud would yield 1,300 billion kilowatts of power, the equivalent of about 500,000 large conventional power stations. This amount of energy could heat the Arctic and provide sky illumination of more than 500 lux, basically eliminating the long polar night. It would also eliminate the differences among the seasons and between the climate at the poles and that at the equator. Counterproposals existed at the time to cool the planet by positioning a sunshade over the equator between 30°N and 30°S—this about forty-five years before the current batch of proposals to manage solar radiation (chapter 8).

Bombs Away

The scientists and cold warriors who meddled with the Earth's atmosphere and near-space environment believed that "they could control everything," even radiation and nuclear fallout.[52] They had supporters in high places, such as Senate majority leader Lyndon B. Johnson, chair of the Preparedness Subcommittee. The launch of *Sputnik 1* in October 1957 diverted the world's attention from the scientific concerns of the ongoing International Geophysical Year and heightened American apprehensions of a "missile gap" and possible national security threats from space. The launch of *Sputnik 2* in November further fueled these fears. Johnson warned in early 1958 that the Russian *Sputniks* were not "play toys" and proclaimed that the very future of the United States depended on its first seizing ownership of space and controlling it for military purposes.

> The testimony of the scientists is this: Control of space means control of the world, far more certainly, far more totally than any control that has ever or could ever be achieved by weapons, or by troops of occupation. From space, the masters of infinity would have the power to control the earth's weather, to cause drought and flood, to change the tides and raise the levels of the sea, to divert the Gulf Stream and change temperate climates to frigid. . . . If, out in space, there is the ultimate position—from which total control of the earth may be exercised—then our national goal and the goal of all free men must be to win and hold that position.[53]

Later that month, the United States launched its first satellite, *Explorer 1*, with a modified Redstone military missile, the Juno 1.

In August 1958, during the extensive series of bomb tests known as Operation Hardtack, the military tested its antiballistic missile and communication disruption capabilities with two high-altitude shots named Teak and Orange. In each test, an army Redstone rocket launched a 3.8-megaton hydrogen bomb warhead. Teak detonated at 48 miles altitude in the mesosphere, and Orange at 27 miles in the stratosphere. Each blast illuminated the night sky as if it were daylight, with the added excitement that due to a malfunction of the missile guidance system, the Teak shot occurred directly over Johnston Island, in the North Pacific, instead of at the planned spot 48 miles downrange. Apparently, the experimenters had no qualms about destroying either themselves or any sensitive or protective layers of the atmosphere.

In Operation Argus, conducted in August and September 1958, just six months after the discovery of the Van Allen radiation belts by the satellites *Explorer 1* and *3*, the U.S. military and the Atomic Energy Commission decided that they should try to destroy or disrupt what had just been discovered. They did this with the full cooperation of astronomer James Van Allen.[54] A specially equipped naval convoy launched and detonated three 1.7-kiloton atomic bombs at altitudes ranging from 125 to 335 miles above the South Atlantic Ocean to "seed" the exosphere with electrons. The participants hyped it as the "greatest scientific experiment of all time" and claimed it was a test of a geophysical theory proposed by Nicholas C. Christophilos of Lawrence Berkeley Laboratory.[55] In scale it was indeed impressive, involving nine ships and 4,500 people, with "nuclear observations" taken by the overflying satellite *Explorer 4*, a barrage of high-altitude five-stage Jason sounding rockets, airplane flights, and ground stations—but there was very little science, apparently. Test results and other documentation remained classified for the next twenty-five years. The military purpose was most likely to see if and how nuclear explosions disrupted communication channels. Since an atmospheric test-ban treaty was then under negotiation, the military was quick to point out that this test was not *in* the atmosphere but "above it."

Other nuclear tests in near space ensued, such as the much larger Starfish explosion of July 1962 above Johnston Island, which disrupted the Van Allen belts and created an artificial magnetic belt and an "aurora tropicalis" visible as far away as New Zealand, Jamaica, and Brazil. Three Soviet high-altitude explosions that year had similar effects. A *New Yorker* cartoon depicted a serious-looking technocrat questioning a colleague in a high-tech laboratory setting: "But how do you *know* destroying the inner Van Allen belt will create havoc until you try it?"[56] It was quite a year for near-space fireworks, with the British, Danes, and Australians issuing formal protests, led by the astronomical community. During the tests, some hotels in the Pacific apparently offered "rainbow" bomb parties on their roofs so guests could watch the light shows.

One of the more bizarre items that crossed Harry Wexler's desk at the U.S. Weather Bureau in 1961 was a technical report simply called "Weather Modification," by M. B. Rodin and D. C. Hess at Argonne National Laboratory. The authors made the reasonable suggestion that applying heat directly to a rain cloud, or to a moist air mass with rain potential, might alter the natural precipitation in a given geographical region by increasing the buoyancy of the cloud or air parcel. This was James Espy's century-old convective theory. The modern twist: they favored using large, hovering nuclear reactors "wherever safety criteria can be met" to deliver the huge amounts of heat required (figure 7.4).[57] Such nuclear-powered aircraft were never built.

Not all space seeding was nuclear. In 1960 the Department of Defense and MIT's Lincoln Laboratory announced a plan to launch 500 million tiny copper wires into an 1,800-mile orbital ring to serve as radio antennae. Since the Earth's

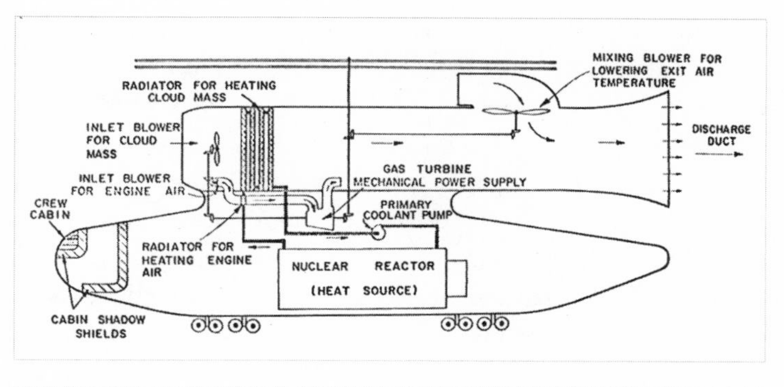

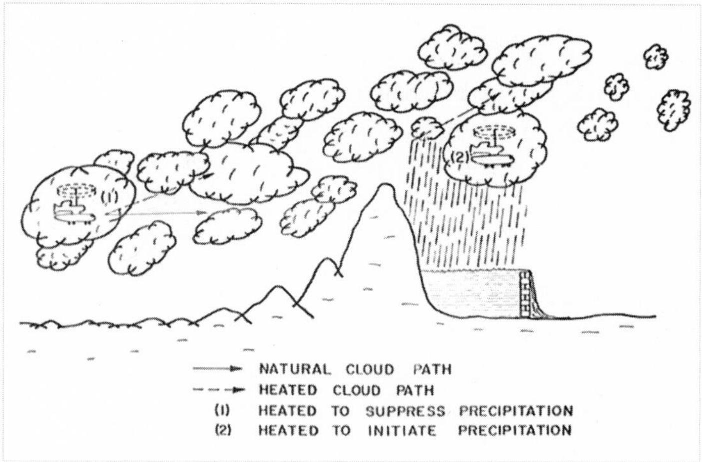

7.4 "Weather Modification": (*above*) schematic drawing of the layout for a hovering-type aircraft equipped with a nuclear heat source (note the lead-lined crew cabin and the little pinwheel blowers for air inlet and mixing); (*below*) nuclear weather modification helicopter in action (1) suppressing rain on one side of the mountain and (2) filling a reservoir on the other. (WEXLER PAPERS)

ionosphere was vulnerable to enemy attack by a thermonuclear detonation, and undersea cables might be cut by a hostile power, the military wanted to be able to guarantee secure worldwide communication channels, regardless of the protests of other nations about space debris or the concerns of astronomers about visual or radio interference. The first launch, in 1961, failed, but two years later the detritus injected by Project West Ford (originally called Project Needles) was used to bounce radio messages across the continent.

This is indeed geoengineering. The experiment effectively created an artificial ionosphere, "better" than the original since it would not be disrupted by magnetic storms or solar flares. Wexler, however, was concerned that the environmental effects of the cloud of needles had not been fully considered, including their effect on the Earth's heat budget, magnetic field, and ozone levels. Astronomers protested bitterly, since the layer of needles interfered with their observations, especially in the new field of radio astronomy.[58] Although the cloud of needles behaved broadly as designed and mostly dispersed after about three years, rendering it useless for radio communication, as of 2010 some copper "needles" are still in orbit. Occasionally, one of them reenters the Earth's atmosphere and flashes briefly as it burns up as an artificial meteor. Astronomers soon will be forced to oppose proposals for solar radiation management, since any attempt to attenuate sunlight will also attenuate starlight (chapter 8).

In February 1962, Wexler was informed of a review by an ad hoc panel at NASA convened to consider the "High Water Experiment," the upcoming release of almost 100 tons of water into the ionosphere. The delivery vehicle was a Saturn test rocket to be launched from Cape Canaveral to an altitude of 65 miles and then destroyed. The panel, chaired by atmospheric scientist William W. Kellogg of the RAND Corporation, concluded, on the basis of some back-of-the-envelope calculations, that "it was unable to predict exactly what would happen following the rupture of the Saturn tanks."[59] They supposed that the water would boil instantly in the vacuum of space and then form ice crystals in a cloud about 6 miles wide and up to 20 miles long that would gradually fall out and dissipate downrange (figure 7.5). Some of the water would also dissociate, forming atomic O and H. Noctilucent clouds should form, and the radio properties of the ionosphere might be affected, with possible disruption to stratospheric ozone. The members of the panel knew that "introducing more H would change *something*" (4), but they could not say what. Nevertheless, they considered the scale of this test, literally a "drop in the bucket," and predicted that "no major change in the atmosphere will take place that will hinder human activities" (1). They also predicted, correctly, that "in fact it may turn out to be hard to detect any effects at all (alas!), after the first few minutes" (1).

Kellogg and the panel were not completely confident that they understood all the factors involved in this experiment and readily admitted to "a good deal of uncertainty." At the time, atmospheric scientists were used to the idea "that on occasion small changes can 'trigger' larger ones, if the conditions in the atmosphere are in a kind of metastable state." Kellogg asked, "Is there such a condition in the upper atmosphere?" (6). He was unable to identify any, and the panel suggested no contingency plans for any "trigger" effects.

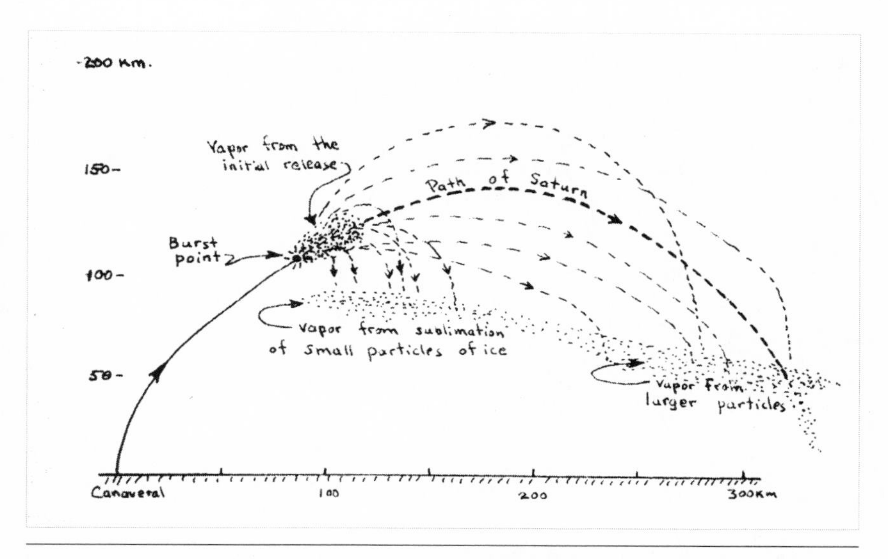

7.5 "High Water Experiment," February 1962: William W. Kellogg's "Sketch showing how the various sizes of ice particles produced from the moving Saturn vehicle would be expected to travel in the upper atmosphere and finally sublime as they fall to lower altitudes," prepared on the basis of an ad hoc NASA panel discussion. (WEXLER PAPERS)

Harry Wexler and the Possibilities of Climate Control

"The subject of weather and climate control is now becoming respectable to talk about."[60] This was Harry Wexler's opening line in his 1962 speech "On the Possibilities of Climate Control." Wexler based his remarks on newly available technical capacities in climate modeling and satellite remote sensing, new scientific insights into the Earth's heat budget and stratospheric ozone layer, and new diplomatic initiatives, notably President John F. Kennedy's 1961 speech at the United Nations proposing "cooperative efforts between all nations in weather prediction and eventually in weather control."[61] Soviet premier Nikita Khrushchev, flush with the success of two space spectaculars carrying Russian cosmonauts into orbit, had also mentioned weather control in his report to the Supreme Soviet in July 1961. Wexler noted that the subject had recently received serious attention from the President's Scientific Advisory Committee, the Department of State, and the National Academy of Sciences Committee on Atmospheric Sciences. The last had recommended increased funding for large-scale cooperative weather control projects and made it part of the proposal that led to the creation of the National Center for Atmospheric Research. For its part, the United Nations, with

Wexler's scientific input through the State Department, had issued a resolution on international cooperation in the peaceful uses of outer space, recommending "greater knowledge of basic physical forces affecting climate and the possibility of large-scale weather modification."[62] Before this, statements about controlling the atmosphere had typically been provided by non-meteorologists: chemists, cloud-seeding enthusiasts, futurists, generals, and admirals.

Wexler was none of the above. He was one of the most influential meteorologists of the first half of the twentieth century, and his career, as revealed in his publications and well-preserved office files, touched every aspect of weather and climate science. He was born in 1911 in Fall River, Massachusetts, and died suddenly of a heart attack in August 1962 at age fifty-one during a working vacation in Woods Hole, Massachusetts. The third son of Russian immigrants Samuel and Mamie (Hornstein) Wexler, Harry was interested in science at an early age, an interest he shared with his childhood friend and future brother-in-law, the noted meteorologist Jerome Namias. Wexler majored in mathematics at Harvard University, graduating magna cum laude in 1932. He then attended MIT, earning his master's degree in 1934 and his doctorate in 1939 under the mentorship of the influential meteorologist Carl-Gustav Rossby. Wexler worked for the U.S. Weather Bureau throughout his career, initially on operationalizing Bergen School techniques for air mass and frontal analysis and later as head of research.

Following the outbreak of war in Europe, Wexler served in the U.S. military's crash program to train a new cadre of weather forecasters. On a leave of absence from the weather bureau, he taught at the University of Chicago as an assistant professor of meteorology. In 1941 he returned to the weather bureau as senior meteorologist in charge of training and research, working to assist in defense preparations. He accepted a commission as captain in the U.S. Army in 1942 and served as the senior instructor of meteorology to the U.S. Army Air Force's Aviation Cadet School at Grand Rapids, Michigan. While in this position, he joined the University Meteorological Committee, established to coordinate military efforts in meteorological training.[63]

After his honorable discharge in January 1946 with the rank of lieutenant colonel, Wexler returned to the weather bureau, becoming chief of the Special Scientific Services division and serving on the Pentagon's Research and Development Board. In this capacity, he encouraged the development of new technologies, including tracing nuclear fallout, airborne observations of hurricanes, sounding rockets, and weather radar. He was a pioneer in the use of electronic computers for numerical weather prediction and general circulation modeling, serving as the weather bureau liaison to von Neumann's IAS meteorology project and playing a more central role than has hitherto been acknowledged. In 1954 he helped institutionalize the U.S. Joint Numerical Weather Prediction Unit, a partnership of the weather bureau with the air force and navy "to produce prognostic weather charts on an operational basis using numerical techniques." A year later, after a successful numerical experiment by Norman Phillips, in which he was able to simulate realistic features of the general circulation of the atmosphere, von Neumann and Wexler argued for the creation of a General Circulation Research Section (later Laboratory) in the weather bureau. The Geophysical Fluid Dynamics Lab in Princeton, New Jersey, and the National Centers for Environmental Prediction in Camp Springs, Maryland, trace their origins to these roots.[64]

Taking up Rossby's call for more information about geophysical interactions between the Northern Hemisphere and the Southern Hemisphere, Wexler accepted the added challenge involved in serving as chief scientist for the U.S. expedition to the Antarctic for the International Geophysical Year (IGY; 1957–1958). By doing so, he could integrate critical new information about both the South Pole and the Southern Hemisphere into a global picture of circulation and dynamics of the entire atmosphere. Wexler also incorporated the results of theoretical work on the influence of rising carbon dioxide levels into the weather bureau's climate-modeling efforts and instituted radiation, ozone, and, notably, carbon dioxide measurements at the Mauna Loa Observatory, which were established under his guidance just prior to the IGY.[65]

Atmospheric observation by rockets and satellites also came under Wexler's purview. He served as chairman of several influential committees on this subject, including the Upper-Atmosphere Committee of the American Geophysical Union, the National Advisory Committee on Aviation's Special Committee for the Upper Atmosphere, and the National Research Council's Space Science Board. Wexler was in charge of the meteorology of the TIROS (Television Infrared Observation Satellite) meteorological satellite program and helped support the first Earth heat budget experiment flown on *Explorer 7*. In 1961 the Kennedy administration appointed Wexler as the lead negotiator for the United States in talks with the Soviet Union concerning the joint use of meteorological satellites. The negotiations expanded into a multinational effort to institute a World Weather Watch (WWW), with Wexler and Soviet academician Victor A. Bugaev as the architects for a new program to be administered by the World Meteorological Organization in Geneva. Formally established in 1963 and still in existence, the World Weather Watch coordinates the efforts of member nations by combining observing systems, telecommunication facilities, and data-processing and forecasting centers to make available meteorological and related environmental information needed to provide efficient weather services in all countries. Wexler was clearly on top of his science, a leader in new techniques and technologies, and a figure of international importance. In other words, he was a meteorological heavyweight.[66]

In 1958 Wexler published a paper in *Science* that examined some of the consequences of tinkering with the Earth's heat budget. He began by describing the two streams of radiant energy and their seasonal and geographic distribution, one stream directed downward and the other upward, which "dominate the climate and weather of the planet Earth."[67] The downward stream of energy consists of the solar radiation absorbed by the Earth's surface and atmosphere after accounting for losses by reflection. The upward stream is infrared radiation emitted to space by the Earth's surface and atmosphere, the latter mostly from atmospheric water vapor, clouds, carbon dioxide, and ozone. Wexler wrote: "In seeking to modify climate and weather on a grand scale it is tempting to speculate about ways to change the shape of these basic radiation curves by artificial means" (1059), especially by changing the reflectivity of the Earth. After a brief examination of possible albedo changes caused by using carbon dust to blacken the deserts and the polar ice caps, Wexler turned to the notion, probably originating with Teller, that detonating ten really "clean" hydrogen bombs in the Arctic Ocean would produce a dense ice cloud in northern latitudes and would likely result in the removal of the sea ice. The balance of the paper in *Science* is an examination of the radiative, thermal, and meteorological consequences of this outrageous act, not only for warming the polar regions but also for the equatorial belt and middle latitudes. Noting perceptively that "the disappearance of the Arctic ice pack would not necessarily be a blessing to mankind" (1062–1063) and implying that a nation like the Soviet Union already had the firepower to try such an experiment, Wexler concluded with a paragraph whose relevance has not been diminished by time: "When serious proposals for large-scale weather modification are advanced, as they inevitably will be, the full resources of general-circulation knowledge and computational meteorology must be brought to bear in predicting the results so as to avoid the unhappy situation of the cure being worse than the ailment" (1063).

In 1962, armed with the latest results from computer models and satellite radiance measurements as applied to studies of the Earth's heat budget, Wexler expanded his study to examine theoretical questions concerning natural and anthropogenic climate forcings, both inadvertent and purposeful. He did this in his lectures to technical audiences: "On the Possibilities of Climate Control," presented at the Boston chapter of the American Meteorological Society, the Traveler's Research Corporation in Hartford, and the UCLA Department of

Meteorology.[68] After reminding his listeners that Kennedy and Khrushchev had made climate modification "respectable" to talk about, Wexler quoted extensively from Zworykin's weather control proposal and von Neumann's response to it.

Wexler discussed the problem of increasing global pollution from industry and reviewed recent developments in atmospheric science, including computing and satellites that led him to believe that manipulating and controlling large-scale phenomena in the atmosphere were becoming distinct possibilities. He cited rising carbon dioxide emissions as an example of indirect control, mentioning the Callendar effect as one of the ways in which humanity was already inadvertently modifying global climate: "We are releasing huge quantities of carbon dioxide and other gases and particles into the lower atmosphere, which may have serious effects on the radiation or heat balance, which determines our present pattern of climate and weather."[69]

Wexler warned that the space age was introducing an entirely new kind of "atmospheric pollution" problem. He was particularly worried that some types of rocket fuel might release chlorine or bromine, "which could destroy naturally occurring atmospheric ozone and open up a 'hole,' admitting passage of harmful ultra-violet radiation to the lower atmosphere" (2). Changes in the upper atmosphere caused by increasing contrails, space experiments gone awry, or the actions of a hostile power could disrupt the ozonosphere, the ionosphere, or even the general circulation and climate on which human existence depends. Wexler felt that it was urgent to use "the most advanced mathematical models of atmospheric behavior" (3) to study the physical, chemical, and meteorological consequences of such interferences. He then explained how the weather bureau was in the process of acquiring new computers and developing new models to "simulate the behavior of the actual atmosphere and examine artificial influences that Man is introducing in greater and greater measure as he contaminates the atmosphere" (4). Both the *Christian Science Monitor* and the *Boston Globe* ran prominent stories on this aspect of his lecture.[70]

Wexler told his audiences that he was concerned with planetary-scale manipulation of the environment that would result in "rather large-scale effects on general circulation patterns in short or longer periods, even approaching that of climatic change."[71] He assured them that he did not intend to cover all possibilities "but just a few . . . *limited primarily to interferences with the Earth's radiative balance on a rather large scale*: I shall discuss in a purely hypothetical framework those atmospheric influences that man might attempt deliberately to exert and also those which he may now be performing or will soon be performing, perhaps in ignorance of its consequences. We are in weather control *now* whether we know it or not" (4).

He was clearly interested in both inadvertent climatic effects—such as might be created by industrial emissions, rocket exhaust gases, or space experiments gone awry—and purposeful interventions, whether peaceful or done with hostile intent. Echoing von Neumann's 1955 warning about technology, Wexler continued: "Even in this day of global experiments, such as the world-wide Argus electron seeding of the Earth's magnetic field at 300 miles height, man and machinery orbiting the Earth at 100 miles seventeen times in one day, and 100 megaton bombs—are we any closer to some idea of the approaches which could lead to an eventual 'solution' [to the problem of climate control]?" (3). He noted "a growing anxiety" in the public pronouncements that "Man, in applying his growing energies and facilities against the power of the winds and storms, may do so with more enthusiasm than knowledge and so cause more harm than good."[72]

Wexler was well aware that any intervention in the Earth's heat budget would change the atmospheric circulation patterns, the storm tracks, and the weather itself, so, as he pointed out, weather and climate control are not two different things. After presenting some twenty technical slides on the atmosphere's radiative heat budget and discussing means of manipulating it, Wexler concluded with a grand summary of highly speculative techniques to heat, cool, or otherwise restructure the atmosphere:

> (a) increase global temperature by 1.7°C [3°F] by injecting a cloud of ice crystals into the polar atmosphere by detonating 10 H-bombs on the Arctic sea ice;
>
> (b) lower global temperature by 1.2°C [2.2°F] by launching a ring of dust particles into equatorial orbit to shade the Earth;
>
> (c) warm the lower atmosphere and cool the stratosphere by injecting ice, water, or other substances into space; and
>
> (d) destroy all stratospheric ozone, raise the tropopause, and cool the stratosphere by up to 80°C [144°F] by an injection of a catalytic de-ozonizer such as chlorine or bromine.[73]

Cutting a Hole in the Ozone Layer

One of the most stunning aspects of Wexler's lectures was his awareness that catalytic reactions of chlorine and bromine could severely damage the ozone layer. Wexler was concerned that inadvertent damage to ozone might occur if increased rocket exhaust polluted the stratosphere or if near-space "seeding" experiments went awry: "The exhausts from increasingly powerful and numerous space rockets will soon be systematically seeding the thin upper atmosphere with large quantities of chemicals it has never possessed before or only in small quantities."[74] He was also concerned that the cold war and the space age might provide rival militaries with both the motivation and the wherewithal to damage the ozone layer. He cited a 1961 study by the Geophysics Corporation of America on possible harm to the Earth's upper atmosphere caused by the oxidizers in rocket fuel. He was also aware that Operations Argus and Starfish, Project West Ford, and Project High Water constituted recent significant interventions in the near-space environment that were accompanied by unknown and unquantified risks.

On the topic of purposeful damage, Wexler turned to the 1934 presidential address to the Royal Meteorological Society, in which the noted geoscientist Sydney Chapman had asked, "Can a hole be made in the ozone layer?"[75] That is, can all or most of the ozone be removed from the column of air above some chosen area? Chapman was thinking of an event that would provide a window for astronomers to extend their observations some hundreds of angstroms farther into the ultraviolet without the interference of atmospheric ozone. Possible health effects of human exposure to shortwave radiation did not appear to Chapman to be an important issue, since the hole he was contemplating would be localized, probably in a remote area (he suggested Chile), and would be short-lived, somewhere between a day and an hour, timed for the benefit of astronomers only. Cutting such a hole, Chapman continued, would require "the discharge of a deozonizing agent" perhaps by airplanes, balloons, or rockets. Chapman proposed two possibilities: a large amount of a one-to-one destructive agent such as hydrogen that would reduce O_3 molecules to O_2 or "some catalyst which, without itself undergoing permanent change, could promote the reduction of large numbers of ozone molecules in succession" (134). Although the choice of the agent would have to be left to the chemists, Chapman concluded that "the project of making a [temporary] hole in the ozone layer [a 90 percent reduction for the benefit of astronomers] does not seem quite impossible of achievement" (135).

In November 1961, Wexler gathered weather bureau staff for a briefing on ozone depletion and circulated this memo, titled "Deozonizer":

> Sydney Chapman proposed making a temporary "hole" in the ozone layer by inserting a substance which could be oxidized by the ozone. He suggested that hydrogen might be dispersed but wondered if there might be a catalyst gas or fine

powder which might perhaps be dispersed in smaller quantities than the 1 to 1 ratio hydrogen would require. Could you or your colleagues suggest suitable agents that might do the job with maximum efficiency consistent with the least weight?[76]

Bill Malkin suggested that Wexler might wish to raise the possibility with the country's national defense research arm, "that serious consideration be given to the possibility of artificially and temporarily altering (up or down) the ozone concentration over an area, as a most effective weapon."[77] Using the radiation model of Syukuro Manabe and F. Möller, Wexler was able to calculate a catastrophic 80°C (144°F) stratospheric cooling that would occur with no ozone layer.[78]

Seeking further advice on how to cut a "hole" in the ozone layer, Wexler turned to chemist Oliver Wulf at Caltech, who suggested that "from a purely chemical viewpoint, chlorine or bromine might be a 'deozonizer.'"[79] Wulf and Wexler exchanged numerous letters between December 1961 and April 1962 and met face-to-face in March, and Wulf met with Chapman in April. All these exchanges point to the conclusion (a stunning one, given the received history of ozone depletion) that chlorine or bromine atoms might act in a catalytic cycle with atomic oxygen to destroy thousands of ozone molecules. For example, Wulf wrote in early January 1962, "chlorine or bromine photosensitized decomposition [of ozone] might come closest to a reaction in which a small amount of added material would cause a relatively large amount of decomposition."[80] Wexler replied immediately, adding that he even had a delivery system in mind "à la West Ford dipoles" but had "no intention of suggesting or backing any such proposal."[81]

Wexler estimated that a 100-kiloton bromine "bomb" would destroy all ozone in the polar regions, and four times that amount would be needed near the equator. In a handwritten note composed in January 1962 he scrawled the following (figure 7.6):

UV decomposes $O_3 \rightarrow O$ in presence of a halogen like Br, Cl.

$O \rightarrow O_2$ and so prevents O_3 from forming.

100,000 tons Br. could theoret[ically] prevent all O_3 north of 65°N from forming.

And in another note (figure 7.7):

$Br_2 \rightarrow 2\,Br$ in sunlight destroys $O_3 \rightarrow O_2 + BrO$

These are essentially the basis of the modern ozone-depleting chemical reactions.

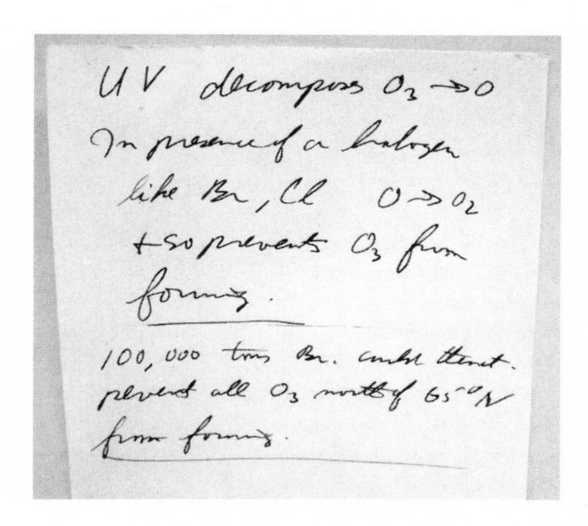

UV decompose $O_3 \to O$
In presence of a halogen
like Br, Cl $O \to O_2$
+ so prevents O_3 from
forming.

100,000 tons Br. could thereby
prevent all O_3 north of 65°N
from forming.

$Br_2 \to 2Br$ in sunlight
destroys $O_3 \to O_2 +$
BrO

7.6 (*top*) Harry Wexler's handwritten note on ozone depletion, January 1962. (WEXLER PAPERS)

7.7 (*bottom*) Harry Wexler's handwritten note on bromine reactions, January 1962. (WEXLER PAPERS)

Wexler's rough note of December 20, 1961, jotted down during a telephone conversation with Wulf, constitutes an ozone-depletion Rosetta Stone.[82] It links Chapman's 1934 speech, Wulf, rocket fuel emissions, ozone-destroying reactions triggered by chlorine and bromine as catalysts, particulates, methane destruction, and an estimate that a minuscule amount of atomic bromine could cause immense harm (figure 7.8).

In the summer of 1962, Wexler accepted an invitation from the University of Maryland Space Research and Technology Institute to present a lecture titled "The Climate of Earth and Its Modifications" and might, under normal circumstances, have prepared his ideas on geoengineering and ozone destruction for publication. However, he was cut down in his prime by a sudden heart attack on August 11, 1962, during a working vacation at Woods Hole. The documents

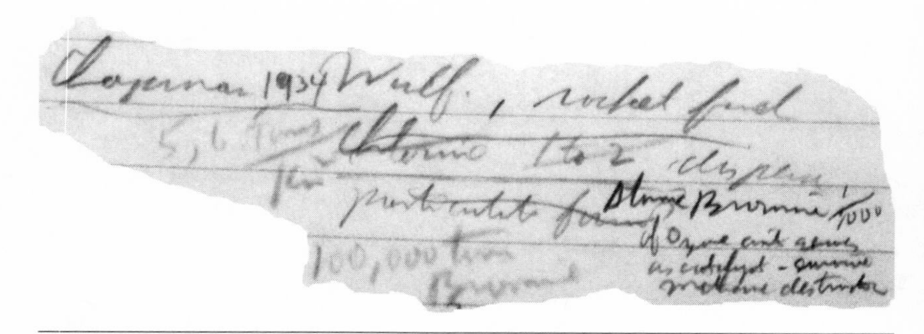

7.8 Harry Wexler's "Rosetta Stone" note, linking Sydney Chapman, Oliver Wulf, rocket fuel, and catalytic ozone-destroying reactions triggered by chlorine and bromine. (WEXLER PAPERS)

relating to his career—from his early work at MIT, his work as liaison to the IAS meteorology project, his research into all sorts of new technologies, to his final speeches on ozone depletion and climate control—headed into the archives, probably not to be seen and certainly not to be reevaluated until today.

The well-known and well-documented supersonic transport (SST) and stratospheric-ozone-depletion issues date only to the 1970s and do not include Wexler's role. The idea that bromine and other halogens could destroy stratospheric ozone was published in 1974, while chlorofluorocarbon production expanded rapidly and dramatically after 1962.[83] Had Wexler lived to publish his ideas, they would certainly have been noticed and could have led to a different outcome and perhaps an earlier coordinated response to the issue of stratospheric ozone depletion. Recently, I have been in correspondence with three notable ozone scientists about Wexler's early work: Nobel laureates Sherwood Rowland and Paul Crutzen and current National Academy of Sciences president Ralph Cicerone. They are uniformly interested and quite amazed by Wexler's insights and accomplishments.

Remarkable, too, is the fact that with all his sophistication and the leading roles he played in the development of computer modeling, satellite monitoring, and many, many other technical fields, Wexler still opened his 1962 lectures by quoting extensively from Zworykin's "Outline of Weather Proposal" (1945) and von Neumann's response to it. A colleague who heard Wexler's lecture in Boston wrote that climate engineering constituted "a delightful area of mental gymnastics. Let's hope the entire world is satisfied to play the game on this plane until the state of meteorological knowledge is truly adequate for big league experimentation."[84] Wexler replied, "I hope that before we get into large experimentation that not only will the state of meteorological knowledge be much more advanced than it is now, but also the state of our socio-political affairs as well."[85] Remember, it was not Paul Crutzen in 2006 but Harry Wexler about fifty years before who first claimed that climate control was now "respectable to talk about," even if he considered it quite dangerous and undesirable.

* * * * *

The possibility of manipulating global climate through planetary-scale engineering is currently being actively debated, although its feasibility and desirability are highly questionable, if not contentious. Most of the debate centers on back-of-the -envelope calculations (which are not good enough) or basic climate models (which are also not good enough). Still, the current crop of geoengineers has yet to acknowledge the checkered history of the subject.

Accounts of the early history of computers in meteorology follow a well-rehearsed script, identifying Vilhelm Bjerknes and Louis Fry Richardson as early pioneers and emphasizing progress after 1946 through the work of a familiar cast of characters and technical breakthroughs. Through the career of Harry Wexler, we can now see that the two histories, the familiar and the (until now) unwritten, are closely interrelated and that climate control is not so much a newcomer in the age of global warming as something that has been up in the air for quite a long time.

The recent history of climate fears, fantasies, and possibilities is positioned firmly between the work of two colleagues, John von Neumann and Harry Wexler. An examination of general climate fears and specific climate fantasies reveals that some were no more than hand-waving proposals, while some were actual field projects. Anchoring this in time were the high hopes that futurists had for new emerging technologies such as digital computers, to provide stunning precision and predictability; nuclear energy, to power continental-scale transformations or violently alter the geophysical status quo; and satellites, to monitor the Earth continuously with eagle eyes and to serve as platforms for active interventions.

Wexler's work on geoengineering in the period 1958 to 1962 applied the results of new computer climate experiments, nuclear tests in near space, and newly available satellite heat budget measurements. His work on ozone destruction, in particular, is notable since it predated the Nobel Prize–winning work of Paul Crutzen, Sherwood Rowland, and Mario Molina by about a decade, although Wexler died before he could publish the results. It is clear that Wexler was well qualified to speak authoritatively about the otherwise "nebulous" subjects of climate, climate change, and climate control. He served on numerous scientific panels and governmental advisory boards, had access to and helped collect global climate data, understood the theoretical issues and their complexity, and promoted and advanced the latest technologies. He warned then, and we might wisely conclude today, that

> [climate control] can best be classified as "interesting hypothetical exercises" until the consequences of tampering with large scale atmospheric events can be assessed in advance. Most such schemes that have been advanced would require colossal engineering feats and contain the inherent risk of irremediable harm to our planet or side effects counterbalancing the possible short-term benefits.[86]

Based on the visionary foundation provided by Vladimir Zworykin and John von Neumann, and the much more speculative megaprojects being proposed at the time, Wexler's prescient work "On the Possibilities of Climate Control" clearly reminds us that we are not the first generation to be involved with or concerned about geoengineering and places the current debate in the context of at least half a century of continuous and usable history.

8

THE CLIMATE ENGINEERS

How can you engineer a system whose behavior you don't understand?

—RON PRINN, QUOTED IN MORTON, "CLIMATE CHANGE"

DURING the unusually hot summer of 1988, with a major heat wave in the American Midwest, Yellowstone National Park in flames, and issues such as ozone depletion in the headlines, climate modeler James Hansen of NASA announced to the world that "global warming has begun."[1] Hansen reported that he was "99 percent certain that the warming trend was not a natural variation but was caused by a buildup of carbon dioxide and other artificial gases in the atmosphere" and that anthropogenic greenhouse warming "is already happening now." He predicted more-frequent episodes of very high temperatures and drought in the next decade and beyond. Hansen later revised his remarks, but his statement remains the starting point for recent widespread concern about global warming. The question was no longer whether human agency had contributed to global change. That question was answered in the affirmative long ago. The more significant questions involved the magnitude and consequences of the global changes being caused by a combination of natural forces and increasing anthropogenic stresses and what was to be done about it.

That summer, the government of Canada, in collaboration with the United Nations Environment Program (UNEP) and the World Meteorological

Organization (WMO), convened a major conference on the topic "The Changing Atmosphere: Implications for Global Security." The conference statement captured the tone and urgency first expressed in the 1950s by Roger Revelle and John von Neumann: "Humanity is conducting an unintended, uncontrolled, globally pervasive experiment, whose ultimate consequences could be second only to a global nuclear war."[2] The conference recommended reductions of carbon dioxide emissions to 20 percent below 1988 levels, to be achieved by 2005. Needless to say, we did not reach this goal, but a process had been put into motion to set new goals and deadlines.

Also in 1988, the WMO and UNEP established the Intergovernmental Panel on Climate Change (IPCC), whose purpose is to provide periodic assessments of "the scientific, technical and socioeconomic information relevant for the understanding of the risk of human-induced climate change."[3] From modest beginnings, the IPCC has emerged as a representative parliamentary body that has gradually acquired status and authority. It has prepared four major assessments to date, the first in 1990 in preparation for the 1992 Earth Summit in Rio de Janeiro. Here the UN Framework Convention on Climate Change (UNFCCC) set an ultimate objective of stabilizing atmospheric concentrations of greenhouse gases at levels that would prevent "dangerous" human interference with the climate system.[4]

Each of the subsequent IPCC reports (1995, 2001, 2007) has expressed a sense of greater urgency about the climate change problem. The IPCC consensus involves six key points:

1. Anthropogenic emissions are changing the composition of the atmosphere, especially by increasing its radiatively active trace gases.
2. This will enhance the greenhouse effect and will result in long-term global warming.
3. Observed changes in climate on decades-to-centuries time scales are consistent with human influence.
4. Models indicate that future warming is likely to be substantial.
5. Both environment and society will be adversely impacted.
6. Avoiding dangerous human influence in the climate system will require substantial early actions, but may not provide direct benefits for several generations.[5]

It is still not clear what "dangerous human influence" in the climate system actually is or how to avoid it, but mitigation, adaptation, and intervention through climate engineering are now on the table. Deindustrialization will also reduce greenhouse gas emissions, as demonstrated by the political and economic collapse of the Soviet Union.

Ratcheting up the sense of urgency, in 2005 Hansen warned that the Earth's climate is nearing an unprecedented "tipping point"—a point of no return that can be avoided only if the "growth of greenhouse gas emissions is slowed" in the next two decades:

> The Earth's climate is nearing, but has not passed, a tipping point beyond which it will be impossible to avoid climate change with far-ranging undesirable consequences. These include not only the loss of the Arctic as we know it, with all that implies for wildlife and indigenous peoples, but losses on a much vaster scale due to rising seas. . . . This grim scenario can be halted if the growth of greenhouse gas emissions is slowed in the first quarter of this century.[6]

According to Hansen, tipping points occur because of amplifying feedbacks, including loss of sea ice, melting glaciers, release of methane in warming permafrost, and growth of vegetation on previously frozen land. These surface and atmospheric changes increase the amount of sunlight absorbed by the Earth and amplify the warming effect of carbon dioxide produced by burning fossil fuels. Hansen's brief statement, widely distributed by the press, clearly struck a cultural nerve. It acknowledged undesirable and inadvertent human influence on the climate system and pointed to a possible remedy. In the interest of impact, however, Hansen avoided complexities. For example, it is highly unlikely that merely slowing the growth of emissions would be a very effective policy. Hansen may be right: we may be approaching the physical tipping point of climate, or, as James Lovelock argued in his book *The Revenge of Gaia*, we may already have passed it, with catastrophic consequences for humanity.[7] More likely, Hansen, Lovelock, and many others are trying to add the weight of their opinions to a second kind of "tipping point," a behavioral change in which humanity decides to live with only clean energy and takes concerted action against harming the climate system. There is also a third "tipping point"—one that has been reached by a handful of geoengineers who are so concerned about climate change that they are proposing purposeful, even reckless, intervention.

The following discussion will define geoengineering, review its recent history, and provide a critique of current proposals and practices by revealing their assumptions and values. It is an occasion to reflect on the precedents that brought us to this point and to identify a "middle path" of mitigation and adaptation located between doing too little and doing too much. It is offered in the hope that the study of a checkered past can help us avoid a checkered future and with the conviction that if we are indeed facing unprecedented challenges, it is good to consider historical precedents.

What Is Geoengineering?

In 1996 Thomas Schelling wrote, "'Geoengineering' is a new term, still seeking a definition. It seems to imply something global, intentional, and unnatural."[8] More than a decade later, the word remains largely undefined and unpracticed. It is not in the *Oxford English Dictionary*, but it did find its way into the *Urban Dictionary*, where it is loosely defined as "the intentional large-scale manipulation of the global environment; planetary tinkering; a subset of terraforming or planetary engineering . . . the last gasp of a dying civilization."[9] Lovelock subscribes to this definition, at least the first part, and further claims that "we became geoengineers soon after our species started using fire for cooking," or perhaps, as geoscientist William Ruddiman has proposed, millennia ago through the practices of extensive deforestation and agriculture.[10]

In the *OED*, an "engineer" is one who contrives, designs, or invents, "a layer of snares"; a constructor of military engines; one whose profession is the designing and constructing of works of public utility.[11] So engineering, by definition, has both military and civilian aspects, elements potentially both nefarious and altruistic (figure 8.1). By analogy, the neologism "geoengineer" refers to one who contrives, designs, or invents at the largest planetary scale possible for either military or civilian purposes—a layer of snares at the global level. Today geoengineering, as an unpracticed art, is still largely "geo-scientific speculation."

"Ecohacking," another term for geoengineering, made the short list for the Oxford Word of the Year 2008. It is loosely defined as "the use of science in very large-scale [planetary scale] projects to change the environment for the better/stop global warming (e.g., by using mirrors in space to deflect sunlight away from Earth)."[12] A recent report issued by the Royal Society of London defines geoengineering as "the deliberate large-scale manipulation of the planetary environment to counteract anthropogenic climate change."[13] But there are significant problems with such definitions. First of all, an engineering practice defined by its scale (geo) need not be constrained by its stated purpose (environmental improvement), by one of its currently proposed techniques (space mirrors), or by one of perhaps many stated goals (to counteract anthropogenic climate change). Nuclear engineers, for example, are capable of building both power plants and bombs; mechanical engineers can design components for both ambulances and tanks; my father, a precision machinist during World War II, milled both aluminum ice cream scoops and one-of-a-kind components for top-secret military projects. So to constrain the essence of something that does not exist by its stated purpose, techniques, or goals is misleading at best.

8.1 A climate engineer. (FLEMING, "THE CLIMATE ENGINEERS")

"Ecohacking" sounds both too small and too electronic to cover the field of geoengineering. We are all ecohackers, as was the first person to cut down a tree with an axe. In traditional English, hackers are literally those who chop up the Earth, or figuratively those who mangle words or sense. In the computer age, "hacker" is slang for an enthusiast who considers programming an end in itself or, more subversively, who seeks to gain unauthorized access to computer files or networks. Hackers typically have "big projects" about which they obsess. One project of the computer climate engineers is to cut off the sunbeams in a simple climate model to "prove" that the Earth will cool and sea ice will grow. Much more sophisticated modelers have shown that the unknown consequences of doing this may be very, very serious. When people propose to cool the Earth by 2°C (3.6°F) using a technical fix, they are overlooking the fact that Earth has not yet warmed 2°C in the past century. So we are really dealing with dangerous speculation about speculation. A more apt term might be "geohacking," which is hopefully harmless enough if the practice is restricted to tinkering with computer models and never "sees the light of day" in the form of potentially dangerous outdoor demonstration projects or planetary-scale tinkering.

Placing his faith firmly in progress, engineer and policy analyst David Keith is of the opinion that scientific understanding grants us increased Archimedean leverage and an "ever greater capability to deliberately engineer environmental processes on a planetary scale."[14] Echoing William Suddards Franklin and his grasshopper of long ago or Ross Hoffman and his misunderstanding of the butterfly effect, Keith maintains that "accurate knowledge of the atmospheric state and its stability could permit leverage of small, targeted perturbations to effect proportionately larger alterations of the atmospheric dynamics."[15] But no matter how great the scientific wizardry, the modern Archimedes still has no place to stand, no acceptable lever or fulcrum, and no way to predict where the Earth will roll if tipped. Failing ultimate control, geoengineering may indeed have the potential to enrage the chaotic "climate beast" of the influential geochemist and oceanographer Wallace Broecker.[16]

Terraforming and Beyond

Geoengineering is a subset of "terraforming," or the engineering of planetary environments. Martyn J. Fogg reviewed the history and some of the technical aspects of "orchestrated planetary change" in his book on this subject, published, curiously, by the Society of Automotive Engineers, a group that one might expect would be most familiar with automobile air-conditioning.[17] He defined "planetary engineering" as "the application of technology for the purpose of influencing the global properties of a planet" and "terraforming" as the process of "enhancing the capacity of an extraterrestrial planetary environment to support life. The ultimate in terraforming would be to create an uncontained planetary biosphere emulating all the functions of the biosphere of the Earth—one that would be fully habitable for human beings."[18]

Fogg described how ecological-engineering techniques might be used someday to implant life on other planets and how geoengineering might be used to ameliorate (or perhaps exacerbate) the currently "corrosive process" of global change on the Earth. He presented order-of-magnitude calculations and the results of some simple computer modeling to assess the plausibility of various planetary-engineering scenarios. He deemed it "rash to proclaim" impossible any scheme that does not "obviously violate the laws of physics." Yet Fogg focused only on possibilities, not on unintended consequences, and left unaddressed questions of whether the schemes are desirable, or even ethical. According to Fogg, geoengineering is not simply, or even primarily, a technical problem because people, their politics, and their infrastructures get in the way. That is, it involves the implications and dangers of attempting to tamper with an immensely complex biosphere on an inhabited planet.

The epigraph of Fogg's book cites Hungarian-born engineer and physicist Theodore von Kármán to the effect that "scientists study the world as it is; engineers create the world that has never been." This quote has an ominous ring, however, when it comes to terraforming, since some "worlds" perhaps should never be. Fogg traced inspiration for the field to Olaf Stapleton's *Last and First Men* (1930), Robert Heinlein's *Farmer in the Sky* (1950), and James Lovelock and Michael Allaby's *The Greening of Mars* (1984). In his "concise history of terraforming," Fogg mentioned the work of naturalists John Ray (English, seventeenth century) and Georges-Louis Leclerc, Comte de Buffon (French, eighteenth century), who looked on the Earth as unfinished, with man taking the role of a junior partner in creation, taming the wilderness as part of a historical progression toward "perfection."[19] From there, Fogg dropped the names of George Perkins Marsh (1801–1882), an American diplomat and naturalist who wrote about replanting forests, channeling rivers, and reclaiming deserts in *Man and Nature* (1864); Vladimir Vernadsky (1863–1945), the Russian mineralogist and geochemist who popularized the notion of the interconnectedness of the "biosphere"; and Pierre Teilhard de Chardin (1881–1955), the French cleric and philosopher who placed the "noosphere," the realm of human thought, in evolutionary succession to the geosphere and the biosphere.

Such expansive antecedents belie recent attempts to restrict the definition of geoengineering to the purposeful and large-scale alteration of the shortwave side of the Earth's energy budget with the intent of affecting climate. In the literature of planetary terraformation, geoengineering is much, much more than that. It comprises macro-scale projects to control not only the supposed relatively simple and straightforward interaction of albedo and temperature but also much more complex and potentially unknowable interactions of Earth system science—involving the lithosphere, the hydrosphere, the atmosphere, the biosphere, and, perhaps most important, society. After all, engineering deals with the technical side of *human* affairs, and the prefix "geo" potentially involves all aspects of the planet, perhaps also its most prominent companions, the Sun and the Moon. Fogg ventured into hyper-speculative territory when he discussed "astroengineering," or modifying the properties of the Sun, by intervening in its opacity, nuclear reactions, mass loss, chemical mixing, and even "accretion into a central black hole." Tellingly, Fogg admitted that "technical difficulties associated with astroengineering will be immense" (457–458).

Ethical Consequences

Most studies have ignored, minimized, or barely mentioned important ethical issues regarding geoengineering.[20] The report of a 2009 study group, composed of prominently placed geoengineering advocates, candidly admits that the most important sociopolitical and ethical constraints on implementing climate engineering were largely outside the expertise of the technically oriented participants and thus beyond the scope of their study.[21] Every engineer has to seek a building permit for every project, to engage the community and the local authorities in discussion, and to obsess (a lot) about design, safety, and cost. A well-engineered project, especially at the "geo" scale, must be based on ethical principles and practices, sound science, technologies and testing methods, economics (not just immediate costs), politics (including legal and diplomatic aspects), and attention to social, cultural, medical, and environmental concerns. However, if it ever comes down to it, who has the right to issue a permit for the intentional manipulation of the global environment? Who does cost-benefit and safety analysis for the planet? Who is liable for any engineering shortcomings or failures? Would climate engineering, by counteracting the effects of greenhouse gas emissions, create moral traps—for example, by reducing incentives to mitigate or by burdening future generations with expensive and unwieldy projects? Where would the global thermostat be located, and who would control it? Could designer geoengineering be practiced at regional levels to address the greatest problems while seeking to avoid a one-size-fits-all planetary fix? What if some group or nation decided, unilaterally, to intervene in a heavy-handed way in planetary processes and the results were viewed as detrimental to a region or even to the globe? Could today's climate control engineering fantasies, if acted out, lead to undesirable consequences and exacerbate international tensions?[22]

In this vein, atmospheric scientist Alan Robock, a leader in modeling efforts to evaluate climate-engineering schemes, recently wrote,

> The reasons why geoengineering may be a bad idea are manifold, though a moderate investment in theoretical geoengineering research might help scientists to determine whether or not it is a bad idea. Small-scale deployments are out of the question until we are sure that known adverse consequences can be avoided. Then there are the [Donald Rumsfeld–like] multiple unknown unknowns that argue against ever undertaking a large-scale deployment.[23]

His list of twenty reasons (subsequently pared down to seventeen) why geoengineering (especially solar radiation attenuation by sulfates) may be a bad idea includes:

> (1) Potentially devastating effects on regional climate, including drought in Africa and Asia, (2) Accelerated stratospheric ozone depletion, (3) Unknown environmental impacts of implementation, (4) Rapid warming if deployment ever stops, (5) Inability to reverse the effects quickly, (6) Continued ocean acidification, (7) Whitening of the sky, with no more blue skies, but nice sunsets, (8) The end of terrestrial optical astronomy, (9) Greatly reduced direct beam solar power, (10) Human error, (11) The moral hazard of undermining emissions mitigation, (12) Commercialization of the technology, (13) Militarization of the technology, (14) Conflicts with current treaties, (15) Who controls the thermostat? (16) Who has the moral right to do this? (17) Unexpected consequences.[24]

Some of these results (1–5) are derived from general circulation model simulations and others (6–9) from back-of-the-envelope calculations; most, however, (10–17) stem from historical, ethical, legal, and social considerations. Robock admits that geoengineering would have certain benefits, including cooling the planet, possibly reducing or reversing sea ice and ice sheet melting and sea level rise, and increasing plant productivity and thus the terrestrial carbon sink.

Most enthusiasts for solar radiation management have overlooked, however, its "dark" side: the scattering of starlight as well as sunlight, which would further degrade seeing conditions for both ground-based optical astronomy and general night sky gazing. A recent article by astronomers Christian Luginbuhl, Constance Walker, and Richard Wainscoat discusses the rapid growth of light pollution from ground-based sources but does not consider aerosol scattering effects that reduce nighttime seeing.[25] Imagine the outcry from professional astronomers and the general public if the geoengineers pollute the stratosphere with a global sulfate cloud; imagine a night sky in which sixth-magnitude stars are invisible, with a barely discernible Milky Way and fewer visible star clusters or galaxies. This would be worse than Project West Ford. It would constitute a worldwide cultural catastrophe.

When contemplating *planetary*-scale engineering, regionally or nationally based technical initiatives are not nearly broad enough. As the Tyndall Centre for Climate Change Research pointed out, the equity issues are likely to be substantial: "There will be winners and losers associated with geo-engineering (as there will be with climate change itself). Should the losers be compensated, and if so how? Where the losses include non-market goods, which may be irreplaceable, how are they to be valued?"[26] The process of discussion and decision making needs to include an interdisciplinary mix of historians, ethicists, policymakers, and a broad and inclusive array of international and intergenerational participants—features that have been sorely lacking in recent meetings, which featured mostly white, Western, scientifically trained, and technocratically oriented males.[27] In fact, the field's current lack of diversity indicates that some of the most critical questions have probably not even been posed! For example, how would geoengineering alter fundamental human relationships to nature? Does this or the other questions posed so far have univocal answers? How do they play out in different cultures? Has anyone considered this? A large-scale environmental technological fix framed as a response to undesired climate change could be seen as an act imposed on the multitude by the will of the few, for the primary benefit of those already in power. Many would undoubtedly interpret it as a hostile or an aggressive act. Isn't geoengineering in the category of "Western solutions to global problems"? Rather than engaging in speculative large-scale climate engineering, isn't it better to reduce the effects of greenhouse gas emissions—by reducing greenhouse gas emissions? Gavin Schmidt, a climate modeler at the NASA Goddard Institute for Space Studies, offered a "rock the boat" analogy to illustrate the point:

> Think of the climate as a small boat on a rather choppy ocean. Under normal circumstances the boat will rock to and fro, and there is a finite risk that the boat could be overturned by a rogue wave. But now one of the passengers has decided to stand up and is deliberately rocking the boat ever more violently. Someone suggests that this is likely to increase the chances of the boat capsizing. Another passenger then proposes that with his knowledge of chaotic dynamics he can counterbalance the first passenger and, indeed, counter the natural rocking caused by the waves. But to do so he needs a huge array of sensors and enormous computational resources to be ready to react efficiently but still wouldn't be able to guarantee abso-

lute stability, and indeed, since the system is untested, it might make things worse. So is the answer to a known and increasing human influence on climate an ever more elaborate system to control the climate? Or should the person rocking the boat just sit down?[28]

Protection, Prevention, and Production

In 1930 Harvard geographer and meteorologist Robert DeCourcy Ward sorted climate intervention strategies into three categories: (1) protection, which is "perfectly passive"; (2) prevention, which is more proactive; and (3) production, which is the most active and aggressive of the three.[29] Today we might call these approaches adaptation, mitigation, and intervention. Ward pointed out that protection from the elements, which started in cave dwellings and tropical huts, now involved heated buildings and, "more and more in the future," buildings "artificially cooled during the heat of summer." As in today's discussions of weather-related natural disasters, Ward cited increasing populations in areas visited by tropical cyclones and the need for "better methods of building," coastal setbacks "beyond the reach of the storm waves," and seawalls and breakwaters for coastal cities. For protection against tornadoes, "the most violent disturbances in the atmosphere," Ward recommended storm cellars and solid steel and concrete buildings. For protection from electrical fields, he touted the Faraday "cage" and the grounded lightning rod. High walls, narrow streets, and covered awnings traditionally provide shady relief in hot climates. Ward noted that in America by 1930, newly built arcades and department stores were providing shelter for shoppers, who tended to frequent them more and perhaps spend more money during periods of inclement weather.

Prevention required more effort and more resources. Planting trees for windbreaks to protect crops and prevent soil erosion was a widespread practice in Ward's day. "Frost-fighting" involved regular observations, forecasts for agricultural regions, and cooperative arrangements among farmers and fruit growers—for example, by flooding the cranberry bogs or lighting smudge pots in orchards. Overall, however, Ward had very few successful examples of prevention on which to draw. Fog dispersal worked on only a very small scale. The electrified sand experiments of L. Francis Warren indicated that clouds could be modified somewhat but not controlled, given the vast scale of the atmosphere.

For Ward the third stage, production, was "the most active and aggressive" and also the least possible. Best known to him was the history of artificial rainmaking—a history of promise and hype. James Espy's theory of lighting huge fires was theoretically sound and demonstrable on a small scale, yet impossible to implement operationally. Ward called Robert Dyrenforth's experiments a "national disgrace" and thought it "highly important that no such occasion should arise again" (13). He called the production of rain for profit to "hoodwink" desperate farmers the work of "pure fakirs." He claimed, perhaps too hastily, that "the speculations of former times have been discarded," and now we know the facts. How could he have known that speculation would *increase* over the next eight decades? Asking "How far can man control his climate?" Ward replied that we can protect against and prevent unwanted weather damage, but "we can not produce rain or change the order of nature." He saw "no hope . . . of our ever being able to bring about any but local modifications of the weather and climate" (18). Citing the opinion of Sir Napier Shaw, Ward concluded, "We are lords of every specimen of air which we can bottle up or imprison in our laboratories [but] in the open air we are practically powerless" (6). These words were written in 1930, before the dawn of cloud physics as a field, before the General Electric Corporation's cloud-seeding experiments, before the fantasies of ultimate control, and before the rise of serious fears of weather and climate warfare in the 1950s and 1960s.

Climate Leverage

The noted Soviet geoscientist Mikhail Ivanovitch Budyko (1920–2001) was deeply concerned about both the enhanced greenhouse effect and the growing problem of waste heat. At a 1961 conference in Leningrad on "problems of climate control," he pointed out that at current and projected rates of growth, the waste heat produced by human energy generation could, in two hundred years, rival that of the Earth's radiation balance, rendering life on Earth "impossible."[30] Cities already generated more than five times more energy than the natural radiation balance, and if thermonuclear power was harnessed, he warned, dangerous temperature levels could be reached within a few decades. The threat of such excessive heat led him to become a strong advocate for learning to control and regulate climate. His colleague, academician M. Ye. Shvets, advanced a proposal to inject 36 million tons of 1-micron dust particles into the stratosphere, which would blanket the Northern Hemisphere within six months. His calculations indicated that such a dust screen would reduce solar radiation by 10 percent and temperatures by 2 to 3°C (3.6 to 5.4°F). Such an intervention was also expected to reduce evaporative losses, increase precipitation, and thus increase water supply.[31]

Budyko found this scheme preferable to other ideas of the time, such as the one to create thermal mountains. In James Black and Barry Tarmy's article "The Use of Asphalt Coatings to Increase Rainfall" (1963), two workers for the Esso Research and Engineering Company in New Jersey argue that "useful amounts of rainfall might be produced economically in arid regions near seas and lakes" by "coating a large area with asphalt to produce thermal updrafts which increase the sea breeze circulation and promote condensation."[32] One acre of petrochemical paving materials, conveniently supplied by Esso, would be needed for every 2 to 3 acres of enhanced rainfall area. The authors cited the ancient Babylonian practice of burning their fields after harvest, supposedly to create a blackened area that would produce extra rainfall for the next crop (but possibly for other reasons), and the early work of Espy on producing rain by large conflagrations. Turning to the recent literature, they cited papers on "man-made tornadoes" by Jean Dessens, who burned an acre-size pool of fuel oil at the rate of 1 ton a minute to create artificial clouds and even a small tornado, and suggested that the weather could be controlled artificially if an inexpensive means could be developed "*to paint the Earth black*" (emphasis added).[33] This sounds very much like the Sherwin Williams paint slogan "Cover the Earth" or perhaps the irreverent bumper sticker "Earth First! We'll Pave the Other Planets Later."

In 1962 Harry Wexler was the first to use the new methods of computer climate modeling and satellite heat budget measurements to warn of the possibilities, dangers, and excesses of "climate control," including ways to destroy the ozone layer either inadvertently or with possible harmful intent. The following year, the Conservation Foundation report *Implications of Rising Carbon Dioxide Content of the Atmosphere*, based largely on the work of Charles David Keeling and Gilbert Plass, predicted climate problems ahead and noted: "As long as we continue to rely heavily on fossil fuels for our increasing power needs, atmospheric CO_2 will continue to rise and the Earth will be changed, more than likely for the worse."[34]

Gordon J. F. MacDonald, professor of geophysics at UCLA, was of the opinion that weather control, even of severe storms such as hurricanes and typhoons, was just the beginning step in an escalating game of environmental and geophysical warfare using climate engineering. He thought that belligerents might, for example, cut a hole in the ozone layer over a target area to let in lethal doses of ultraviolet radiation, manipulate the Arctic ice sheet to cause climatic changes or massive tidal waves, trigger earthquakes from a distance, and in general manipulate or "wreck" the planetary environment and its geophysics on a strategic scale. MacDonald developed his perspective as a high-level government adviser, Pentagon confidant, chair of the National Academy of Sciences Panel on Weather and

Climate Modification, and member of the President's Science Advisory Committee (PSAC) in the Johnson administration.[35]

In 1965 the PSAC issued a report titled *Restoring the Quality of Our Environment,* which contained 104 recommendations about pollution of air, soil, and waters. Appendix Y of this report, the work of a subcommittee on atmospheric carbon dioxide chaired by Roger Revelle, is now widely cited as the first official government statement on global warming. It pointed out that "carbon dioxide is being added to the earth's atmosphere by the burning of coal, oil, and natural gas at the rate of 6 billion tons a year. By the year 2000 there will be about 25 percent more carbon dioxide in our atmosphere than at present." Increases in atmospheric CO_2 resulting from the burning of fossil fuels could modify the Earth's heat balance to such an extent that harmful changes in climate could occur. The subcommittee also explored the possibilities of deliberately bringing about "countervailing climatic changes." One ill-conceived suggestion involved increasing the Earth's solar reflectivity by dispersing buoyant reflective particles over large areas of the tropical sea at an annual cost of about $500 million. The subcommittee pointed out that this technology, which was not excessively costly, might also inhibit hurricane formation. No one thought to consider the side effects of particles washing up on tropical beaches or choking marine life or the negative consequences of intervening in hurricanes. And no one thought to ask if the local inhabitants would be in favor of such schemes. Another speculation involved modifying high-altitude cirrus clouds to counteract the effects of increasing atmospheric carbon dioxide. The subcommittee failed to mention the most obvious option: reducing fossil fuel use.[36]

In 1968 Joseph O. Fletcher (b. 1920) of the RAND Corporation published a review of the known patterns and causes of global climate change. In addition to natural causes, the main influencing factors seemed to be the side effects of industrial civilization: carbon dioxide emissions, smog and dust pollution, and waste heat. As Wexler had argued in 1962, purposeful climate modification was also a theoretical possibility, but Fletcher was beginning to argue that it was now becoming a necessity. He reported on recent activities in the Soviet Union aimed at climate control, none of them very promising, and asked: "What can be done to speed progress" in this field?[37] Fletcher's prescription was that climate science must follow what he considered an inevitable four-stage progression: observation, understanding, prediction, and control. Global observations were being conducted or planned at the time using new satellite platforms and large-scale field research campaigns, while theoretical groups were forming around increased computing resources and new mathematical models of atmospheric and oceanic circulation. Fletcher thought that "an inevitable result" of all this would be "the development of a more sophisticated theory to explain climatic change which, in turn will trigger an avalanche of 'climatic experiments' testing the predictions of the improved theory of climate" (22). Is scientific progress linear? Can it be managed?

The following year, Fletcher issued a report on "managing climatic resources" in which he came to the "inescapable conclusion" that due to rising population, greater vulnerabilities, and the irreversible damage being done to the climate system, "purposeful management of global climatic resources and control of the planet's climate would eventually become necessary to prevent undesirable changes."[38] Citing a recent upsurge of research in weather modification and climate control, he thought that humanity had reached a technological threshold at which it was already "within man's engineering capacity" to influence the global system by altering patterns of thermal forcing. He considered it feasible to carry out climate-influencing schemes such as creating large inland seas, deflecting ocean currents, seeding clouds extensively, and (the reverse of today's sentiment) even removing the Arctic pack ice. Then, as now, he left unresolved the huge and complex economic, sociological, legal, and political problems that such intervention would generate.

Fletcher stated, in no uncertain terms, that "an increase in CO_2 causes global warming" (2). Referring to the work of Guy Stewart Callendar and Gilbert Plass, who attributed the warming of the previous century of approximately 0.5°C (0.9°F) to this cause, he warned that a future warming of three times this amount or even more could be possible by the year 2000 (this did not happen) and could bring about "important changes of global climate during the next few decades" (this might yet happen) (3). Another, longer-term problem that he highlighted, echoing Budyko, was that heat pollution from energy generation could grow to rival the energy provided by the Sun. Still, Fletcher hedged his bets by pointing to the strong negative feedback of increased low-level cloudiness; the assumed enormous capacity of colder ocean water to absorb carbon dioxide; the 30 percent increase in turbidity, or "global dimming," due to air pollution and aircraft condensation trails; and the overall complexity of the climate system, which renders specific cause-and-effect estimates very uncertain.

After reviewing the complex patterns of past climate change and the workings of the global climate "machine," Fletcher concluded that the most important outstanding problem was developing a quantitative understanding of the general circulation, especially oceanic heat transport and ocean–atmosphere heat exchange. (Note that computer modeling was still in its infancy in 1969 and the El Niño–Southern Oscillation [ENSO] had not yet been identified, although aspects of the El Niño–La Niña system were known.) Fletcher also discussed feedbacks that acted as "triggers" of climate change and provided the example of the dramatic warming of the Arctic, identified and measured by 1940, which, had it continued, could have resulted in "a new and stable climatic regime" in which the Arctic Ocean became ice-free.

From climate "triggers," Fletcher moved on to a discussion of the possibilities of deliberately influencing climate. Here he followed the theoretical lead of Russian scientist M. I. Yudin, who sought to identify critical "instability points" for intervening in the development of cyclones, by changing either their winds or steering currents or their heat budget.[39] Using back-of-the-envelope calculations that have become de rigueur among geoengineers, Fletcher estimated that it would take only sixty C-5 aircraft to conduct cloud-seeding operations over the entire Arctic Basin and to exert "enormous thermal leverage" by creating or dissipating clouds, influencing the reflectivity of the Arctic pack ice with soot or carbon black, or even changing the course of ocean currents with macro-engineering projects.

Fletcher again presented his four-stage model of what he called "progress toward climate control": "We must observe how nature behaves before we can understand why, we must understand before we can predict, and we must be able to predict the outcome before we undertake measures for control."[40] He warned, however, that while modern technology was already capable of influencing the global climate system or "heat engine" by altering patterns of thermal forcing, the consequences of such acts could not be adequately predicted. The situation was pretty much the same then as it is now. Geoengineers tend to argue linearly, in a mythical orderly series from science, to engineering, to a public discussion with other "citizens," who can then be educated on the wonders of science and the possibilities of engineering. Prefiguring later optimism, Fletcher thought that an improved observational system, combining ground stations and satellite surveillance, paleoclimatic reconstructions, much faster computers, and better models, would resolve the problems and allow simulations to be performed in enough detail "to evaluate the consequences of specific climate modification acts." He estimated that this capability would be available by 1973, but close to four decades later it is still a desideratum (for some).

Having spent most of his time on technical speculations, Fletcher turned briefly to what he called "international cooperation" for the management of global climatic resources, basing his comments on his assumption that purposeful climate modification deserved the attention of scientific and government leaders. Repeating the opening lines of Wexler's lecture (could Fletcher have been in the audience in 1962?), he invoked John F. Kennedy's statement to the United Nations regarding "further co-operative efforts between all nations in weather prediction and eventually in weather control" (21). Fletcher also cited a joint congressional resolution of April 1, 1968, to the effect that the United

States would be a full participant in the World Weather Watch, which certainly involved observation and prediction, if not understanding and control, and would take steps to support "the theoretical study and evaluation of inadvertent climate modification and the feasibility of intentional climate modification" (22). While the WWW is still functioning and there have been numerous integrated assessments of climate change, recall that even as Fletcher was writing this piece, Project Popeye and Operation Motorpool were under way in the jungles of Southeast Asia, giving a black eye to schemes for the intentional modification of the environment.

Budyko included a section on climate modification in his book *Climatic Changes* (1974). Noting how difficult it had been to control urban air pollution, he predicted that it would be even more difficult to prevent an increase in the carbon dioxide content of the atmosphere and a growth in waste heat release. Agreeing with Fletcher, he concluded that "in the near future climate modification will become necessary in order to maintain current climatic conditions."[41] Budyko was quite skeptical of plans to remove the polar ice, rehydrate Africa, or redirect ocean currents, commenting that it remained "quite unclear how they may influence climate" (239). He was more favorable, however, toward the possibility of triggering instabilities in large-scale atmospheric flows.

Budyko's preferred technique—one discussed by the National Academy of Sciences in 1992 and still under discussion—involved increasing the aerosol content of the lower stratosphere using aircraft or rocket delivery systems. In a back-of-the-envelope calculation, he estimated that a 2 percent reduction in direct solar radiation and a 0.3 percent decrease in total radiation were needed to cool the Earth by several degrees. This could be accomplished by generating an artificial cloud of 600,000 tons of sulfuric acid, the result, under favorable circumstances and assumptions, of burning some 100,000 tons of sulfur per year. Budyko considered this to be a "negligible" quantity compared with other anthropogenic and natural sources. He wrote that "such amounts [of sulfur] are not at all important in environmental pollution" (240), with the important exceptions of the unfavorable effects of such injections on the ozone layer and on agricultural activity, which required further study. Budyko was aware that current simplified theories were inadequate to specify all the possible changes in weather conditions resulting from modifications of the aerosol layer of the stratosphere. Obviously, he believed then, and it holds true today, that deliberate climate modification would be premature before the consequences could be calculated with confidence.

In 1974 William Kellogg and Stephen Schneider published an article in *Science* titled "Climate Stabilization: For Better or Worse?" One of their major concerns was food and water shortages ravaging Africa and whether climatologists could do anything to alleviate this situation while simultaneously cooling the planet. Noting that human activities were increasingly pushing on certain "leverage points" that control the heat balance of the system, they admitted that, as yet, there was no comprehensive theory that could explain—much less predict—temperature trends or rainfall patterns. They drove this point home by listing as a "cause" of climate change the behavior of the climate system as described by Edward Lorenz: "An interactive system as complex as the oceans and atmosphere can have long-period self-fluctuations, even with fixed external inputs."[42] According to Lorenz, chaotically forced internal fluctuations with timescales longer than the thirty- to forty-year interval used to define a climatological average might easily be misinterpreted or confused with climatic changes forced by external variations. This fundamental property of complex systems has vexed those seeking to attribute climate change to any one factor.[43]

Nevertheless, imagining a future in which climate changes could be forecast, Kellogg and Schneider laid out three basic (but not morally equivalent) options: (1) do nothing, (2) alter our patterns of land and sea use in order to lessen the impact of climate change, or (3) anticipate climate change and implement schemes to control it. As they noted, the third option would be extremely contentious and would inevitably generate conflict, for the atmosphere is a highly complex and interactive resource common to all nations. The second option is related to the "middle path."

What if one nation developed the skills to predict climate? This would dramatically change international economic market strategies and might lead to pressure for climate control. What if, after purposeful manipulation, climatic cause-and-effect linkages could be traced? Accusations would abound, and nations might use perceived damages as an excuse for hostility. Given the immense costs of miscalculation (or perception of miscalculation), who then would decide and who would implement climate modification and control schemes? Kellogg and Schneider noted with some irony, but prophetically, "We have the impression that more schemes will be proposed for climate control than for control of the climate controllers."[44] They ended their article by calling for interdisciplinary studies of climate change and its consequences for society. These sorts of studies have subsequently been pursued by the IPCC, but so far with little attention given to geoengineering fantasies.

In 1977 Cesare Marchetti used the term "geoengineering" to refer to the capture and injection of carbon dioxide into the ocean in down-welling currents.[45] He identified the Mediterranean undercurrent at Gibraltar as a likely candidate, with the capacity to sequester all the carbon dioxide emissions of Europe. Today, geoengineers discuss carbon capture and sequestration (CCS) and solar radiation management techniques at their meetings. Also in 1977, the National Academy of Sciences looked at a variety of ideas to reduce global warming, should it ever become dangerous, and concluded that investing in renewable energy was more practical than climate engineering.[46] That same year, Freeman Dyson estimated the scale and cost of an emergency program to plant fast-growing trees to control the carbon dioxide in the atmosphere from fossil fuel burning. He later suggested transporting and dispersing sulfates into the stratosphere using smokestack emissions from burning high-sulfur coal in power plants. Recently, he proposed dumping snow in Antarctica to reduce sea levels.[47] These wild ideas, not taken seriously, were intended as illustrations of how to buy time for society to switch to non-carbon-based energy sources.

In a 1983 report for the National Research Council on "changing climate," Thomas Schelling wrote that "technologies for global cooling, perhaps by injecting the right particles into the stratosphere, perhaps by subtler means, [might] become economical during coming decades."[48] Economics however, was not the most important dimension. Echoing von Neumann's 1955 warning, Schelling wrote that climate control, like nuclear weapons, could become "more a source of international conflict than a relief" (470) if several nations possessed the technology and if they disagreed on the optimum climatic balance. He cited the possibility that one nation might view landfalling hurricanes as disasters, while another might see them as providing necessary water for crops. Concerning interventions that might last for decades or centuries, Schelling predicted that future environmental agendas might well change, as they had in the past and that "CO_2 may not . . . dominate discussion of anthropogenic climate change as it does now" (470). "It is difficult to know what will still look alarming 75 years from now" (482)—that is, after 2050. Also, in 1983 the idea of nuclear winter emerged. A major nuclear war would certainly inject smoke and dust into the stratosphere, yet no one in his right mind would consider such a holocaust an offset to global warming.[49]

Growing concern about anthropogenic global warming led Stanford Solomon Penner, director of the Center for Energy and Combustion Research at the University of California–San Diego, and his associates to suggest in 1984 that the heating from a doubling of CO_2 could be offset if commercial airlines would fly at an altitude of 8 to 20 miles for a ten-year period and tune their engines to emit more particulates to increase the Earth's albedo. A major problem with this suggestion, beyond polluting the stratosphere (which concerned Wexler in 1962), was that commercial aircraft rarely fly at or above 8 miles (although military aircraft do).[50] About this time, studies by cloud physicists indicated that an increase in the amount or brightness of marine stratocumulus clouds in the lower atmosphere might provide significant offsets to global warming. One possible mechanism would be through adding cloud condensation nuclei from emissions of sulfur dioxide; several hundred coal-fired power plants might do the job.

In 1989 James Early, a scientist at Lawrence Livermore National Laboratory, revisited the issue of space mirrors and linked space manufacturing fantasies with environmental issues in his wild speculations on the construction of a solar shield "to offset the greenhouse effect."[51] His back-of-the-envelope calculations indicated that a massive shield some 1,250 miles in diameter would be needed to reduce incoming sunlight by 2 percent. He estimated that an ultra-thin shield, possibly manufactured from lunar materials using nano-fabrication techniques, might cost "from one to ten trillion dollars." Launched from the Moon by an unspecified "mass driver," the shield would reach a "semi-stable" orbit at the L_1 point 1 million miles from the Earth along a direct line toward the Sun, where it would perch "like a barely balanced cart atop a steep hill, a hair's-width away from falling down one side or the other."[52] Here it would be subjected to the solar wind, harsh radiation, cosmic rays, and the buildup of electrostatic forces. It would have to remain functional for "several centuries," which would entail repair missions. It would also require an active positioning system to keep it from falling back to the Earth or into the Sun. In other words, it was not feasible. Early did not indicate what a guidance system might look like for a 5-million-square-mile sheet of material possibly thinner than kitchen plastic wrap, with a mass close to 1 billion kilograms (2.2 billion pounds in Earth gravity). He alluded to the enormous scale and costs of this project and its "major undefined systems," while disingenuously declaring it to be a simpler project, "much smaller in size and scale," than controlling the temperatures on *other* planets of the solar system. By this "logic," even controlling the temperature of the entire solar system would be "simple" compared with galactic-scale engineering!

National Academy, 1992

The publication of the National Academy of Sciences report *Policy Implications of Greenhouse Warming: Mitigation, Adaptation, and the Science Base* (1992) is well within the memory of the current generation of climate engineers. The massive report, whose synthesis panel was chaired by Daniel J. Evans, former governor and U.S. senator from Washington State, examined what was known about greenhouse gases and their climatic effects and then presented geoengineering as one of the cheapest mitigation options, at least in its direct costs. One of the controversial aspects was the report's conclusion that "assumed gradual changes in climate" would produce impacts "that will be no more severe, and adapting to them will be no more difficult, than for the range of climates already on the Earth and no more difficult than for other changes humanity faces."[53] Another problem was the report's narrow focus on cost-effectiveness and the assumed ease of implementing remedial policies. These include fantastic geoengineering schemes conflated with energy-switching and efficiency options under the catch-all category "mitigation."

Here are the National Academy's geoengineering options, notable for their impracticability:

- *Space mirrors.* Place 50,000 mirrors, each 40 square miles in area, in Earth orbit to reflect incoming sunlight.
- *Stratospheric dust.* Use guns, rockets, or balloons to maintain a dust cloud in the stratosphere to increase the reflection of sunlight.
- *Stratospheric bubbles.* Place billions of aluminized, hydrogen-filled balloons in the stratosphere to provide a reflective screen.
- *Low-stratospheric dust, particulates, or soot.* Use aircraft delivery systems or fuel additives to maintain a cloud of dust, particulates, or soot in the lower stratosphere to reflect or intercept sunlight.
- *Cloud stimulation.* Burn sulfur in ships or power plants to form sulfate aerosols in order to stimulate additional low marine clouds to reflect sunlight.
- *Laser removal of atmospheric chlorofluorocarbons.* Use up to 150 extremely powerful lasers, consuming up to 2 percent of the world power supply, to break up CFCs in the lower atmosphere.[54]
- *Ocean biomass stimulation.* Fertilize the oceans with iron to stimulate the growth of CO_2-absorbing phytoplankton.
- *Reforestation.* Plant 3 percent of the entire U.S. surface area (100,000 square miles) with fast-growing trees to sequester 10 percent of U.S. carbon dioxide emissions. (433–464)

Washington insider Robert A. Frosch—a vice president of General Motors Research Labs, former deputy director of the Advanced Research Projects Agency, former assistant secretary of the navy for research and development, and former administrator of NASA—spearheaded the geoengineering aspects of the study. His enthusiastic promotion of climate engineering was seen as a rationale for GM and other corporations to argue against cutting carbon dioxide emissions. At the time, Frosch said,

> I don't know why anybody should feel obligated to reduce carbon dioxide if there are better ways to do it. When you start making deep cuts, you're talking about spending some real money and changing the entire economy. I don't understand why we're so casual about tinkering with the whole way people live on the Earth, but not tinkering a little further with the way we influence the environment.[55]

Yale economist William Nordhaus, also a contributor to the National Academy study, used geoengineering scenarios in his dynamic integrated climate economy (DICE) model to calculate the balance between economic growth (or decline) and climate change. Defining geoengineering as "a hypothetical technology that provides *costless* mitigation of climate change" (emphasis added), he came to the controversial conclusion that "geoengineering produces major benefits, whereas emissions stabilization and climate stabilization are projected to be worse than inaction."[56] At one point, he referred to the scale of his global economic projections as "mind-numbing," but he could well have applied this description to his overall conclusions regarding the potential for a geoengineering solution. Stephen Schneider later wrote: "As a member of that panel, I can report that the very idea of including a chapter on geoengineering led to serious internal and external debates. Many participants (including myself) were worried that even the thought that we could offset some aspects of inadvertent climate modification by deliberate modification schemes could be used as an excuse to continue polluting."[57] In fact, it was precisely in this way—as an alternative to reducing emissions—that geoengineering discussions found their way into the twenty-first century.

Such sentiments echoed the dismal opinions of economists at the time on pollution solutions. In 1991, for example, World Bank economist Lawrence Summers (who later resigned as president of Harvard University following a no-confidence vote of the faculty and now directs the White House's National Economic Council) wrote, in what he assumed would remain a private, and what he later deemed a sarcastic, memo: "Shouldn't the World Bank be encouraging MORE migration of the dirty industries to the LDCs [less developed countries]. . . . I think the economic logic behind dumping a load of toxic waste in the lowest wage country is impeccable and we should face up to the fact that . . . under populated countries in Africa are vastly UNDER-polluted."[58] The outrage generated when this memo became public in 1992, just before the first Earth Summit in Rio de Janeiro, motivated José Lutzenberger, Brazil's secretary of the environment, to respond to Summers:

> Your reasoning is perfectly logical but totally insane. . . . Your thoughts [provide] a concrete example of the unbelievable alienation, reductionist thinking, social ruthlessness and the arrogant ignorance of many conventional "economists" concerning the nature of the world we live in. . . . If the World Bank keeps you as vice president it will lose all credibility. To me it would confirm what I often said . . . the best thing that could happen would be for the Bank to disappear.[59]

"Insane," "reductionist," "ruthless," "arrogant"—such modifiers suit most geoengineering proposals quite well. Nordhaus wrote in 2007 that geoengineering is, at present, "the only economically competitive technology to offset global warming."[60]

A Naval Rifle System

Frosch called his proposal to bombard the stratosphere using an array of 350 naval guns "designer volcanic dust put up with Jules Verne methods" (figure 8.2).[61] He envisioned each $1 million, 16-inch gun being able to fire 1 ton of sulfate or aluminum oxide into the stratosphere about every ten minutes. Each barrel would need replacement after 1,000 to 1,500 shots. Thus a single cannon would have a useful life of less than two weeks, and a total of 300,000 cannon would be needed for a forty-year program! The naval guns had been designed in 1939 and were first put into service in 1943, so they would have to be updated. The cost of ammunition for 400 million shots was estimated at $4 trillion, the barrels would be $300 billion, the firing stations $200 billion, and the personnel costs $100 billion—for a total of $5 trillion over forty years. This system could deliver dust to the stratosphere for about $14 a pound, and each pound was expected to mitigate 45 tons of carbon emissions.[62] Balloon delivery systems were estimated to cost $36 a pound and sounding rockets, $45.

Frosch was aware that damaging side effects could result, such as stratospheric ozone destruction, widespread drought, or unacceptable atmospheric haze, but he did not emphasize that. Instead, he reassured his readers that "the rifle system appears to be inexpensive, to be relatively easily managed, and to require few launch sites" (460). He concluded that "the rifles could be deployed at sea or in military reservations where the noise of the shots and the fallback of expended shells could be managed" (817–819). What Frosch forgot to take into account was the lower tropospheric air pollution generated by the bombardments. If, for example, each 650-pound explosive charge contained pure

8.2 Shooting dust into the stratosphere to offset global warming, one proposal by the National Academy of Sciences in its report published in 1992. Nobel laureate Paul J. Crutzen revived the idea in 2006. (CARTOON BY JOHN IRELAND, IN *GEOGRAPHICAL MAGAZINE*, MAY 1992)

nitroglycerine ($C_3H_5N_3O_9$), it would generate about 380 pounds of carbon dioxide when fired, so 400 million cannon shots would produce about 76 million tons of carbon dioxide. This calculation does not take into account other gaseous by-products, such as smoke or nitrous oxide, nor does it consider the carbon emissions involved in manufacturing or transporting 300,000 cannon barrels, each of which is over 65 feet long and weighs over 130 tons.[63] Could such a long-term and violent bombardment be sustained without any accidents or other side effects? Is declaring war on the stratosphere the best mitigation strategy? The authors of the 2009 Novim Group report on geoengineering seem to think so and discuss, apparently without a sense of irony, the possibility of opening fire on the ozone layer with M1 tank guns loaded with aerosols.[64]

Ocean Iron Fertilization

The other scheme hatched at the time was ocean iron fertilization (OIF). "Give me half a tanker of iron, and I'll give you an ice age," biogeochemist John Martin (a Colby College graduate) reportedly quipped in a Dr. Strangelove accent at a conference at Woods Hole in 1988.[65] Martin and his colleagues at Moss Landing Marine Laboratories proposed that iron was a limiting nutrient in certain ocean waters and that adding it stimulated explosive and widespread phytoplankton growth. They tested their iron deficiency, or "Geritol," hypothesis in bottles of ocean water, and subsequently experimenters added iron to the oceans in a dozen or so ship-borne "patch" experiments extending over hundreds of square miles. OIF worked, just like pouring Miracle-Gro on your tomatoes. Was it possible that the blooming and die-off of phytoplankton, fertilized by the iron in natural dust, was the key factor in regulating atmospheric carbon dioxide concentrations during glacial–interglacial cycles? Dust bands in ancient ice cores encouraged this idea, as did the detection of natural plankton blooms by satellites.[66]

Enter the geoengineers. Could OIF speed up the biological carbon pump to sequester carbon dioxide, and was it a solution to global warming? Because of this possibility, Martin's hypothesis received widespread public attention. What if entrepreneurs or governments could turn patches of ocean soupy green and claim that the carbonaceous carcasses of the dead plankton sinking below the waves constituted biological "sequestration" of undesired atmospheric carbon? Or could plankton blooms increase the production of dimethyl sulfate (DMS) and cool the Earth by making marine clouds slightly more reflective? Several companies—Climos, Planktos (now out of the business), the aptly named GreenSea Ventures, and the Ocean Nourishment Corporation—have proposed entering the carbon-trading market by dumping either iron or urea into the oceans to stimulate both plankton blooms and ocean fishing. The scientific consensus, however, supported by diplomatic negotiations, held that more research was needed to evaluate risks and benefits before anyone should even think of selling carbon offsets from ocean iron fertilization. Some of the key questions that are as yet unanswered include the amount and fate of carbon from a bloom, how long it would remain sequestered, and, most important, how all this could be verified. If the commercial companies are going to try to sell an artificial and beneficial "rain" of ocean phytoplankton, then all the caveats and all the verification and attribution challenges of artificial rainmaking apply. It is similar to the relationship between cloud physics and commercial cloud seeding; as Kenneth Coale, director of Moss Landing, pointed out, "iron experiments are about how nature works; commercial ocean seeding is about getting nature to work for us."[67]

Totally unresolved issues related to all large-scale OIF projects include possible damage to the ocean food web and the world's fishing industry caused by disturbing marine ecosystems, production of biological "dead zones," pollution of the deep ocean by the buildup of iron compounds, possible destruction of stratospheric ozone, and generation of undesirable greenhouse gases such as methane and nitrous oxide. OIF projects could be undertaken unilaterally by a rogue state or a group out to make a point; and if the fertilization ever stopped, the carbon dioxide would immediately begin to return to the atmosphere.[68] Regarding OIF, chemist Whitney King wrote in 1994,

> No engineer would consider designing a building which has a less than a 1 percent chance of standing up and the potential of wiping out a whole city if it falls. Yet this is probably a good analogy for our state of understanding the carbon cycle and the role iron plays in controlling it. So why has the iron fertilization theory gained so much attention? I would suggest that we enjoy the thought of being able to control global climate.[69]

Artificial Trees or Lackner Towers

Klaus Lackner of the Earth Institute at Columbia University, collaborating with Tucson, Arizona–based Global Research Technologies, envisions a world filled with millions of inverse chimneys, some of them more than 300 feet high and 30 feet in diameter, inhaling up to 30 billion tons of carbon dioxide from the atmosphere every year (the world's annual emissions) and sequestering it in underground or undersea storage areas. Picture in your mind's eye Al Gore's *An Inconvenient Truth* movie logo of a smokestack apparently exhaling a Katrina-like hurricane; now run the smokestack in reverse and imagine millions of such giant planetary vacuum cleaners or, more accurately, air filters. Lackner prefers a different, somewhat greener metaphor: a forest of artificial trees covered in CO_2-absorbing artificial leaves. Note, however, that trees sequester carbon, not carbon dioxide, and they return oxygen to the atmosphere. Unlike Lackner forests, trees also provide shade, habitat, and food for squirrels, birds, and other living things. Although Lackner says he is not a geoengineer, but merely interested in compensating for current emissions, he envisions his devices being enlisted in the "fight against climate change."[70] Others hope someday to attain negative global carbon dioxide emissions, but this would entail immense storage problems, similar to nuclear waste disposal.[71]

Lackner has built a demonstration unit in which a filter filled with caustic and energy-intensive sodium hydroxide can absorb the carbon dioxide output of a single car. He admits, however, that this system is not safe or practical, so he is currently looking into proprietary "ion-exchange resins" with undisclosed energetic and environmental properties. Of course, the capture, cooling, liquefaction, and pumping of 30 billion tons of atmospheric carbon dioxide would require an astronomical amount of energy and infrastructure, and it is not at all certain that the Earth has the capacity for safe long-term storage of such a large amount of carbon.

Let us look briefly at the chemical energetics that make air capture of carbon and its sequestration (CCS) such a non-starter (with the possible exception of selected local sites, such as North Sea oil rigs, where pumping carbon dioxide back into the wells makes more sense than venting it). First there is the mass problem. Combustion of coal (mainly carbon-12) results in carbon dioxide waste products of molecular weight 44. Thus a mole of carbon burned yields energy, plus a mole of carbon dioxide waste, a mass gain of 267 percent. Lackner did not acknowledge this mass problem when he wrote, "Ultimately *carbon* extraction must be matched by *carbon dioxide* capture and storage. For every *ton* of carbon pulled from the ground, another *ton* of carbon must be taken out of the mobile carbon pool" (emphasis added).[72] In actuality, for every ton of carbon used, 3.67 tons of carbon dioxide have to be captured and stored. There are other problems as well. Liquefying such a huge amount of CO_2 requires immense pressures, on the order of 80 to 100 atmospheres. We are still not done, since the liquid CO_2 must be piped to injection sites. Imagine all the pipelines needed for the pipe dream of erecting 3 million large Lackner towers worldwide, or up to 100 million smaller, container-size units.

The image of a Lackner tower, resembling a huge flyswatter, was superimposed over New York's Central Park in a photomontage created by a Canadian film company.[73] Left unmentioned was the fact that Manhattan alone would require hundreds of these scrubbers for its resident population of 1.5 million people (not counting visitors), so about one in every ten large structures in the city would resemble a giant flyswatter. The cost of the land to build them, energy to run them, piping to drain them, and makeup and maintenance of their "idealized" filters were not specified, making Lackner's proposal at present untenable. Oh, by the way—each of the large Lackner towers would cost at least $20 million.

Lackner's dream of carbon sequestration found support from his Columbia University colleague Wallace Broecker and science writer Robert Kunzig in their book, *Fixing Climate* (2008). Their overall thesis is that carbon dioxide capture and storage represents the equivalent of sewage treatment, which modern societies deem a necessity. They quoted Harrison Brown, the Caltech geochemist, eugenicist, futurist, and role model for the current presidential science adviser, John Holdren. In 1954 Brown imagined feeding a hungry world by increasing the carbon dioxide concentration of the atmosphere to stimulate plant growth:

> We have seen that plants grow more rapidly in an atmosphere that is rich in carbon dioxide.... If, in some manner, the carbon-dioxide content of the atmosphere could be increased threefold, world food production might be doubled. One can visualize, on a world scale, huge carbon-dioxide generators pouring the gas into the atmosphere.... In order to double the amount in the atmosphere, at least 500 billion tons of coal would have to be burned—an amount six times greater than that which has been consumed during all of human history. In the absence of coal ... the carbon dioxide could be produced by heating limestone.[74]

Recall that Nils Ekholm and Svante Arrhenius had suggested in the first decade of the twentieth century that, facing the return of an ice age, atmospheric carbon dioxide might be increased artificially by opening up and burning shallow coal seams—a process that would also fertilize plants. Broecker and Kunzig end their book with just such a fantasy:

> Our children and grandchildren, having stabilized the CO_2 level at 500 or 600 ppm [parts per million], may decide, consulting their history books, that it was more agreeable at 280 ppm. No doubt our more distant descendents will choose if they can to avert the next ice age; perhaps, seeing an abrupt climate change on the horizon, they will prevent it by adjusting the carbon dioxide level in the greenhouse. By then they will no longer be burning fossil fuels, so they would have to deploy some kind of carbon dioxide generator, shades of Harrison Brown, to operate in tandem with the carbon dioxide scrubbers. (232)

Lackner reportedly agreed, adding that capturing, storing, and releasing carbon dioxide may one day be possible. Can you imagine a world in which Lackner's carbon dioxide scrubbers and the Ekholm–Arrhenius/Brown carbon dioxide generators would operate in tandem as a kind of planetary thermostat? "Trying to see that far into the future is crazy, of course" (232) (Broecker and Kunzig's words, not mine).

How can we comprehend such proposals? It may be common for people living in close proximity to the megalopolis—for example, just off Broadway—to see the sky as an open sewer and try to "fix it." Bus exhaust, steam vents, fumes, and foul odors serve as constant reminders that something is indeed wrong with the sky. Fly into a major city near or after sunset, and you will see, on approach, streams of red taillights and white headlights of opposing traffic. Stand on the street, perhaps daring to cross, and you will see the grim visages of oncoming drivers or the tailpipes and exhaust plumes of the passing traffic. In such a dense infrastructure, when almost every building is air-conditioned, it is not hard to imagine a future in which every tenth building might in fact be a giant outdoor air filter or an inverse chimney. To put it simply, when you see pipes sticking out everywhere, it is not hard to imagine more pipes—good pipes correcting emissions from the bad pipes.

Today's city dwellers, especially the influential ones, do not choose to spend much time on crowded, dangerous, and uncomfortable streets. Instead, they shuttle between microtopian environments, from air-conditioned vehicles to air-conditioned buildings, and even to air-conditioned shopping malls and sports arenas. Near Washington, D.C., certainly a city known for both its international influence and its need for air-conditioning, the power brokers always wear business suits to signal their status and their unlimited access to HVAC. The tunnels running under Capitol Hill and into the Library of Congress symbolize this, as does Pentagon City, across the Potomac River in nearby Virginia, which is known for its underground warrens where not only can the air be conditioned, but Muzak can be pumped in and the homeless can be kept out. This is the type of "environment" in which most of our decision makers operate.

Recycling Ideas

In the twenty-first century, geoengineers have convened several meetings, regularly exchange views on Google Groups, and continue to hatch, nurture, and recycle their ideas. In September 2001, the U.S. Climate Change Technology Program quietly held an invitational conference on "response options to rapid or severe climate change." Sponsored by a White House that was officially skeptical about greenhouse warming, the meeting gave new status to the control fantasies of the climate engineers. According to one participant, however, "If they had broadcast that meeting live to people in Europe, there would have been riots"—a comment indicative of a much more robust green movement in Europe, which at the time still hoped the United States might sign the Kyoto Protocol.

Two years later, the Pentagon released a controversial report titled "An Abrupt Climate Change Scenario and Its Implications for United States National Security." The report explained how global warming might lead to rapid and catastrophic global cooling through mechanisms such as the slowing of North Atlantic deep-water circulation—and recommended that the government "explore geoengineering options that control the climate."[75] Such actions would have to be studied carefully, of course, given their potential to exacerbate conflict among nations. A symposium sponsored in 2004 by the Tyndall Centre for Climate Change Research in Cambridge, England, set out to "identify, debate, and evaluate" possible but highly controversial options for the design and construction of engineering projects for the management and mitigation of global climate change.

"Russian Scientist Suggests Burning Sulfur in Stratosphere to Fight Global Warming," read the headlines from Moscow in November 2005. The article referred to a letter from the prominent scientist Yuri Izrael, the head of the Global Climate and Ecology Institute, to President Vladimir Putin warning that global warming required immediate action and suggesting burning thousands of tons of sulfur in the stratosphere as a remedy. Izrael said his plan was based on the idea of putting aerosols into the atmosphere at an altitude of 8 to 12 miles to create a reflective layer that would lower the heating effect of solar radiation: "In order to lower the temperature of the Earth by 1–2 degrees we need to pump about 600,000 tons of aerosol particles. To do that, we need to burn from 100–200,000 tons of sulfur. And we do not have to burn the sulfur there, we can simply use sulfur-rich aircraft fuel." It seems that Izrael did not make his own calculations, but simply dusted off Penner's 1984 idea and used the exact figures of his countryman Budyko, published in 1974.[76]

A 2006 editorial on geoengineering by Nobel laureate Paul Crutzen contained a proposal that was very similar to Budyko's, although framed by different environmental and policy concerns.[77] He came to similar conclusions, too, except his calculations indicated that ten times more sulfur would be needed, between 1 and 2 *million* tons per year. Budyko had pegged the geoengineering sulfur input at about one-ten-thousandth of "that due to man's activity," while Crutzen had it as "*only* 2–4 percent of the current input" (emphasis added).[78] Crutzen wrote that albedo enhancement was not the best solution to global warming "by far," but he still recommended that the Budyko/National Academy notions of using artillery guns, balloons, or aircraft to inject sulfates or other particles into the stratosphere "might again be explored and debated."[79] He suggested that the attack on the upper atmosphere might be conducted from "remote tropical island sites or from ships" (213), but nowhere indicated that he had considered the need to consult residents of the tropics for their opinions on this. Whether from Budyko, Penner, Izrael, or Crutzen, the idea of purposeful stratospheric pollution, for whatever purpose, is extremely grating to modern sensibilities. Nevertheless, there have been several more workshops in recent years. NASA-Ames and the Carnegie Institution convened one in 2006 on the Phaethon-like topic "managing solar radiation." Participants could not help but laugh when a meeting coordinator apologized for not being able to control the temperature of the room. Ad hoc meetings on climate engineering pop up on a regular basis. In 2007 one was held at the American Academy of Arts and Sciences, another at MIT in 2009, and a large gathering on "climate intervention" in 2010 at the Asilomar campus in Pacific Grove, California. The topics under discussion at these meetings have been far from mainstream, involving more speculation than science.[80]

Geoengineering does not have a widespread following. In 2006 Ralph Cicerone wrote: "Ideas on how to engineer the Earth's climate, or to modify the environment on large scales . . . do not enjoy broad support from scientists. Refereed publications that deal with such ideas are not numerous nor are they cited widely."[81] The situation has not changed substantially since then. According to a 2008 report by the Tyndall Centre, geoengineering proposals have not advanced beyond the outline/concept stage and are best confined to computer model simulations, since small-scale field experiments would be inconclusive and global experiments would be far too risky and socially unacceptable. Recently, atmospheric scientist Richard Turco, founding director of the UCLA Institute of the Environment and one of the authors of "Nuclear Winter" (1983), called many geoengineering plans "preposterous" and "mind-boggling." He saw "no evidence" that technological quick fixes to the climate system would be as cheap or as easy as their proponents claim, and he said that many of them "wouldn't work at all" and could not be field-tested without unacceptable, even unpredictable, risks. Embarking on such projects, he said, "could be foolhardy."[82]

Late in 2009, Izrael and his colleagues reported on what they called a geoengineering field experiment in Russia to study solar radiation passing through aerosol layers. Citing Crutzen's 2006 editorial, they made the dubious and self-referential claim that "injection of reflecting aerosol submicron particles into the stratosphere can be an *optimal* option to compensate warming" (emphasis added).[83] The experimenters then proceeded to experiment, not in the stratosphere but near the ground. In several tests, a military helicopter burning "metal-chloride pyrotechnic" flares and a military truck spraying an "overheated vapor-gas mixture of individual fractions of petroleum products" generated thick toxic clouds of smoke (266). The petroleum device was not unlike Irving Langmuir and Vincent Schaefer's smoke screen generator of World War II. Ground-based solar radiation measurements then showed what everyone already knows—a thick cloud obscures the Sun: "Possible changes in the irradiance are estimated in this case rather approximately. The irradiance reduction in this case was about 28%" (269). Note the experimenters' use of the terms "possible" and "approximately"; a reduction of sunlight of 28 percent, sustained globally, would devastate life on Earth. Another experimental trial also yielded inconclusive results. Cloudy weather with sky clearing made it difficult for the researchers to detect a "possible change in the solar radiation caused by the artificial aerosol sample passing over the instrument complex against the background of natural changes" (269). Nevertheless, for this team, inconclusive small-scale experiments near the ground were seemingly a sufficient proof of concept: "Based on the experimental results obtained in our work, it is shown how it is principally possible to *control* solar radiation passing through artificially created aerosol formations in the atmosphere with different optical thickness" (272). We can only hope these Russian experimenters are not in charge of managing solar radiation for the globe.

An editorial cartoonist for the *New York Times* captured the essence (and the absurdity) of one of the proposed techniques (figure 8.3). Two overheated polar bears are feverishly trying to pump sulfur into the air, but they seem to be having trouble keeping their hose erect, especially if their ice floe shrinks any further. And whose warships are those in the distance? Do they carry Frosch/Crutzen sulfate cannons, or are they trying to stop the geoengineering? Russian opinion has long favored an open Arctic Ocean, and some scientists, including Budyko, believe that the beneficial effects of global warming might "pep up" cold regions and allow more grain and potatoes to be grown, making the country wealthier.[84] Better check with Vladimir Putin before we screw (with) the Arctic.

Naval artillery is only one of the many "manly" ways to declare "war" on global warming by using military equipment. The cartoon alludes to a proposal by Edward Teller's protégé Lowell Wood to attach a long hose to a nonexistent but futuristic military High Altitude Airship (a Lockheed-Martin–Defense Department stratospheric super blimp now on the drawing board with some twenty-five times the volume of the Goodyear blimp) to "pump" reflective particles into the stratosphere. According to Wood, "Pipe it up; spray it out!"[85] Wood has worked out many of the details—except for high winds, icing, and accidents, since the HAAs are likely to wander as much as 100 miles from

8.3 "Screwing (with) the Planet," as interpreted by Henning Wagenbreth. (© *NEW YORK TIMES*, OCTOBER 24, 2007)

their assigned stations. If the geoengineers cannot keep it up, however, imagine a 25-mile phallic "snake" filled with 10 tons of sulfuric acid ripping loose, writhing wildly, and falling out of the sky. Carol Cohn said it best in her classic article "Sex and Death in the Rational World of Defense Intellectuals": "The dominant voice of militarized masculinity and decontextualized rationality speaks so loudly in our culture, it will remain difficult for any other voices to be heard . . . until that voice is delegitimated."[86] The geoengineers have been playing such games with the planet since computerized general circulation models were first developed. While this kind of research will undoubtedly continue, it should remain indoors between consenting adults. What must be aired are the underlying assumptions.

A Royal Society Smoke Screen

The Royal Society of London recently dedicated a special issue of its venerable *Philosophical Transactions* to the topic "Geoscale Engineering to Avert Dangerous Climate Change."[87] The journal bills itself as "essential reading for mathematicians, physicists, engineers and other physical scientists," which is noteworthy, since climate engineering is not solely or even essentially a technical problem and none of the eleven papers in the special issue were written by historians, ethicists, or other humanists or social scientists. Editors Brian Launder, an engineer, and Michael Thompson, an applied mathematician/solar physicist, began by blaming China and India for their soaring greenhouse gas emissions, praising the developed world (at least the European Union) for struggling to meet its carbon-reduction targets, and wondering if the day may come when geoengineering solutions are "*universally* perceived to be less risky than doing nothing" (emphasis added). Only a few of the articles did what the editors promised: subject macro-engineering options to "critical appraisal by acknowledged experts in the field." Most of the articles had been recycled from the 2004 Tyndall Centre meeting on climate engineering and were written by advocates standing to benefit directly from any increase in funding.

Survey articles by Stephen Schneider and James Lovelock questioned, in broad brushstrokes, the validity and overall viability of the geoengineering enterprise. Schneider briefly reviewed the fifty-year history of schemes to modify large-scale environmental systems or control climate. He pointed out that schemes are typically presented as cost-effective alternatives or as ways to buy time for mitigation, but he expressed doubts that they would work as planned or that they would be socially feasible, given the potential for transboundary conflicts if negative climatic events occur during geoengineering activities.[88]

Lovelock, invoking a metaphor he has long used, posed as a "geophysiologist," or planetary physician, and diagnosed the Earth as having a fever induced by the parasite *Disseminated primatemia* (the superabundance of humans). As treatment, he recommended a low-carbon diet combined with nuclear medicine. He likened geoengineering to crude planetary surgery, as practiced by the butcher/barber surgeons of old. While the patient would definitely survive, the parasites had a much lower probability: "Our ignorance of the Earth system is overwhelming. . . . Planetary scale engineering might be able to combat global warming, but as with nineteenth century medicine, the best option may simply be kind words and letting Nature take its course."[89] Lovelock is a freethinker who advocates nuclear power, imagines dystopian futures caused by climate change, and has had Michael Mann's "hockey stick" graph pinned on the wall above his desk for a number of years. He and Chris Rapley have recently proposed their own geoengineering fix for the "pathology of global warming," specifically, a vast array of vertical pipes placed in the oceans to bring colder, nutrient-rich water to the surface to spur the growth of carbon dioxide–absorbing plankton. But many worry that the idea might interfere with fishing, disrupt whale populations, and release more carbon dioxide into the atmosphere than it captures.[90] Most recently Lovelock has supported "biochar," the conversion of massive amounts of agricultural "waste" into non-biodegradable charcoal and its subsequent burial.[91] This surely qualifies for Nathaniel Hawthorne's Hall of Fantasy, since it would mark the end of composting and would generate massive amounts of the known carcinogen benzo[a]pyrene. Its practitioners risk the fate of Hawthorne's Dr. Cacaphodel, "who had wilted and dried himself into a mummy by continually stooping over charcoal furnaces and inhaling unwholesome fumes during his researches."[92]

In the *Philosophical Transactions* special issue on geoengineering, two teams of oceanographers examined ocean iron fertilization field experiments and model studies to gauge whether this technique can "become a viable option to sequester CO_2." Victor Smetacek and S. W. A. Naqvi impugned the current "apparent consensus against OIF [as] premature." They praised vague but possibly positive side effects of the widespread use and commercialization of this technique (more krill may mean more whales), while they minimized discussion of any negative side effects, such as disruption of the ocean food chain or the creation of anoxic dead zones. Without providing any details, they offered the hollow reassurance that "negative effects of possible commercialization of OIF could be controlled by the establishment of an international body headed by *scientists* to supervise and monitor its implementation" (emphasis added).[93] Scientists typically have little or no training in history, ethics, or public policy, while global climate change is a human problem, not merely a scientific issue.

The article by John Latham and colleagues rehearsed the idea of seeding marine stratus clouds with seawater to increase their albedo and possibly make them more persistent. They concluded, to no one's surprise, that it might—just might—work. A companion piece by Steven Salter and colleagues pointed out that an armada of robotic spray ships plying the high seas would be needed and that their spray would make the clouds brighter by introducing so many cloud condensation nuclei that the cloud droplets would be much smaller and more numerous. This "overseeding" technique was attempted using silver iodide in the 1950s as a means to prevent rain. Thus the worldwide array of brighter clouds proposed by Latham and Salter might produce less rain than unaltered clouds, with unknown environmental consequences. It looks like the international body of scientists mentioned by the oceanographers will be busy monitoring this technique too.[94]

Ken Caldeira and Lowell Wood offered perhaps the most disingenuous paper by using an "idealized" (read: relatively simple) climate model in which they turned down the sunlight at the top of the atmosphere by using various aerosols. They did not specify where this magic knob might actually be located, but every undergraduate student in atmospheric science knows that the "knob" is built into the models as an indication of the climate's sensitivity to solar insolation. Wonder of wonders, when the sunlight is turned down, the planet cools; and when the sunlight is turned down over the Arctic Circle, the Arctic cools and parameterized sea ice grows. By focusing on physics rather than on the complexities of atmospheric science or ecology, and by tuning their model assumptions, they concluded that their "engineered high CO_2 climate" could be made to emulate a perhaps more desirable but presently unattainable low CO_2 climate. Caldeira and Wood used back-of-the-envelope calculations to push forward their case for military hardware with unspecified failure rates delivering unspecified aerosols into the stratosphere with unknown environmental consequences. They ignored the recent, more sophisticated modeling work of Alan Robock, Luke Oman, and Georgiy Stenchikov indicating that stratospheric aerosols injected at high latitudes would soon be carried by the winds as far south as 30°N, interfering with the Asian summer monsoon. Since stratospheric aerosols would not stay confined above the Arctic Circle, the "yarmulke plan" of Caldeira and Wood is physically impossible. Their non-sequitur conclusion: "Implementing insolation modulation appears to be feasible." Their most honest admission: "Modeling of climate engineering is in its infancy."[95]

The article in the volume with the greatest integrity, by the most sophisticated team of modelers, and the one that offered a fresh and rather sobering assessment of the consequences of injecting sulfate aerosols into the stratosphere was by Philip Rasch and his colleagues. Their simulations indicated that while the Northern Hemisphere might cool overall after such an intervention, significant and undesirable reductions in precipitation could occur over vulnerable areas such as North Africa and India, possibly leading to drought conditions and damage to agricultural productivity. Such climate engineering would also cause significant changes in the overall spectrum of solar radiation, with more biologically damaging ultraviolet-B radiation reaching the Earth's surface, with negative consequences likely for human health and biological populations. The worldwide sulfate haze would also reduce direct-beam solar radiation and increase diffuse sky radiation with unwelcome aesthetic effects, interfere with optical astronomy, dramatically reduce the capacity for generation of solar power, and probably cause unwanted stresses on plant ecosystems and crops. Rasch and his colleagues also warned of increased ozone depletion attributable to the presence of additional sulfate particles in the stratosphere. A related article in *Science* by Simone Tilmes, Rolf Müller, and Ross Salawitch supported this conclusion: "An injection of sulfur large enough to compensate for surface warming caused by the doubling of atmospheric CO_2 would strongly increase the extent of Arctic ozone depletion during the present century for cold winters and would cause a considerable delay, between 30 and 70 years, in the expected recovery of the Antarctic ozone hole."[96] So much for Crutzen's proposal.

In 2009 oceanographer John Shepherd and I were on a panel presenting testimony to the U.S. Congress on the governance of geoengineering. He introduced a recent study that he chaired for the Royal Society of London with the comment "geoengineering is no magic bullet." I immediately thought, "It is no bullet at all" and we would be better off not shooting our ordnance at the atmosphere.[97] The published report recommends, sensibly, that nations make increased efforts toward mitigating and adapting to climate change, but it also supports further research and development of geoengineering, including appropriate observations, development and use of climate models, and (more ominously) "carefully planned and executed experiments," including small- to medium-scale experiments both in the laboratory and in field trials.[98]

Field Tests?

In his 2008 testimony to the British House of Commons, Launder spotlighted his recent editorship of the *Philosophical Transactions* special issue on geoengineering and urged the government to go beyond paper studies and "earmark" a portion of its budget for a program of field tests leading to possible geo-scale deployment. The response of mainstream engineers, however, was lukewarm. In the opinion of Britain's Royal Academy of Engineering, "All the current proposals have inherent environmental, technical and social risks and none will solve all the problems associated with energy and climate change." The academy recommended that the government "stay well informed" but treat geoengineering with caution.[99] Geographer Dan Lunt, from the University of Bristol, and others pointed out that the missing dimension in all of this was a large-scale program to determine the efficacy, side effects, practicality, economics, and ethical implications of geoengineering, a kind of ethical, legal, social implications (ELSI) approach common in other controversial fields. If American geoengineers are seeking funding, a single agency with deep pockets—for example, the Department of Energy, NASA, or even the Department of Defense or Homeland Security—is not the way to go. Neither is a private company in which commercial goals may overwhelm scientific objectivity. This field needs enhanced public input and open peer review, such as that provided by the National Science Foundation.

Recently, atmospheric scientist William Cotton pointed out the relationship between weather engineering and climate engineering, along with their systematic problems and structural differences. In weather modification experiments, the scientific community requires "proof" that cloud seeding has increased precipitation. Following an intervention, such proof would include "strong physical evidence of appropriate modifications to cloud structures and highly significant statistical evidence"—that is, effects that exceed the natural background variability of the atmosphere. But intervention is not control. In 1946 Kathleen Blodgett at General Electric told Irving Langmuir that intervening in or modifying a cloud was a far cry from controlling its subsequent motion and growth or the characteristics of its precipitation. Having experienced the promise and hype of cloud seeding, and after having worked for fifty years in this field, Cotton admitted, "We cannot point to strong physical and statistical evidence that these early claims have been realized."[100] He went on to note that proof of success in climate engineering would be far harder to establish than in weather engineering. In fact, it would be impossible, for several reasons: climate models are not designed to be predictive, so there is no forecast skill; global climate experiments cannot be randomized or repeated and cannot be done without likely collateral damage; climate variability is very high, so the background-noise-to-signal ratio is overwhelming; and climate change is slow to develop because of built-in thermal lags due to oceans and ice sheets. What all this adds up to is that experimental "results" could not be established even within the experimenters' life spans. Did I mention the chaotic behavior of the climate system? That alone would overwhelm any attribution of experimental interventions by climate engineers. Cotton warned that in times of drought or climate stress, politicians would emerge with the need to demonstrate that they were doing something, that they were in control of the situation, even if they only enacted what he called political placebos.

The Middle Course

In 1983 Thomas Schelling outlined four basic policy choices for responding to carbon dioxide–induced climate change:

1. Reduce its production.
2. Adapt to increasing carbon dioxide and changing climate.
3. Remove it from the atmosphere.
4. Modify climate, weather, and hydrology.

The first two options, practiced worldwide, with foresight and moderation, constitute the "middle course." "Mitigation" properly refers to a complex array of initiatives involving primarily decarbonizing and increasing the efficiency of the energy supply, afforestation and the prevention of further deforestation, and other efforts aimed at reducing anthropogenic emissions and concentrations of radiatively active trace gases. "Adaptation," or climate resilience, involves collective means taken to avoid, cope with, or reduce the adverse impacts of climate change, both on humans and on all living creatures and ecosystems. The first climate migrants in prehistorical times were adapting to the onset of an ice age. Ward's categories of prevention and protection, from 1930, are close matches. Some mitigation efforts, however, involving proposed carbon capture and sequestration can indeed be massive in scale, such as ocean iron fertilization and a worldwide array of Lackner towers, and deserve the same caveats as direct climate intervention schemes.

In a 2008 book, Gabrielle Walker and Sir David King surveyed the problems of global warming and some of the technological and political "solutions," or at least responses that might arise. They discussed the oft-cited "stabilization wedges" of Princeton professors Stephen Pacala and Robert Socolow, which offered the hopeful vision of stabilizing atmospheric carbon dioxide levels (but not necessarily the climate system) using existing technologies.[101] Pacala and Socolow rightly emphasized efficiency first—in electricity generation, passenger vehicle transport, shipping, and other end-use sectors—followed by new renewable energy sources and as-yet-unproven carbon capture and storage. Walker and King wrote that stabilizing atmospheric carbon dioxide levels at 450 parts per million would require implementing the following "wedges" immediately:

> Double the fuel economy of two billion cars, halve the annual average distance traveled by two billion cars, cut carbon emissions from buildings and appliances by one-quarter, capture and store carbon dioxide from 800 gigawatts of coal power plants and 1600 gigawatts of natural gas power plants, build two million 1-megawatt wind turbines (about 50 times more than exist today), stop all felling of tropical forests and plant 740 million acres of new trees in the tropics, double the current amount of nuclear power, quadruple the amount of natural gas used to generate electricity . . . , increase the use of biofuels in vehicles to fifty times today's

level, use low-tillage farming methods on all the world's cropland, and increase the global area of solar panels by a factor of seven hundred. (92–93)

Note that all these wedges involve large geometric factors: 2-, 50-, even 700-fold increases (or decreases) in current practices, with unspecified costs or other considerations. The global back-of-the-envelope nature of these suggestions, the sheer scale of the challenge, and the lack of fine-grained analysis regarding local and regional implementation strategies had led some, geoengineers included, to wonder if it can be done efficiently—or at all. At the 2007 American Academy meeting on geoengineering, Socolow was musing aloud about adding a geoengineering wedge to his portfolio; I suggested that this was probably a premature move.

The middle course is Phaethon's ideal path, as advised by his father, Helios, to spare the whip, hold tight the reins, and "keep within the limit of the middle zone," neither too far south or north, nor too high or low: "the middle course is safest and best."[102] Inviting people from different cultures and diverse walks of life to take action on climate change involves defining a middle course, not a path of least resistance but one between doing too little about climate change (which has been the case recently with much of U.S. policy) and doing too much (which would most certainly be the case if the climate engineers have free rein to turn the planet into a machine). For that matter, it is also possible for well-meaning social engineers to attempt to do too much, by promoting overly aggressive or one-size-fits-all approaches to energy-climate-environment issues. We have yet to demonstrate that economic prosperity can exist or that development can proceed without the use of fossil fuels, although it seems we must indeed do so in the interest of long-term sustainability. Many minds are currently working on plans for expanding the middle course, but these should not include taking up Phaethon's reins and repeating his mistake.

* * * * *

Paper, even that provided by the Patent Office, lies still for anything to be written on it. Recall the 1880 patent of Daniel Ruggles "for producing rain fall by conveying and exploding explosive agents within the cloud realm"; the 1887 patent of J. B. Atwater "to destroy or disrupt tornadoes"; the 1892 patent of Laurice Leroy Brown for a tower to transport and detonate explosives automatically for "aiding rainfall"; and the patent awarded in 1918 to John Graeme Balsillie for ionizing a volume of air and switching the electrical polarity of clouds, "by means of suitable ray emanations." Not to slight the modern era, recall also the 2003 promise by Earthwise Technologies to clear the air and enhance rainfall in Laredo,

Texas, using patented "ionization towers." Now, are you ready for the "Welsbach Patent" to offset global warming? It was granted in 1991 to inventors David B. Chang and I-Fu Shih of Hughes Aircraft Company. Chemtrail conspiracy theorists, suspicious of the U.S. government, are certain that the military is using this technique to seed the lower stratosphere with microscopic particles of aluminum and barium oxide emitted in jet aircraft exhaust.[103] Climate engineers may point out that the early patents were just fantasies, while Welsbach seeding would actually "work." This depends on what you mean by making something "work" in more than a narrow technical sense.

Earlier modification plans always were couched in the context of the pressing issues and available technologies of their eras: James Espy wanted to purify the air and make rain for the East Coast, General Robert Dyrenforth set out to solve the problem of drought in the West, and L. Francis Warren hoped to clear airports in the 1920s, while the Russians and Americans vied over militarizing weather and climate control throughout the cold war. In 1971 climatologist Hubert Lamb wrote that the greatest pending climate emergency might be the overuse of the natural water supply in Central Asia and elsewhere. In 1991 Michael MacCracken at Lawrence Livermore National Laboratory turned his attention to geoengineering the climate as a response to global warming; that same year, Ralph Cicerone and his colleagues proposed injecting alcane gases (ethane and propane) into the ozone layer as a possible way to heal the damage being caused by chlorine compounds.[104] Each generation, it seemed, has had its own leading issues for investing in technologies of control.

Ideas about fixing the sky are seemingly endless, and wild new ideas arrive daily in my in-box. Recently, an engineer claiming space science credentials proposed to shuttle tanks of liquid nitrogen to high altitudes to chill the air and form high, thick contrail-like clouds—this with no analysis of the energetics of the process, either in producing sufficient quantities of liquid nitrogen (which is energy-intensive), in delivering it to altitude (which is both polluting and energy-intensive), or in understanding the radiative consequences of artificial cirrus haze (which can serve to warm the atmospheric layers beneath it). Ideas about stirring up the deep-ocean column with Mixmasters, laying down gigantic plastic bags of liquefied CO_2 on the ocean floor, tilling "biochar" charcoal into the soil, blowing bubbles to make the oceans more reflective, or perhaps shooting each individual CO_2 molecule into space have all been floated as trial balloons. Just yesterday, my in-box had a proposal to flood the Sahara Desert and the Australian Outback to plant mega-forests of eucalyptus trees. My personal favorite involves super-nanotechnology of the future and would entail adding tiny "shock absorbers" to each carbon–oxygen bond in CO_2 molecules. This would serve to keep the molecular bonds from vibrating and rotating freely, thus preventing properly retrofitted CO_2 molecules from acting as strong infrared absorbers and emitters; in lay terms, it would stop them from behaving like greenhouse gases. Imagine the boost to industry and "American competitiveness" in developing high-tech and "green-collar jobs" to manufacture and install more than 10^{26} such submolecular devices worldwide! Of course this is nonsense, but in the coming years you will see many such proposals to "fix the sky." They will be couched in the language of possibility and will convey a sense of unprecedented urgency. But now you know the precedents, the checkered history of weather and climate control.

In 2009 I participated in an America's Climate Choices workshop on geoengineering. At the meeting, convened by the National Academy of Sciences at the request of Representative Alan Mollohan (D-West Virginia), the dominant voices were those of scientists and social scientists interested in a full assessment of the possibilities and dangers of geoengineering; influential policymakers were the primary audience. Unlike earlier meetings, advocates for particular technological fixes were not in the ascendant. I consider this an encouraging development as we seek more nuanced perspectives.

In my presentation, the only one representing the history of science and technology, I pointed out how climate engineers mistakenly claim that they are the "first generation" to propose climate control and how commercial and military interests have inevitably influenced what scientists and engineers have considered purely technical issues. Geoengineering, like climate change, involves, quintessentially, socio-technical hybrid issues. As the American Meteorological Society is recommending, any enhanced research on the scientific and technological potential for geoengineering the climate system must be accompanied by a comprehensive study of its historical, ethical, legal, and social implications, an examination that integrates international, interdisciplinary, and intergenerational issues and perspectives and includes lessons from past efforts to modify weather and climate. History can provide scholars in other disciplines with detailed studies of past interventions by rainmakers and climate engineers as well as structural analogs from a broad array of treaties and interventions. Only in such a coordinated fashion, in which researchers and policymakers participate openly, can the best options emerge to promote international cooperation, ensure adequate regulation, and avoid the inevitable adverse consequences of rushing forward to fix the sky. I repeated this message in my 2009 testimony to the U.S. House of Representatives, Committee on Science and Technology.[105]

In his *Critique of Pure Reason*, the philosopher Immanuel Kant wrote: "The whole interest of my reason . . . is concentrated in the three following questions:

1. What can I know?
2. What should I do?
3. What may I hope?"[106]

These general questions are of immense theoretical, practical, and moral import. Here we conclude by applying them to weather and climate control.

- *What can I know?* We know that climate is nebulous, complex, and unpredictable. We know that it is always changing, on all temporal and spatial scales; and we know few of the turbulent details: what the weather will be next week or if a sudden and disruptive climate change looms in the near or distant future. We know that humans, especially the "Takers," have perturbed the climate system through agriculture, by the burning of fossil fuels, and by the sum total of many additional practices. We do not know the ultimate outcome of all this, but we strongly suspect that it may not be good. We know that weather and climate control has a checkered history, rooted in hubris and populated with charlatans and sincere but misguided scientists; we also know that most plans for weather and climate control were speculative responses to the urgent problems of the day and were based on then-fashionable cutting-edge technologies—cannon, chemicals, electric discharges, airplanes, H-bombs, space probes, computers—with much of it military in origin. We know that those who understand the climate system best are most humbled by its complexity and are among the least likely to claim that they have simple, safe, or cheap ways to "fix" it. We also know that many weather and climate engineers thought they were the "first generation" to think about these things and, since they faced "unprecedented" problems, were somehow exempt from historical precedents. On the contrary, they were critically in need of historical precedents.
- *What should I do?* We should all be asking this question and working together to implement the most reasonable, just, and effective answers. My colleagues at the Climate Institute are eloquent supporters of middle course solutions, but they also advocate responsible geoengineering research, while educating and gently correcting the speculators.[107] Some have asked if the risk of geoengineering is worse than the risk of global warming. I think that it just might be, especially if we neglect the historical precedents and cultural implications. We should cultivate a healthy dose of humility, even awe, before the complexities of nature (and human nature). Do not propose simplistic technical solutions to complex socioeconomic problems; do not even propose simplistic socioeconomic solutions. Do not claim credit for unverifiable results. Adopt the Hippocratic prescription for a planetary fever: "to help, or at least to do no harm."

Practice mitigation and adaptation in a pluralistic world with many climes and many cultures. Perhaps as a first ethical approximation, we could follow Kant's categorical imperative: "Act only on a maxim that you can at the same time will to become a universal law."[108]

■ *What may I hope?* We can hope that fears and anxieties that freeze us into inaction or that tempt us to do too much might be overcome and that a middle course of climate mitigation and adaptation might emerge—amenable to all, reasonable, practical, equitable, and effective.

NOTES

Introduction

1. National Academy of Sciences, "Geoengineering Options."
2. Crutzen, "Albedo Enhancement."
3. Fleming, "Global Climate Change and Human Agency."
4. Glacken, *Traces on the Rhodian Shore*, 501–705; Fleming, *Historical Perspectives on Climate Change*, 11–32.
5. Fleming, *Meteorology in America*, 44, 99.
6. Jordan, "Art of Pluviculture," 81–82; Spence, *Rainmakers*.
7. Ekholm, "On the Variations of the Climate," 61.
8. Fleming, *Callendar Effect*, xiii.
9. Wexler, "U.N. Symposium."
10. President's Science Advisory Committee, *Restoring the Quality*, 127; Weart, "Climate Modification Schemes."
11. Fleming and Jankovic, eds., *Klima*.
12. Curry, Webster, and Holland, "Mixing Politics and Science," 1035.
13. Weinberg, "Can Technology Replace Social Engineering?"
14. Emerson, *Conduct of Life*, 86–87.

1. Stories of Control

1. Bulfinch, *Age of Fable*, 62; Ovid, *Metamorphoses*, 1.750–779, 2.1–400.
2. Emanuel, *What We Know About Climate Change*, 53.
3. Milton, *Paradise Lost*, bk. 10, lines 649–650.
4. Dante, *Divine Comedy: Inferno*, canto 4, lines 146–148.
5. Catlin, *North American Indians*, 1:152–153.
6. Quinn, *Ishmael*, 80–81.
7. Verne, *Purchase of the North Pole*, 143.
8. Twain, *American Claimant*, opening section, "The Weather in This Book."
9. Hahn, *Wreck of the South Pole, or the Great Dissembler*, 48.
10. Griffith, *Great Weather Syndicate*, 6.
11. Cook, *Eighth Wonder*, 55–56.
12. Gratacap, *Evacuation of England*, 54.
13. Train and Wood, *Man Who Rocked the Earth*, 11.
14. England, *Air Trust*, 17–21.
15. Wilson, "Rain-Maker," 503.
16. Mergen, *Weather Matters*, 233.
17. Roberts, *Jingling in the Wind*, 3–6.
18. Nash, *Rainmaker*, 60–61.
19. Quoted in "From the Footlights," n.p.
20. [McGavin], "Darren's Theatre Page," n.p.
21. Kael, "Reviews A–Z," s.v. "The Rainmaker."
22. Liebling, *Just Enough Liebling*, 283.
23. Clarke, "Man-Made Weather," 74.
24. Bellow, *Henderson the Rain King*, 201.
25. Vonnegut, *Fates Worse Than Death*, 26.
26. Bohren, "Thermodynamics," n.p.

2. Rain Makers

1. Bacon, *Works*, vol. 8, *Great Instauration*, 23–24.
2. Ibid., vol. 5, *New Atlantis*, 399.
3. Ibid., vol. 8, *New Organon*, 115.
4. Fleming, "Meteorology," 184–188.
5. Butterfield, *Origins of Modern Science*, viii.
6. Merchant, *Death of Nature*, 193.
7. Cohen, *Revolution in Science*; Shapin, *Scientific Revolution*.
8. Fleming, *Meteorology in America*, 23–54, 66–73, 78–81, 95–106.
9. Espy, *Third Report on Meteorology*, 100.
10. Espy, *Philosophy of Storms*, 492–493.
11. *Congressional Globe*, 25th Cong., 3d sess., December 18, 1838, 39–40.

12. Quoted in Meyer, *Americans and Their Weather*, 87.
13. Espy, *Philosophy of Storms*, 492.
14. Quoted in Harrington, "Weather Making," 51–52.
15. Espy, *Second Report on Meteorology*, 14–19.
16. Quoted in Espy, *Fourth Meteorological Report*, 35–36.
17. Leslie, "Rain King," 11.
18. Meyer, *Americans and Their Weather*, 89.
19. Hawthorne, "Hall of Fantasy," 204.
20. Le Maout, *Effets du canon*, 13.
21. Quoted in Le Maout, *Lettre à M. Tremblay*, 5.
22. Le Maout, *Effets du canon*, 11.
23. Le Maout, *Encore le canon*, 13.
24. "Caius Marius," in *Plutarch's Lives*, 3:220.
25. Humphreys, *Rain Making*, 30.
26. Powers, "Rain-Making," 52.
27. Stone, "Rain-Making by Concussion," 52.
28. U.S. House of Representatives, *Production of Rain by Artillery-Firing*, 5.
29. Ruggles, "Method of Precipitating Rainfalls," 1; "Novel Method of Precipitating Rainfalls," 342.
30. Ruggles, *Memorial*, 1.
31. Van Bibber, "Rain Not Produced," 405.
32. "How About That Patent for Rain-Making?" *Farm Implement News*, October 22, 1891, clipping in NOAA Central Library.
33. Williams, "Bizarre & Unusual Will of Robert St. George Dyrenforth," 12.
34. Dyrenforth, *Report of the Agent*, 8–10.
35. *Farm Implement News*, September 1891, n.p., clipping in NOAA Central Library.
36. Dyrenforth, *Report of the Agent*, 16–17.
37. *Farm Implement News*, September 1891, n.p.
38. Dyrenforth, *Report of the Agent*, 25.
39. Quoted in Hering, "Weather Control," 181–182.
40. Dyrenforth, *Report of the Agent*, 32.
41. "Government Rainmaking," 309–310.
42. *Scientific American Supplement*, October 17, 1891, 13160.
43. Clarke, "Ode to Pluviculture," 260.
44. Curtis, "Rain-Making in Texas," 594.
45. Quoted in *Scientific American Supplement*, October 17, 1891, 13160.
46. Hering, "Weather Control," 182.
47. Blake, "Can We Make It Rain?" 296–297.
48. Blake, "Rain Making," 420.
49. Moore, "Famine," 45–46.
50. "Another Rain Controller," 113.
51. Harrington, "Weather Making," 47.
52. Brown, "Tower and Dynamite Detonator."

3. Rain Fakers

1. Hering, *Foibles and Fallacies of Science*, 240.
2. Seneca, *Naturales Quaestiones*, quoted in Frazer, "Some Popular Superstitions," 142.
3. White, *History of the Warfare*, 1:323–372.
4. Arago, *Meteorological Essays*, 219.
5. Lomax, *Bells and Bellringers*, 20.
6. Arago, *Meteorological Essays*, 211–213.
7. Fitzgerald, "At War with the Clouds," 629–636; Hering, "Weather Control," 185.
8. Cerveny, *Freaks of the Storm*, 36.
9. Abbe, "Hail Shooting in Italy," 358.
10. *Owning the Weather* (Greene).
11. Humphreys, *Rain Making*, 1.
12. News clipping, undated, Franklin Papers.
13. Franklin, "Weather Control," 496–497.
14. Franklin, "Much-Needed Change of Emphasis," 452.
15. Franklin, "Weather Control," 496–497.
16. Franklin, "Weather Prediction and Weather Control," 378.
17. Abbe, "Cannonade Against Hail Storms," 738–739.
18. Caldwell, "Some Kansas Rain Makers," 309.
19. Spence, *Rainmakers*, 52–63.
20. "Current Notes," 192.
21. "Kansas All Right Now She Has Ten Rainmaking Outfits Ready for Service," *St. Louis Republic*, April 25, 1894, 12
22. "A Successful Rainmaker: How Clayton B. Jewell Coaxes Moisture from Cloudless Skies," *Columbus Enquirer-Sun*, August 12, 1894, 6.
23. Tate and Tate, *Good Old Days Country Wisdom*, 95–96.
24. "Unfortunate Rain-maker," 735.
25. Quoted in T. E. Murtaugh, "Rainmaker C. M. Hatfield as Seen at Close Range," *Los Angeles Examiner*, March 19, 1904, reprinted in *San Antonio Daily Express*, April 2, 1905, 8.
26. Moore, "Fake Rainmaking," 153.
27. Patterson, "Hatfield the Rainmaker," n.p.
28. "Fake Rainmaker," 84.
29. Canada, House of Commons, *Official Report of Debates*, 562
30. Tuthill, "Hatfield the Rainmaker," 107–110.
31. Jenkins, *Wizard of Sun City*, 5–6.
32. Patterson, "Hatfield the Rainmaker," n.p.; Spence, *Rainmakers*, 79–99.
33. *Seattle Post Intelligencer*, July 17, 1920, reprinted in *Bulletin of the American Meteorological Society* 1 (1920): 80–82.
34. "Rain Maker Asked to Turn Off Faucets," *Wyoming State Tribune*, May 23, 1921, 2.
35. "Rainmaker Makes Hit: Dr. Hatfield Biggest Hero in Italy Today," *Morning Oregonian*, September 2, 1922, 4; "Rainmaker Fails," 591.
36. Carpenter, "Alleged Manufacture of Rain," 377.
37. Liebling, *Honest Rainmaker*, 9.

38. "Sykes Sells Sunshine," 40; Spence, *Rainmakers*, 128–131.
39. Liebling, *Honest Rainmaker*, 38.
40. "Belmont Park to Test Rainmaker's Magic," *New York Times*, September 10, 1930, 23; "Rainmaker Fails in Test," *New York Times*, September 16, 1930, 11; Liebling, *Honest Rainmaker*, 47–48.
41. Goodstein, "Tales in and out of 'Millikan's School,'" n.p.
42. Fleming, "Sverre Petterssen," 79.
43. Petterssen, *Weathering the Storm*, 181–182.
44. DuBridge interview.
45. Roger Hammond to Irving Langmuir, February 3, 1950, copy in Vonnegut Papers.
46. Willard Haselbach, "'Rain Maker of the Rockies': History's Biggest Weather Experiment Underway," *Denver Post*, April 22, 1951, 17A.
47. "Brief History of Artificial Weather"; Changnon, "Paradox of Planned Weather Modification," 29.
48. James Y. Nicol, "Weather's Miracle Man," *Star Weekly Magazine*, April 6, 1957, 10–11.
49. Kobler, "Stormy Sage," 69–70.
50. Bundgaard and Cale, "Irving P. Krick."
51. "Biography," Wilhelm Reich Museum, http://www.wilhelmreichmuseum.org/biography.html (accessed January 20, 2010).
52. "Cloudbuster."
53. "Goodbye Chemtrails."
54. Webb County Commissioners, "Official Minutes," April 14, 2003.
55. Raul Casso, Webb County, Texas, chief of staff (2003), in Berler, KGNS-TV broadcasts.
56. Berler, KGNS-TV broadcasts.
57. Guerra, "Stormy Weather," 6.
58. McAdie, "Natural Rain-makers," 77.
59. Carpenter, "Alleged Manufacture of Rain," 376–377.
60. Jordan, "Art of Pluviculture."
61. Humphreys, *Rain Making*, vii.
62. National Academy of Sciences, *Critical Issues*, 3–4.
63. Fleming, "Fixing the Weather"; Fleming, "The Pathological History."

4. Foggy Thinking

1. McAdie, "Control of Fog," 36.
2. Shakespeare, *Hamlet*, act 2, scene 2; Coleridge, *Rime of the Ancient Mariner*, 12; Doyle, *Study in Scarlet*, 27.
3. Abbe, "Tugrin Fog Dispeller," 17.
4. Williams, *Climate of Great Britain*, cited in Jankovic, *Reading the Skies*, 1.
5. Fleming, *Meteorology in America*, 26–27.
6. Lodge, "Electrical Precipitation," 34.

7. Balsillie, "Process and Apparatus for Causing Precipitation."
8. Bancroft to Warren, August 23, 1920, Bancroft Papers.
9. Warren to Bancroft, May 21, 1921, Bancroft Papers.
10. Chaffee, "Second Report on Dust Charging."
11. Warren to Bancroft, September 19, 1921, Bancroft Papers.
12. "Fliers Bring Rain with Electric Sand," *New York Times*, February 12, 1923, 3.
13. "Wright Sees Sand Rip Clouds Away," *New York Times*, February 18, 1923, E1.
14. Smith, "Dr. Warren—Rain Maker."
15. Warren to Bancroft, July 11, 1922, Bancroft Papers.
16. Talman, "Can We Control the Weather?"
17. "Rain-making Not Feasible Says U.S. Weather Bureau," U.S. Department of Agriculture press release, March 21, 1923, *New York Tribune* and *New York Times*, clippings in Bancroft Papers.
18. Ibid.
19. *Cornell Daily Sun*, March 24, 1923, clipping in Bancroft Papers.
20. McFadden, "Is Rainmaking Riddle Solved?" 30.
21. Warren, *Facts and Plans*, 26.
22. "Miracle."
23. Warren, *Facts and Plans*, 7.
24. "Dispersal of Clouds Here Accomplished," *Hartford Courant*, June 19, 1926, 1; "The Rainmaker Comes to Town," *Hartford Courant*, June 20, 1926, D1; "Rainmaker Successful for 2d Time," *Hartford Courant*, August 25, 1926, 1.
25. Warren to Bancroft, telegram, June 18, 1926, Bancroft Papers.
26. "Rainmaker's Work in City Is Completed," *Hartford Courant*, October 22, 1926, 1.
27. Warren, *Facts and Plans*, 17.
28. Warren to Bancroft, April 7, 1927, May 4, 1928, and February 21, 1929, Bancroft Papers.
29. Servos, "Wilder D. Bancroft," 4. Spence calls this episode "scientific," but it was not (*Rainmakers*, 103–115).
30. Quoted in Houghton and Radford, "On the Local Dissipation," 5.
31. Bergeron, "On the Physics"; Bergeron, "Some Autobiographic Notes."
32. Houghton and Radford, "On the Local Dissipation," 13–26.
33. Houghton Papers.
34. "Fog Broom," n.p.
35. Bowles and Houghton, "Method for the Local Dissipation," 48–51; "Controlled Weather," 205; Wylie, *M.I.T. in Perspective*, 80–81.
36. Houghton to N. McL. Sage, July 18, 1940, Office of the President, Institute Archives, MIT.
37. Houghton and Radford, "On the Local Dissipation," 6.
38. Quoted in Fleming, *Callendar Effect*, 51.
39. Humphreys, *Rain Making*; Brunt, "Artificial Dissipation"; Ogden, "Fog Dispersal."
40. "FIDO Conference Program"; Banks, *Flame over Britain*.
41. Records of the Petroleum Warfare Department, British National Archives; Banks, *Flame over Britain*.

42. "FIDO Conference Program."
43. Banks, *Flame over Britain*, 148–150.
44. Williams, *Flying Through Fire*, 20.
45. Banks, *Flame over Britain*.
46. Clarke, "Man-Made Weather," 188.
47. Ogden, "Fog Dispersal," 34.
48. Fleming, *Callendar Effect*, 56–59.
49. Ibid., 59–60.
50. Ibid., 60; Ogden, "Fog Dispersal," 38.
51. Gregg, "Address at the Dedication Ceremonies."
52. "U.S. Weather Chief Hails Achievement of Air Conditioning," *Chicago Tribune*, July 12, 1934, 4.
53. Devereaux, "Meteorological Service of the Future," 217.
54. Mindling, "Raymete and the Future."
55. Devereaux, "Meteorological Service of the Future," 218.

5. Pathological Science

1. Rosenfeld, *Quintessence of Irving Langmuir*; Suits and Martin, "Irving Langmuir"; Fleming, "Pathological History," 8–12.
2. Langmuir, "Pathological Science."
3. Fleagle, "Second Opinions," 100.
4. Quoted in Langmuir, "Pathological Science," 11–12.
5. Associated Press, "Panel Finds Misconduct by Controversial Fusion Scientist," July 18, 2008, http://www.cbc.ca/news/story/2008/07/18/fusion-misconduct.html.
6. Schaefer, autobiography, chap. 6, Schaefer Papers.
7. Langmuir, "Growth of Particles," 167–174; Smoke Generation (1940–1947), Schaefer Papers.
8. Schaefer interview; Schaefer to C. Guy Suits, October 10, 1946, Schaefer Papers.
9. Lambright, *Weather Modification*.
10. Gathman, "Method of Producing Rainfall," 1; Gathman, *Rain Produced at Will*, 27–28.
11. Sanford, "Rain-Making," 490–491.
12. "Who Owns the Clouds?" 43.
13. McAdie, "Natural Rain-makers," 80.
14. "Clouds Sprayed," 418.
15. Grunow, "Der Künstliche Regen," 602.
16. Veraart, *Meer zonneschijn*.
17. Quoted in Kramer and Rigby, "Selective and Annotated Bibliography," 191–192.
18. Byers, "History of Weather Modification," 5–6.
19. Schaefer Laboratory Notebook, July 12, 1946, Schaefer Papers.
20. Schaefer interview.
21. Langmuir Laboratory Notebook, July 1946, Langmuir Papers.

22. Schaefer interview.
23. Schaefer Laboratory Notebook, July 31, 1946, Schaefer Papers.
24. Schaefer to Suits, October 10, 1946, Schaefer Papers.
25. Croy, "Rainmakers," 214.
26. Press release, November 13, 1946, GE Archives.
27. "Scientist Creates Real Snowflakes," *New York Times*, November 14, 1946, 33.
28. Press release, November 13, 1946, GE Archives; Schaefer, "Production of Ice Crystals"; Byers, "History of Weather Modification."
29. Schaefer Laboratory Notebook, November 13, 1946, Schaefer Papers.
30. Suits to V. H. Fraenckel, November 13, 1946, Schaefer Papers.
31. Press release, November 14, 1946, GE Archives.
32. *New York Times*, November 15, 1946, 24; *Boston Globe*, November 15, 1946, 1.
33. Schaefer Papers.
34. General Electric Corporation, *56th Annual Report*, 27.
35. Fleming, "Fixing the Weather," 177.
36. Goldstein to GE, November 18, 1946, copy in Schaefer Papers.
37. Goldston, "Legal Entanglements"; Langmuir, "Summary of Results."
38. Havens, "History of Project Cirrus," 13.
39. "Interim Report."
40. "Project Cirrus" (1952), 13.
41. "Law Asked to Bar Suits Against the Rainmakers," *New York Times*, January 13, 1949, 25.
42. "Many Legal Entanglements."
43. "Project Cirrus" (1950), 287.
44. Schaefer interview; Droessler, "Federal Government Activities," 253.
45. Doubleday, "Air Force Activities"; Byers, "History of Weather Modification," 16–17.
46. News clippings, including *New York Times*, October 12, 1947, 24; *Albany Times-Union*, October 12, 1947, n.p.; *Christian Science Monitor*, October 14, 1947, 10; *Los Angeles Times*, October 14, 1947, 1, Schaefer Papers.
47. Larry Murray, "Hurricane Study Only Begun, Says Schaefer," October 17, 1947, clipping in Schaefer Papers.
48. Langmuir, "Growth of Particles," 183–185.
49. Enders to Suits, November 13, 1947, with clipping enclosed, "Dry Ice and a Crazy Hurricane," Schaefer Papers.
50. Schaefer, "Preliminary Report."
51. Langmuir, "Growth of Particles," 185.
52. Richard Gray, "U.S. Government Aims to Tame Hurricanes," *Daily Telegraph*, August 2, 2008, http://www.telegraph.co.uk.
53. Vonnegut, "Nucleation of Ice," 593–595.
54. Vonnegut interview.
55. Vonnegut Laboratory Notebook, November 1946, Vonnegut Papers.
56. Vonnegut interview.
57. Press release, October 21, 1947, GE Archives.
58. Vonnegut interview.

59. Havens, Juisto, and Vonnegut, *Early History*.
60. Vonnegut interview.
61. Langmuir, "Report on Evaluation," 23.
62. Petterssen, *Weathering the Storm*, 293–294.
63. Langmuir, "Seven Day Periodicity"; Brier, "7-Day Periodicities."
64. Suits to Langmuir, August 26, 1949, copy in Vonnegut Papers.
65. Press release, January 2, 1950, GE Archives.
66. Langmuir, "Abstract of Remarks, Oct. 1950," GE Archives.
67. Hosler, "Weather Modification"; Steinberg, *Slide Mountain*, 106–134, 191–194.
68. Langmuir, note card, "TV Aug. 24, 8:33 am, Dave Garroway, Today," after 1953, Langmuir Papers.
69. "Review of Savannah Hurricane."
70. "Rain Making Ineffective," 375; Reichelderfer, "Letter," 38.
71. "Scientist Would Move Rain Tests to South Pacific," *Albuquerque Tribune*, April 29, 1955, 50, clipping in Schaefer Papers.
72. Langmuir, "Report on Evaluation," 23.
73. "Scientist Would Move Rain Tests," 50.
74. Langmuir, "Production of Rain"; Langmuir, "Report on Evaluation," 23; "Langmuir Predicts Hurricane Prevention," *Washington Post*, August 25, 1955, 2.
75. Elliott, "Experience of the Private Sector"; "City Flip-Flop on Rainmaking," *Daily News*, November 5, 1951, clipping in Schaefer Papers; Landsberg, "Memorandum for the Record."
76. *Science of Superstorms* (British Broadcasting Corporation).
77. Ibid., cited in Richard Gray, "How We Made the Chernobyl Rain," *Sunday Telegraph*, April 22, 2007, n.p.
78. *Science of Superstorms* (British Broadcasting Corporation).
79. Ibid.

6. Weather Warriors

1. *Owning the Weather* (Greene).
2. Stephan Farris, "Ice Free," *New York Times*, July 27, 2008, MM 20.
3. Fuller, *Weather and War*; Fuller, *Thor's Legions*.
4. Fleming, "Sverre Petterssen," 75–83.
5. Polybius, *Universal History*, bk. 8; Rossi, *Birth of Modern Science*.
6. Hacker, "Military Patronage"; Mendelsohn, "Science, Scientists."
7. McNeill, *Pursuit of Power*; Fleming, "Distorted Support."
8. Gillispie, *Science and Polity*.
9. U.S. Army Medical Department, "Regulations," 227.
10. Fleming, "Storms, Strikes, and Surveillance"; see also Whitnah, *History of the United States Weather Bureau*, 22–42; and Hawes, "Signal Corps."
11. Bates and Fuller, *America's Weather Warriors*, 16–26; Fleming, "Distorted Support," 51–53.

12. Fleming, "Fixing the Weather and Climate," 176.
13. Ibid., 175–176.
14. Teller, *Memoirs*, 253.
15. Simpson and Simpson, "Why Experiment?"; news clippings in Schaefer Papers.
16. "Weather Control Called 'Weapon,'" *New York Times*, December 10, 1950, 68.
17. Suits, "Statement on Weather Control."
18. V. H. Fraenckel to C. Guy Suits, telegram, 1947; W. H. Milton Jr. to C. Guy Suits, September 2, 1947, Schaefer Papers.
19. R. E. Evans to Vincent Schaefer, December 10, 1947, Schaefer Papers.
20. Vincent Schaefer to C. J. Brasefield, May 4, 1948; Vincent Schaefer to Michael J. Ference Jr., May 13, 1948, and reply May 14, 1948, Schaefer Papers.
21. Kobler, "Stormy Sage," 70.
22. Byers, "History of Weather Modification," 13.
23. Petterssen, *Weathering the Storm*, 295.
24. Droessler, *Federal Government Activities*, 254.
25. Byers, "History of Weather Modification," 25–27.
26. Petterssen, *Cloud and Weather Modification*.
27. Act of Congress, August 13, 1953 (67 Stat. 559), as amended July 9, 1956 (70 Stat. 509); U.S. Library of Congress, Congressional Research Service, *Weather Modification*.
28. Orville, "Weather Made to Order?" 26.
29. Nate Haseltine, "Cold War May Spawn Weather-Control Race," *Washington Post and Times Herald*, December 23, 1957, A1.
30. Quoted in "Weather Weapon," 54.
31. Fedorov, "Modification of Meteorological Processes," 391.
32. Quoted in Arthur Krock, "An Inexpensive Start at Controlling the Weather," *New York Times*, March 23, 1961, 32.
33. U.S. National Oceanic and Atmospheric Administration, "Hurricane Research Division History"; Willoughby et al., "Project STORMFURY."
34. Gentry, "Hurricane Modification," 497.
35. U.S. Navy, "Technical Area Plan," 1.
36. Simpson interview.
37. *Science of Superstorms* (British Broadcasting Corporation).
38. Simpson interview; MacDonald, "Statement"; Fleming, "Distorted Support."
39. Jack Anderson, "Air Force Turns Rainmaker in Laos," *Washington Post*, March 18, 1971), F7; Seymour M. Hersh, "Rainmaking Is Used as a Weapon by U.S.," *New York Times*, July 3, 1972, 1.
40. Cobb et al., *Project Popeye*.
41. *Science of Superstorms* (British Broadcasting Corporation).
42. Fleming, "Pathological History," 13.
43. Fuller, *Air Weather Service Support*, 30–32.
44. "Motorpool."
45. Shapley, "Weather Warfare," 1059–1061; Westmoreland, *Soldier Reports*, 342.
46. Appleman et al., *Fourth Annual Survey Report*.

47. Chary, *History of Air Weather Service Weather Modification*, abstract.
48. National Academy of Sciences, *Weather and Climate Modification*, 24.
49. Fleming, "Pathological History," 13.
50. "Dienbienphu Push Renewed by Reds," *New York Times*, April 23, 1954, 2.
51. *Contributions to the History of Dien Bien Phu*, 201.
52. Doel and Harper, "Prometheus Unleashed."
53. *Science of Superstorms* (British Broadcasting Corporation).
54. Handler to Pell, July 25, 1972, in U.S. Senate, *Prohibiting Military Weather Modification*, 153.
55. Quoted in U.S. House of Representatives, *Prohibition of Weather Modification*, 5; Munk, Oreskes, and Muller, "Gordon James Fraser MacDonald."
56. "Hearing on Senate Resolution 281"; *Congressional Record*, July 11, 1973, 233303–5.
57. Seymour M. Hersh, "U.S. Admits Rain-Making from '67 to '72 in Indochina," *New York Times*, May 19, 1974, 1; U.S. House of Representatives, *Weather Modification as a Weapon*.
58. Gromyko to United Nations secretary-general, August 7, 1974, in U.S. House of Representatives, *Weather Modification as a Weapon*, 11–12.
59. U.S. House of Representatives, *Weather Modification as a Weapon*, 510–513; Juda, "Negotiating a Treaty," 28.
60. United Nations, "Convention on the Prohibition."
61. United Nations, *Multilateral Treaties Deposited*, 667.
62. Goldblat, "Environmental Convention," 57.
63. *Owning the Weather* (Greene).
64. Chamorro and Hammond, *Addressing Environmental Modification*.
65. United Nations, *Multilateral Treaties Deposited*, 667.
66. Conway, "World According to GARP," 131–147.
67. Marshall, "Greening of the National Labs," 25.
68. Stapler, "New Face of the Pacific," n.p.
69. Cohen, "DoD News Briefing," n.p.

7. Fears, Fantasies, and Possibilities of Control

1. Tuan, *Landscapes of Fear*, 6.
2. von Neumann, "Can We Survive Technology?"
3. Harper, *Weather by the Numbers*; see also Phillips, "General Circulation"; Smagorinsky, "Beginnings of Numerical"; Thompson, "History of Numerical"; and Nebeker, *Calculating the Weather*.
4. Zworykin, "Outline of Weather Proposal," 8.
5. von Neumann to Zworykin, October 24, 1945, in Zworykin, "Outline of Weather Proposal."
6. Athelstan F. Spilhaus, "Comments on Weather Proposal," November 6, 1945, in Zworykin, "Outline of Weather Proposal."

7. Waldemar Kaempffert, "Julian Huxley Pictures the More Spectacular Possibilities That Lie in Atomic Power," *New York Times*, December 9, 1945, 77; "Blasting Polar Ice," *New York Times*, February 2, 1946, 11.
8. "Sarnoff Predicts Weather Control and Delivery of the Mail by Radio," *New York Times*, October 1, 1946, 1.
9. "Talk of the Town," *New Yorker*, October 12, 1946, 23.
10. MacCracken, "On the Possible Use of Geoengineering."
11. "Storm Prevention Seen by Scientist," *New York Times*, January 31, 1947, 16.
12. "Weather to Order," *New York Times*, February 1, 1947, 14.
13. Wexler, "Trip Report—Princeton, N.J., Oct. 14–15, 1946," box 2, Wexler Papers.
14. Harald Sverdrup to Francis Reichelderfer, June 2, 1946, box 2, Wexler Papers.
15. "$28,000,000 Urged to Support M.I.T.," *New York Times*, June 15, 1947, 46.
16. Mann, "War Against Hail," 8.
17. Hoffman, "Controlling Hurricanes"; see also Hoffman, "Controlling the Global Weather"; and Hoffman, Leidner, and Henderson, "Controlling the Global Weather."
18. Shachtman, "NASA Funds Sci-Fi Technology."
19. Quoted in Hoffman, Leidner, and Henderson, "Controlling the Global Weather," 1.
20. *Science of Superstorms* (British Broadcasting Corporation).
21. *Owning the Weather* (Rosen).
22. Mark Schleifstein, "Bill Gates of Microsoft Envisions Fighting Hurricanes by Manipulating the Sea," *Times-Picayune*, July 15, 2009, http://www.nola.com.
23. Lenin, *Materialism*.
24. Burke, "Influence of Man," 1036, 1049–1050.
25. Zikeev and Doumani, *Weather Modification*.
26. Markin, *Soviet Electric Power*, 133.
27. Rusin and Flit, *Man versus Climate*, 174.
28. Rusin and Flit, *Methods of Climate Control*, 2.
29. Adabashev, *Global Engineering*, 161.
30. Shaler, "How to Change," 728–729.
31. Riker, *Conspectus of Power*.
32. Borisov, "Radical Improvement of Climate."
33. Quoted in Adabashev, *Global Engineering*, 192.
34. Borisov, *Can We Control the Arctic Climate?* 6–7.
35. Takano and Higuchi, "Numerical Experiment."
36. Rusin and Flit, *Methods of Climate Control*, 25.
37. Gaskell, *Gulf Stream*, 152.
38. Rusin and Flit, *Methods of Climate Control*, 26.
39. Mackenzie, *Flooding of the Sahara*, 285.
40. "Algerian Inland Sea."
41. Verne, *L'Invasion de la mer*.
42. G. A. Thompson, "Proposes to Turn Sahara into a Sea," *New York Times*, October 15, 1911, C1; Thompson, "Plan for Converting the Sahara."
43. Gall, *Das Atlantropa-Projekt*; Voigt, *Atlantropa*.

44. Johnson, "Climate Control."
45. U.S. Army Air Forces, "Memorandum Report"; Oberth, *Die Rakete*; Oberth, *Wege zur Raumschiffahrt*.
46. "Space Mirror."
47. U.S. Army Air Forces, "Memorandum Report."
48. "Space Mirror."
49. Humphreys, "Volcanic Dust"; Humphreys, *Physics of the Air*.
50. Hoyle and Lyttleton, "Effect of Interstellar Matter"; Krook, "Interstellar Matter."
51. Rusin and Flit, *Man versus Climate*, 60–63.
52. Bruno, "Bequest of the Nuclear Battlefield," 259.
53. "Text of Johnson's Statement on Status of Nation's Defenses and Race for Space," *New York Times*, January 8, 1958, 10.
54. Fleming, "What Counts as Knowledge?"
55. Walter Sullivan, "Called 'Greatest Experiment,'" *New York Times*, March 19, 1959, 1; Christofilos, "Argus Experiment," 869.
56. *New Yorker*, May 26, 1962, 31.
57. Rodin and Hess, "Weather Modification."
58. Lovell and Ryle, "Interference to Radio Astronomy."
59. Kellogg, "Review of Saturn High Water Experiment," 1.
60. Wexler, "On the Possibilities of Climate Control," 1.
61. Kennedy, "Address to the United Nations," n.p.
62. United Nations, "International Co-operation in the Peaceful Uses of Outer Space."
63. Wexler Papers; Yalda, "Harry Wexler."
64. U.S. Joint Numerical Weather Prediction Unit, "Facts Sheet"; Washington, "Computer Modeling."
65. Price and Pales, "Mauna Loa Observatory."
66. European Space Agency, "Harry Wexler."
67. Wexler, "Modifying Weather on a Large Scale," 1059.
68. Wexler, "On the Possibilities of Climate Control."
69. Wexler, "Further Justification for the General Circulation," 1; see also Fleming, *Calendar Effect*.
70. Robert C. Cowen, "Space Fuel: Weather Maker?" *Christian Science Monitor*, January 17, 1962, 5; Sumner Barton, "Space Taint Could Twist Weather," *Boston Globe*, January 21, 1962, A5.
71. Wexler, "On the Possibilities of Climate Control," 1.
72. Wexler, "Further Justification for the General Circulation," 1.
73. Wexler, "On the Possibilities of Climate Control," table 1.
74. Wexler, "Further Justification for the General Circulation," 2.
75. Chapman, "Gases of the Atmosphere," 133.
76. Wexler, "Deozonizer" memorandum.
77. Malkin to Wexler, November 22, 1961, Wexler Papers.
78. Manabe and Möller, "On the Radiative Equilibrium."
79. Wulf to Wexler, December 15, 1961, Wexler Papers.
80. Wulf to Wexler, January 2, 1962, Wexler Papers.

81. Wexler to Wulf, January 5, 1962, Wexler Papers.
82. Wexler to Wulf, telephone call notes, December 20, 1961, Wexler Papers.
83. Crutzen, "Influence of Nitrogen Oxides"; Molina and Rowland, "Stratospheric Sink."
84. C. N. Touart to Wexler, January 11, 1962, Wexler Papers.
85. Wexler to Touart, January 19, 1962, Wexler Papers.
86. Wexler, "U.N. Symposium," 2.

8. The Climate Engineers

1. Philip Shabecoff, "Global Warming Has Begun, Expert Tells Senate," *New York Times*, June 24, 1988, 1.
2. *Changing Atmosphere*, 292.
3. Intergovernmental Panel on Climate Change, "IPCC History"; Bolin, *History of the Science and Politics*.
4. United Nations, "Framework Convention on Climate Change."
5. MacCracken, "Working Toward International Agreement."
6. Hansen, "Tipping Point?" n.p.
7. Lovelock, *Revenge of Gaia*.
8. Schelling, "Economic Diplomacy," 303.
9. *Urban Dictionary*, s.v. "geoengineering," http://www.urbandictionary.com.
10. Lovelock, *Vanishing Face of Gaia*, 139; Ruddiman, *Plows, Plagues, and Petroleum*.
11. *Oxford English Dictionary*, s.v. "engineer."
12. "OUP Blog."
13. Royal Society of London, *Geoengineering*, 77.
14. Keith, "Engineering the Earth System."
15. Keith, "Geoengineering the Climate," 269.
16. Broecker, *Fossil Fuel CO_2*, cover illustration.
17. Fogg, *Terraforming*.
18. Quoted in "Terraforming Information Pages."
19. Glacken, *Traces on the Rhodian Shore*, 415–426, 672–681.
20. For example, Keith and Dowlatabadi, "Serious Look at Geoengineering"; Flannery et al., "Geoengineering Climate"; and Teller, Wood, and Hyde, "Global Warming and Ice Ages."
21. Blackstock et al., *Climate Engineering Responses*, v.
22. Just such a scenario is in Dyer, *Climate Wars*, 190–193.
23. Robock, "20 Reasons," 17–18
24. Robock, "Need for Organized Research."
25. Luginbuhl, Walker, and Wainscoat, "Lighting and Astronomy."
26. Tyndall Centre for Climate Change Research, "Potential of Geo-engineering Solutions," paragraph 2.13.
27. An interesting article on ethics, but with no mention of geoengineering, is Gardiner, "Ethics and Global Climate Change." The work of philosophers Dale Jamieson and

Martin Bunzl also comes to mind. I was personally involved in sessions on geoengineering history, ethics, and policy at the American Geophysical Union (2005), at the American Association for the Advancement of Science (2007), and elsewhere, but I would suggest that more work on this subject is essential.

28. [Gavin Schmidt], "Geo-engineering in Vogue . . . ," n.p.
29. Ward, "How Far Can Man Control His Climate?"
30. Danko, "Budyko," 179–180; Gal'tsov, "Conference on Problems of Climate Control"; Budyko, "Heat Balance of the Earth."
31. Shvets, "Control of Climate Through Stratospheric Dusting."
32. Black and Tarmy, "Use of Asphalt Coatings," 557.
33. Dessens, "Man-made Tornadoes," 13.
34. Eichhorn, *Implications of Rising Carbon Dioxide*, 14; Fleming, "Gilbert N. Plass."
35. MacDonald, "How to Wreck the Environment"; Munk, Oreskes, and Muller, "Gordon James Fraser MacDonald."
36. President's Science Advisory Committee, *Restoring the Quality of Our Environment.*
37. Fletcher, *Changing Climate*, 20.
38. Fletcher, *Managing Climatic Resources*, 2.
39. Yudin, "Possibilities for Influencing."
40. Fletcher, *Managing Climatic Resources*, 18.
41. Budyko, *Climatic Changes*, 236.
42. Kellogg and Schneider, "Climate Stabilization," 1165.
43. Santer et al., "Towards the Detection and Attribution."
44. Kellogg and Schneider, "Climate Stabilization," 1171.
45. Marchetti, "On Geoengineering."
46. National Academy of Sciences, *Energy and Climate.*
47. Dyson, "Can We Control the Carbon Dioxide?"; Dyson and Marland, "Technical Fixes"; "Question of Global Warming."
48. Schelling, "Climatic Change," 469.
49. Turco et al., "Nuclear Winter"; Badash, *Nuclear Winter's Tale.*
50. Penner, Schneider, and Kennedy, "Active Measures."
51. Early, "Space-Based Solar Shield."
52. Tyson, "Five Points of Lagrange," n.p.
53. National Academy of Sciences, *Policy Implications*, 657.
54. Stix, "Removal of Chlorofluorocarbons."
55. Dan Fagin, "Tinkering with the Environment," *Newsday*, April 13, 1992, 7.
56. Nordhaus, "Optimal Transition Path," 1317–1318.
57. Schneider, "Earth Systems Engineering," 418.
58. Summers, "Memo," n.p.
59. Quoted in ibid.
60. Nordhaus, "Challenge of Global Warming," n.p.
61. Fagin, "Tinkering with the Environment," 7.
62. National Academy of Sciences, *Policy Implications*, 451–452.
63. I thank Gregory Cushman for illuminating discussions on tropospheric pollution caused by cannon fire.

64. Blackstock et al., *Climate Engineering Responses*, 47.
65. The quote is widely cited and is based on Martin and Fitzwater, "Iron Deficiency," 341–343.
66. Martin, "Glacial–Interglacial CO_2 Change"; Martin, Gordon, and Fitzwater, "Case for Iron."
67. Buesseler et al., "Ocean Iron Fertilization"; quoted at Woods Hole ocean engineering meeting, September 2007.
68. *Oceanus*.
69. King, "Can Adding Iron?" 134.
70. Lackner, "Capture."
71. Keith, "Why Capture?"
72. Lackner, "Submission." I thank my colleague Thomas Shattuck for an enlightening discussion of chemical energetics.
73. Broecker, *Fossil Fuel CO_2*, fig. 59.
74. Brown, *Challenge of Man's Future*, 142, quoted in Broecker and Kunzig, *Fixing Climate*, 228.
75. Schwartz and Randall, "Abrupt Climate Change," n.p.
76. "Russian Scientist Suggests Burning Sulfur in Stratosphere to Fight Global Warming," *MosNews*, November 30, 2005, n.p.
77. Crutzen, "Albedo Enhancement."
78. Budyko, *Climatic Changes*, 241.
79. Crutzen, "Albedo Enhancement," 212.
80. U.S. National Aeronautics and Space Administration, *Workshop Report*; Kintisch, "Tinkering"; Kintisch, "Giving Climate Change a Kick."
81. Cicerone, "Geoengineering," 221.
82. "No Quick or Easy Technological Fix"; compare with "Five Ways to Save the World."
83. Izrael et al., "Field Experiment," 265.
84. Budyko, "Global Climate Warming," n.p.
85. Wood, "Stabilizing Changing Climate."
86. Cohn, "Sex and Death," 717–718.
87. Launder and Thompson, eds., "Geoscale Engineering."
88. Schneider, "Geoengineering" (2008).
89. Lovelock, "Geophysiologist's Thoughts," quoted in "Medicine for a Feverish Planet: Kill or Cure?" *Guardian*, September 1, 2008.
90. Lovelock and Rapley, "Ocean Pipes," 403.
91. Vince, "One Last Chance."
92. Hawthorne, "Great Carbuncle," 128.
93. Lampitt et al., "Ocean Fertilization"; Smetacek and Naqvi, "Next Generation."
94. Latham et al., "Global Temperature Stabilization"; Salter, Sortino, and Latham, "Sea-going Hardware."
95. Caldeira and Wood, "Global and Arctic Climate Engineering"; Robock, Oman, and Stenchikov, "Regional Climate Responses."

96. Rasch et al., "Overview of Geoengineering"; Tilmes, Müller, and Salawitch, "Sensitivity of Polar Ozone Depletion," 1201.
97. U.S. House of Representatives, *Geoengineering*.
98. Royal Society of London, *Geoengineering*, ix.
99. Royal Academy of Engineering, "Submission."
100. Cotton, "Weather and Climate Engineering," 1.1.
101. Walker and King, *Hot Topic*; Pacala and Socolow, "Stabilization Wedges."
102. Bulfinch, *Age of Fable*, 63.
103. Chang and Shih, "Stratospheric Welsbach Seeding."
104. Lamb, "Climate-Engineering Schemes"; MacCracken, "Geoengineering the Climate"; Cicerone, Elliott, and Turco, "Reduced Antarctic Ozone Depletion"; Cicerone, Elliott, and Turco, "Global Environmental Engineering."
105. American Meteorological Society, "Policy Statement on Geoengineering"; U.S. House of Representatives, *Geoengineering*.
106. Kant, *Critique of Pure Reason*, 690.
107. MacCracken, "Beyond Mitigation."
108. Kant, *Grounding for the Metaphysics of Morals*, xvii.

BIBLIOGRAPHY

Archival and Manuscript Collections

Bancroft, Wilder D. Papers, No. 14-8-135. Division of Rare and Manuscript Collections. Cornell University Library, Ithaca, N.Y.

Bush, Vannevar. Papers. Manuscript Division. Library of Congress, Washington, D.C.

Callendar, Guy Stewart. Archive. University of East Anglia, Norwich, England. Also *The Papers of Guy Stewart Callendar*. Edited by James Rodger Fleming and Jason Thomas Fleming. Digital ed. Boston: American Meteorological Society, 2007.

DuBridge, Lee Alvin. Interview with Judith R. Goodstein, February 19, 1981. Part I, Caltech Archives. http://oralhistories.library.caltech.edu/59/01/OH_DuBridge_1.pdf.

Franklin, William Suddards. Papers. Franklin file. MIT Museum, Cambridge, Mass.

GE Archives. General Electric Corporation. Archival Records and News Bureau Binders. Schenectady Museum, Schenectady, N.Y.

Houghton, Henry G. Papers. Institute Archives and Special Collections, MC 242. Massachusetts Institute of Technology, Cambridge, Mass.

Langmuir, Irving. Papers. Manuscript Division. Library of Congress, Washington, D.C.

Office of the President. Institute Archives and Special Collections, AC4, Box 114.4. Massachusetts Institute of Technology, Cambridge, Mass.

Records of the Defense Research and Development Board. Office of Secretary of Defense. RG-330. National Archives and Records Administration, College Park, Md.

Records of Headquarters. United States Air Force (Air Staff). RG-341. National Archives and Records Administration, College Park, Md.

Records of the Petroleum Warfare Department and Its Successor Bodies. SUPP 15 and AVIA 22/2303, May 15, 1942. British National Archives, Kew, England.

Records of the U.S. Weather Bureau. RG-27. National Archives and Records Administration, College Park, Md.

Schaefer, Vincent J. Interview with Earl Droessler, May 8–9, 1993. Tape Recorded Interview Project, American Meteorological Society and University Corporation for Atmospheric Research.

——. Papers. M. E. Grenander Department of Special Collections and Archives, University at Albany, State University of New York, Albany, N.Y.

Simpson, Robert H. Interview with Edward Zipser, September 6, 1989. Tape Recorded Interview Project, American Meteorological Society and University Corporation for Atmospheric Research.

Vonnegut, Bernard. Interview with B. S. Havens, February 12, 1952. Transcript in Vonnegut Papers.

——. Papers. M. E. Grenander Department of Special Collections and Archives, University at Albany, State University of New York, Albany, N.Y.

Wexler, Harry. Papers. Manuscript Division. Library of Congress, Washington, D.C.

Printed Sources

Abbe, Cleveland. “Cannonade Against Hail Storms.” *Science* 14 (1901): 738–739.

——. “Hail Shooting in Italy.” *Monthly Weather Review*, August 1907, 358.

——. “The Tugrin Fog Dispeller.” *Monthly Weather Review*, January 1899, 17.

Adabashev, Igor. *Global Engineering*. Moscow: Progress Publishers, 1966.

Advisory Committee on Weather Control. *Final Report*. Washington, D.C.: Government Printing Office, 1958.

“An Algerian Inland Sea.” *Nature* 16 (1877): 353–354.

American Meteorological Society. “Policy Statement on Geoengineering the Climate System.” June 2009. http://www.ametsoc.org/POLICY/2009geoengineeringclimate_amsstatement.html.

“Another Rain Controller.” *Scientific American*, August 21, 1880, 113.

Appleman, Herbert S., Laurence D. Mendenhall, John C. Lease, and Robert I. Sax. *Fourth Annual Survey Report on the Air Weather Service Weather-Modification Program* [FY 1971]. Air Weather Service Technical Report 244. Scott Air Force Base, Ill.: Military Airlift Command, 1972.

Arago, François. *Meteorological Essays*. Translated by Edward Sabine. London: Longmans, 1855.

Arrhenius, Svante. *Worlds in the Making: The Evolution of the Universe*. Translated by H. Borns. New York: Harper, 1908.

Bacon, Francis. *The Works of Francis Bacon*. Translated and edited by James Spedding, Robert Leslie Ellis, and Douglas Denon Heath. 15 vols. Boston: Houghton Mifflin, ca. 1900. http://onlinebooks.library.upenn.edu/webbin/metabook?id=worksfbacon.

Badash, Lawrence. *A Nuclear Winter's Tale: Science and Politics in the 1980s*. Cambridge, Mass.: MIT Press, 2009.

Balsillie, John Graeme. "Process and Apparatus for Causing Precipitation by Coalescence of Aqueous Particles Contained in the Atmosphere." U.S. Patent 1,279,823, September 24, 1918. Copy in Bancroft Papers.

Banks, Sir Donald. *Flame over Britain: A Personal Narrative of Petroleum Warfare*. London: Sampson Low, Marston, 1946.

Barber, Charles William. *An Illustrated Outline of Weather Science*. New York: Pitman, 1943.

Bates, Charles C., and John F. Fuller. *America's Weather Warriors, 1814–1985*. College Station: Texas A&M University Press, 1986.

Bellow, Saul. *Henderson the Rain King*. New York: Viking, 1959.

Bergeron, Tor. "On the Physics of Cloud and Precipitation." UGGI 5e assemblée générale, Lisbon, September 1933. *Procés verbaux des séances de l'Association de Météorologie* 2 (1935): 25.

——. "Some Autobiographic Notes in Connection with the Ice Nucleus Theory of Precipitation Release." *Bulletin of the American Meteorological Society* 59 (1978): 390–392.

Berler, Richard "Heatwave." KGNS-TV broadcasts on weather control. Laredo, Tex., 2003.

Black, James F., and Barry L. Tarmy. "The Use of Asphalt Coatings to Increase Rainfall." *Journal of Applied Meteorology* 2 (1963): 557–564.

Blackstock, Jason J., David S. Battisti, Ken Caldeira, Douglas M. Eardley, Jonathan I. Katz, David W. Keith, Aristides A. N. Patrinos, Daniel P. Schrag, Robert H. Socolow, and Steven E. Koonin. *Climate Engineering Responses to Climate Emergencies*. Novim, 2009. http://arxiv.org/ftp/arxiv/papers/0907/0907.5140.pdf.

Blake, Lucien I. "Can We Make It Rain?" *Science* 18 (1891): 296–297.

——. "Rain Making by Means of Smoke Balloons." *Scientific American*, December 31, 1892, 420.

Bohren, Craig. "Thermodynamics: A Tragicomedy in Several Acts." Paper presented at the Gordon Conference on Physics Research and Education: Thermal and Statistical Physics, Plymouth State College, Plymouth, N.H., June 13, 2000.

Bolin, Bert. *A History of the Science and Politics of Climate Change: The Role of the Intergovernmental Panel on Climate Change*. Cambridge: Cambridge University Press, 2007.

Borisov, Petr Mikhailovich. *Can Man Change the Climate?* Moscow: Progress Publishers, 1973.

——. *Can We Control the Arctic Climate?* Translated by E. R. Hope. Ottawa: Defence Scientific Information Service, 1968. [Translated from *Priroda* 12 (1967): 63–73]

——. "Radical Improvement of Climate in the Planet's Polar and Moderate Latitudes." Patent Application 7337. USSR Council of Ministers, Committee for Inventions and Discoveries. Moscow, 1957.

Bowles, Edward L., and Henry G. Houghton. "A Method for the Local Dissipation of Natural Fog." *American Philosophical Society, Miscellanea* 1 (1935): 48–51.

"A Brief History of Artificial Weather Modification." June 10, 1955. Draft prepared for the use of the Advisory Committee on Weather Control. Copy in Schaefer Papers.

Brier, G. W. "7-Day Periodicities in May, 1952." *Bulletin of the American Meteorological Society* 35 (1954): 118–121.

Broecker, Wallace S. *Fossil Fuel CO_2 and the Angry Climate Beast*. New York: Eldigio Press, 2003.

Broecker, Wallace S., and Robert Kunzig. *Fixing Climate: What Past Climate Changes Reveal About the Current Threat—and How to Counter It*. New York: Hill and Wang, 2008.

Brown, Harrison. *The Challenge of Man's Future*. New York: Viking, 1954.

Brown, Laurice Leroy. "Tower and Dynamite Detonator." U.S. Patent Application 473,820, April 26, 1892. Copy in NOAA Central Library, Silver Spring, Md.

Bruno, Laura A. "The Bequest of the Nuclear Battlefield: Science, Nature, and the Atom During the First Decade of the Cold War." *Historical Studies in the Physical and Biological Sciences* 33 (2003): 237–260.

Brunt, D. "The Artificial Dissipation of Fog." *Journal of Scientific Instruments* 16 (1939): 137–140.

Budyko, Mikhail Ivanovitch. *Climatic Changes*. 1974. Translated from the Russian. Washington, D.C.: American Geophysical Union, 1977.

——. "Global Climate Warming and Its Consequence." Blue Planet Prize 1998. http://www.ecology.or.jp/special/9902e.html.

——. "Heat Balance of the Earth and the Problem of Climate Control." Library of Congress, Aerospace Information Division, "Climate Control."

Buesseler, Ken O., Scott C. Doney, David M. Karl, Philip W. Boyd, Ken Caldeira, Fei Chai, Kennieth H. Coale, et al. "Ocean Iron Fertilization: Moving Forward in a Sea of Uncertainty." *Science* 319 (2008): 162.

Bulfinch, Thomas. *The Age of Fable, or Beauties of Mythology*. Boston: Tilton, 1855.

Bundgaard, Robert C., and Richard E. Cale. "Irving P. Krick, 1906–1996." *Bulletin of the American Meteorological Society* 78 (1997): 278–279.

Burke, Albert E. "Influence of Man upon Nature—The Russian View: A Case Study." In William L. Thomas Jr., ed., *Man's Role in Changing the Face of the Earth*, 1035–1051. Chicago: University of Chicago Press, 1956.

Butterfield, Herbert. *The Origins of Modern Science, 1300–1800*. London: Bell, 1949.

Byers, Horace R. "History of Weather Modification." In Wilmot N. Hess, ed., *Weather and Climate Modification*, 3–44. New York: Wiley, 1974.

Caldeira, Ken, and Lowell Wood. "Global and Arctic Climate Engineering: Numerical Model Studies." *Philosophical Transactions of the Royal Society* A 366 (2008): 4039–4056.

Caldwell, Martha B. "Some Kansas Rain Makers." *Kansas Historical Quarterly* 7 (1938): 306–324.

Canada. House of Commons. *Official Report of Debates* (March 26, 1906), 74: 560–567. 10th Parliament, 2nd sess. Ottawa.

Carpenter, Ford Ashman. "Alleged Manufacture of Rain in Southern California." *Monthly Weather Review*, August 1918, 376–377.

Catlin, George. *North American Indians*. 2 vols. Edinburgh: Grant, 1909; Scituate, Mass.: DSI Digital Reproduction, 2000.

Cerveny, Randall S. *Freaks of the Storm: From Flying Cows to Stealing Thunder: The World's Strangest True Weather Stories*. New York: Thunder's Mouth Press, 2006.

Chaffee, E. Leon. "Second Report on Dust Charging." September 1, 1921. Typescript in Bancroft Papers.

Chamorro, Susana Pimiento, and Edward Hammond. *Addressing Environmental Modification in Post–Cold War Conflict*. Occasional Papers. Edmonds, Wash.: Edmonds Institute, 2001. http://www.edmonds-institute.org/pimiento.html.

Chang, David B., and I-Fu Shih. "Stratospheric Welsbach Seeding for Reduction of Global Warming." U.S. Patent 5,003,186, March 26, 1991.

The Changing Atmosphere: Implications for Global Security [in English and French]. Conference proceedings, June 27–30, 1988. Geneva: World Meteorological Organization, 1989. http://www.cmos.ca/ChangingAtmosphere1988e.pdf.

Changnon, Stanley A., Jr. "The Paradox of Planned Weather Modification." *Bulletin of the American Meteorological Society* 56 (1975): 27–37.

Chapman, Sydney. "The Gases of the Atmosphere." Presidential Address. *Quarterly Journal of the Royal Meteorological Society* 60 (1934): 127–142.

Chary, Henry A. *A History of Air Weather Service Weather Modification, 1965–1973*. Air Weather Service Technical Report 247. Scott Air Force Base, Ill.: Military Airlift Command, 1974.

Christofilos, N. C. "The Argus Experiment." *Journal of Geophysical Research* 64 (1959): 869.

Cicerone, Ralph J. "Geoengineering: Encouraging Research and Overseeing Implementation." *Climatic Change* 77 (2006): 221–226.

Cicerone, Ralph J., Scott Elliott, and Richard P. Turco. "Global Environmental Engineering." *Nature* 356 (1992): 472.

———. "Reduced Antarctic Ozone Depletion in a Model with Hydrocarbon Injections." *Science* 254 (1991): 1191–1194.

Clarke, Arthur C. "Man-Made Weather." *Holiday*, May 1957, 74.

Clarke, F. W. "An Ode to Pluviculture; or, The Rhyme of the Rain Machine." *Life*, November 5, 1891, 260. http://www.history.noaa.gov/art/rainmachine.html.

Clement, Hal. *The Nitrogen Fix*. New York: Ace Books, 1980.

"Cloudbuster." http://www.youtube.com/watch?v=8L3tKddZFic.

"Clouds Sprayed with Dry Ice Made to Yield Rain." *Popular Mechanics*, September 1930, 418.

Cobb, James T., Jr., Sheldon D. Elliot Jr., H. E. Cronin, Clifford D. Kern, and W. A. Livingston. *Project Popeye: Final Report*. China Lake, Calif.: Naval Weapons Center, 1967.

Cohen, I. Bernard. *Revolution in Science*. Cambridge, Mass.: Harvard University Press, 1985.

Cohen, William S. "DoD News Briefing." April 28, 1997. http://www.defenselink.mil/transcripts/transcript.aspx?transcriptid=674.

Cohn, Carol. "Sex and Death in the Rational World of Defense Intellectuals." *Signs* 12 (1987): 687–718.

Coleridge, Samuel Taylor. *The Rime of the Ancient Mariner*. 1798. New York: Caldwell, 1889.

Contributions to the History of Dien Bien Phu. Vietnamese Studies, no. 3. Hanoi: Xunhasaba, March 1965.

Conway, Erik. "The World According to GARP: Scientific Internationalism and the Construction of Global Meteorology, 1961–1980." In Margaret Vining and Barton C. Hacker, eds., *Science in Uniform, Uniforms in Science: Historical Studies of American Military and Scientific Interactions*, 131–147. Lanham, Md.: Scarecrow Press, 2007.

"Controlled Weather." *Bulletin of the American Meteorological Society* 15 (1934): 205.

Cook, William Wallace. *The Eighth Wonder: Working for Marvels*. New York: Street and Smith, 1907.

Cotton, William R. "Weather and Climate Engineering." In Jost Heintzenberg and Robert J. Charlson, eds., *Clouds in the Perturbed Climate System: Their Relationship to Energy Balance, Atmospheric Dynamics, and Precipitation*, 339–367. Strugmann Forum Report, vol. 2. Cambridge, Mass.: MIT Press, 2009.

Croy, Homer. "The Rainmakers." *Harper's*, September 1946, 213–220.

Crutzen, Paul J. "Albedo Enhancement by Stratospheric Sulfur Injections: A Contribution to Resolve a Policy Dilemma? An Editorial Essay." *Climatic Change* 77 (2006): 211–220.

——. "The Influence of Nitrogen Oxides on the Atmospheric Ozone Content." *Quarterly Journal of the Royal Meteorological Society* 96 (1970): 320–325.

"Current Notes." *American Meteorological Journal* 11 (1894): 192.

Curry, J. A., P. J. Webster, and G. J. Holland. "Mixing Politics and Science in Testing the Hypothesis That Greenhouse Warming Is Causing a Global Increase in Hurricane Intensity." *Bulletin of the American Meteorological Society* 87 (2006): 1025–1036.

Curtis, George E. "Rain-Making in Texas." *Nature* 44 (1891): 594.

Dando, William A. "Budyko, Mikhail Ivanovitch." In John E. Oliver, ed., *Encyclopedia of World Climatology*, 179–180. Dordrecht: Springer, 2005.

Dante Alighieri. *The Divine Comedy*. 1321. Translated by Henry F. Cary. The Harvard Classics 20. New York: Collier, 1909. http://www.literaryaccess.com/20.

Dessens, Jean. "Man-made Tornadoes." *Nature* 193 (1962): 13–14.

Devereaux, W. C. "A Meteorological Service of the Future." *Bulletin of the American Meteorological Society* 29 (1939): 212–221.

Doel, Ronald E., and Kristine C. Harper. "Prometheus Unleashed: Science as a Diplomatic Weapon in the Lyndon B. Johnson Administration." *Osiris* 21 (2006): 66–85.

"Donald Duck, Master Rain Maker." *Walt Disney's Comics and Stories*. Poughkeepsie, N.Y.: K. K. Publications [Dell Comics], September, 1953.

Doubleday D. C. "Air Force Activities in Cloud Physics Research Program." March 31, 1948. Records of the Defense Research and Development Board, National Archives.

Doyle, Sir Arthur Conan. *A Study in Scarlet*. 1887; New York: Penguin, 2001.

Droessler, Earl G. *Federal Government Activities in Weather Modification and Related Cloud Physics*. Final Report of the United States Advisory Committee on Weather Control. Vol. 2. Howard T. Orville, chairman. Washington, D.C., 1957.

Dyer, Gwynne. *Climate Wars*. Toronto: Random House Canada, 2008.

Dyrenforth, Robert St. George. *Report of the Agent of the Department of Agriculture for Making Experiments in the Production of Rainfall*. U.S. Senate, Ex. Doc. 45, 52nd Cong., 1st sess., 1892.

Dyson, Freeman J. "Can We Control the Carbon Dioxide in the Atmosphere?" *Energy* 2 (1977): 287–291.

Dyson, Freeman J., and Gregg Marland. "Technical Fixes for the Climatic Effects of CO_2." In W. P. Elliott and Lester Machta, eds., *Workshop on the Global Effects of Carbon Dioxide from Fossil Fuels*, 111–118. Rep. CONF-770385. Washington D.C.: Department of Energy, 1979.

Early, James T. "Space-Based Solar Shield to Offset Greenhouse Effect." *Journal of the British Interplanetary Society* 42 (1989): 567–569.

Eichhorn, Noel D. *Implications of Rising Carbon Dioxide Content of the Atmosphere*. New York: Conservation Foundation, 1963.

Ekholm, Nils. "On the Variations of the Climate of the Geological and Historical Past and Their Causes." *Quarterly Journal of the Royal Meteorological Society* 27 (1901): 1–61.

Elliott, Robert D. "Experience of the Private Sector." In Wilmot N. Hess, ed., *Weather and Climate Modification*, 45–89. New York: Wiley, 1974.

Emanuel, Kerry. *What We Know About Climate Change*. Cambridge, Mass.: MIT Press, 2007. [Originally published as "Phaeton's Reins: The Human Hand in Climate Change." *Boston Review*, January–February 2007. http://bostonreview.net/BR32.1/emanuel.php]

Emerson, Ralph Waldo. *The Conduct of Life*. 1860. Vol. 6 of *The Works of Ralph Waldo Emerson*. Boston: Fireside Edition, 1909.

England, George Allan. *The Air Trust*. St. Louis: Wagner, 1915.

Espy, James Pollard. *Fourth Meteorological Report*. U.S. Senate, Ex. Doc. 65, 34th Cong., 3rd sess., 1857.

——. *The Philosophy of Storms*. Boston: Little, Brown, 1841.

——. *Second Report on Meteorology to the Secretary of the Navy*. U.S. Senate, Ex. Doc. 39, 31st Cong., 1st sess., 1851.

——. *Third Report on Meteorology, with Directions for Mariners, etc.* U.S. Senate, Ex. Doc. 39, 31st Cong., 1st sess., 1851.

European Space Agency. "Harry Wexler: The Father of Weather Satellites." http://www.esa.int.

"A Fake Rainmaker." *Monthly Weather Review*, February 1906, 84–85.

Farm Implement News, September–October 1891. Clippings in NOAA Central Library, Silver Spring, Md.

Fedorov, Ye. K. "Modification of Meteorological Processes." In Wilmot N. Hess, ed., *Weather and Climate Modification*, 387–409. New York: Wiley, 1974.

"FIDO Conference Program." May 30, 1945. Callendar Papers.

Fitzgerald, William G. "At War with the Clouds." *Appleton's Magazine*, July–December 1905, 629–636.

Five Ways to Save the World. Produced by Jonathan Barker. Directed by Jonathan Barker, Cecilia Hue, and Anna Abbott. London: British Broadcasting Corporation, 2006.

Flannery, Brian P., Haroon Keshgi, Gregg Marland, and Michael C. MacCracken. "Geoengineering Climate." In Robert G. Watts, ed., *Engineering Response to Global Climate Change*, 379–427. Boca Raton, Fla.: CRC Press, 1997.

Fleagle, Robert G. "Second Opinions on 'Pathological Science.'" *Physics Today*, March 1990, 110.

Fleming, James Rodger. *The Callendar Effect: The Life and Work of Guy Stewart Callendar (1898–1964), the Scientist Who Established the Carbon Dioxide Theory of Climate Change*. Boston: American Meteorological Society, 2007.

——. "The Climate Engineers: Playing God to Save the Planet." *Wilson Quarterly* 31 (2007): 46–60.

———. "Distorted Support: Pathologies of Weather Warfare." In Margaret Vining and Barton C. Hacker, eds., *Science in Uniform, Uniforms in Science: Historical Studies of American Military and Scientific Interactions*, 49–60. Lanham, Md.: Scarecrow Press, 2007.

———. "Fixing the Weather and Climate: Military and Civilian Schemes for Cloud Seeding and Climate Engineering." In Lisa Rosner, ed., *The Technological Fix: How People Use Technology to Create and Solve Problems*, 175–200. New York: Routledge, 2004.

———. "Gilbert N. Plass: Climate Science in Perspective." *American Scientist* 98 (2010): 60–61.

———. "Global Climate Change and Human Agency: Inadvertent Influence and 'Archimedean' Interventions." In James Rodger Fleming, Vladimir Jankovic, and Deborah R. Coen, eds., *Intimate Universality: Local and Global Themes in the History of Weather and Climate*, 223–248. Sagamore Beach, Mass.: Science History Publications/USA, 2006.

———. *Historical Perspectives on Climate Change*. New York: Oxford University Press, 1998.

———. "Meteorology." In Brian S. Biagre, ed., *A History of Modern Science and Mathematics*, 3:184–217. New York: Scribner, 2002.

———. *Meteorology in America, 1800–1870*. Baltimore: Johns Hopkins University Press, 1990.

———. "The Pathological History of Weather and Climate Modification: Three Cycles of Promise and Hype." *Historical Studies in the Physical Sciences* 37 (2006): 3–25.

———. "Storms, Strikes, and Surveillance: The U.S. Army Signal Office, 1861–1891." *Historical Studies in the Physical and Biological Sciences* 30 (2000): 315–332.

———. "Sverre Petterssen, the Bergen School, and the Forecasts for D-Day." *History of Meteorology* 1 (2004): 75–83. http://www.meteohistory.org.

———. "What Counts as Knowledge?" Paper presented at the Asilomar International Conference on Climate Intervention Technologies, Monterey, Calif., March 22–26, 2010, http://www.colby.edu/2010asilomar.

Fleming, James Rodger, and Vladimir Jankovic, ed. *Klima*. *Osiris* 26 (2011).

Fletcher, Joseph O. *Changing Climate*. RAND Report No. P-3933. Santa Monica, Calif.: RAND Corporation, 1968.

———. *Managing Climatic Resources*. RAND Report No. P-4000-1. Santa Monica, Calif.: RAND Corporation, 1969.

"Fog Broom." *Time*, July 30, 1934. http://www.time.com.

Fogg, Martyn J. *Terraforming: Engineering Planetary Environments*. Warrendale, Pa.: Society of Automotive Engineers, 1995.

Franklin, William S. "A Much-Needed Change of Emphasis in Meteorological Research." *Monthly Weather Review*, October 1918, 449–453.

———. "Weather Control." *Science* 14 (1901): 496–497.

———. "Weather Prediction and Weather Control." *Science* 68 (1928): 377–378.

Frazer, James George. *The Golden Bough: A Study in Magic and Religion*. London: Macmillan, 1890.

———. "Some Popular Superstitions of the Ancients." In *Garnered Sheaves: Essays, Addresses, Reviews*. London: Macmillan, 1931.

"From the Footlights." May 2003. http://www.footlightsdc.org/news2003/may03.htm.

Fuller, John F. *Air Weather Service Support to the United States Army: Tet and the Decade After*. Air Weather Service Historical Study 8. Scott Air Force Base, Ill.: Military Airlift Command, 1979.

———. *Thor's Legions: Weather Support to the US Air Force and Army, 1937–1987*. Boston: American Meteorological Society, 1990.

———. *Weather and War*. Scott Air Force Base, Ill.: Military Airlift Command, 1974.

Gall, Alexander. *Das Atlantropa-Projekt: Die Geschichte einer gescheiterten Vision: Hermann Sörgel und die Absenkung des Mittelmeeres*. Frankfurt: Campus, 1998.

Gal'tsov, A. P. "Conference on Problems of Climate Control." Library of Congress, Aerospace Information Division, "Climate Control."

Gardiner, Stephen. "Ethics and Global Climate Change." *Ethics* 114 (2004): 555–600.

Gaskell, Thomas Frohock. *The Gulf Stream*. London: Cassell, 1972.

Gathman, Louis. "Method of Producing Rainfall." U.S. Patent 462,795, November 10, 1891.

———. *Rain Produced at Will*. Chicago, 1891.

General Electric Corporation. *56th Annual Report and Yearbook*. Schenectady, N.Y., 1947.

Gentry, R. Cecil. "Hurricane Modification." In Wilmot N. Hess, ed., *Weather and Climate Modification*, 497–521. New York: Wiley, 1974.

Gillispie, Charles C. *Science and Polity in France at the End of the Old Regime*. Princeton, N.J.: Princeton University Press, 1980.

Glacken, Clarence J. *Traces on the Rhodian Shore: Nature and Culture in Western Thought from Ancient Times to the End of the Eighteenth Century*. Berkeley: University of California Press, 1967.

Goldblat, Jozef. "The Environmental Convention of 1977: An Analysis." In Arthur H. Westing, ed., *Environmental Warfare: A Technical, Legal, and Policy Appraisal*, 53–64. Philadelphia: Taylor and Francis, 1984.

Goldston, Eli. "Legal Entanglements for the Rain-maker." *Case and Comment* 54 (1949): 3–6.

"Goodbye Chemtrails, Hello Blue Skies! The Do-It-Yourself Kit for Sky Repair." http://www.angelfire.com/ak5/energy21/cloudbuster.htm.

Goodstein, Judith. "Tales in and out of 'Millikan's School.'" Pauling Symposium, Oregon State University, Corvallis, 1995. http://oregonstate.edu/dept/Special_Collections/subpages/ahp/1995symposium/goodstei.html.

"Government Rainmaking." *Nation*, October 22, 1891, 309–310.

Gratacap, Louis P. *The Evacuation of England: The Twist in the Gulf Stream*. New York: Bretano, 1908.

Gregg, Willis R. "Address at the Dedication Ceremonies of the Air Conditioned House at the Century of Progress, at Chicago, Ill., on July 11, 1934." Copy in Records of the U.S. Weather Bureau.

Griffith, George. *The Great Weather Syndicate*. London: Bell, 1906.

———. *The World Peril of 1910*. London: White, 1907.

Grunow, Johannes. "Der Künstliche Regen des Holländers Veraart." *Die Umschau* 34 (1930): 602.

Guerra, Maria Eugenia. "Stormy Weather, Maelstrom of Opinions, Soliloquies of Sarcasm, and Chicanery Hallmark Nix for Provaqua Project." *Laredos: A Journal of the Borderlands* 9 (2003): 6–10.

Hacker, Barton C. "Military Patronage and the Geophysical Sciences in America: An Introduction." *Historical Studies in the Physical and Biological Sciences* 30 (2000): 1–5.

Hahn, Charles Curtz. *The Wreck of the South Pole, or the Great Dissembler*. New York: Street and Smith, 1899.

Hansen, James. "The Tipping Point?" *New York Review of Books*, January 12, 2006. http://www.nybooks.com/articles/18618.

Harper, Kristine C. *Weather by the Numbers: The Genesis of Modern Meteorology*. Cambridge, Mass.: MIT Press, 2008.

Harrington, Mark W. "Weather Making, Ancient and Modern." *National Geographic*, April 25, 1894, 35–62.

Havens, Barrington S., comp. *History of Project Cirrus*. G.E. Report No. RL-756. Schenectady, N.Y.: Research Publication Services, The Knolls, 1952.

Havens, Barrington S., James E. Juisto, and Bernard Vonnegut. *Early History of Cloud Seeding*. Socorro: Langmuir Laboratory, New Mexico Institute of Mining and Technology; Atmospheric Sciences Research Center, State University of New York at Albany; and Research and Development Center, General Electric Company, 1978.

Hawes, Joseph M. "The Signal Corps and Its Weather Service, 1870–1890." *Military Affairs* 30 (1966): 68–76.

Hawthorne, Nathaniel. "The Great Carbuncle: A Mystery of the White Mountains." 1837. In *The Celestial Railroad and Other Stories*, 127–142. New York: New American Library, 1963.

——. "The Hall of Fantasy." 1843. In *Mosses from an Old Manse*, 1:199–215. Boston: Ticknor and Fields, 1854.

"Hearing on Senate Resolution 281." *Bulletin of the American Meteorological Society* 53 (1972): 1185–1191.

Hering, Daniel W. *Foibles and Fallacies of Science: An Account of Scientific Vagaries*. New York: Van Nostrand, 1924.

——. "Weather Control." *Scientific Monthly*, February 1922, 178–185.

Hoffman, Ross N. "Controlling the Global Weather." *Bulletin of the American Meteorological Society* 83 (2002): 241–248.

——. "Controlling Hurricanes." *Scientific American*, October 2004, 68–75.

Hoffman, Ross N., Mark Leidner, and John Henderson. "Controlling the Global Weather." NASA Institute for Advanced Concepts report, ca. 2001. www.niac.usra.edu/files/library/meetings/annual/jun02/715Hoffman.pdf.

Hosler, Charles. "Weather Modification and Science and Government." Typescript and personal communication, March 22, 2005.

Houghton, Henry G., and W. H. Radford. *On the Local Dissipation of Natural Fog*. Papers in Physical Oceanography and Meteorology 6, no. 3. Cambridge, Mass.: MIT and Woods Hole Oceanographic Institution, 1938.

Hoyle, Fred, and Raymond A. Lyttleton. "The Effect of Interstellar Matter on Climatic Variation." *Proceedings of the Cambridge Philosophical Society* 35 (1939): 405–415.

Humphreys, William Jackson. *Physics of the Air*. Philadelphia: Lippincott, 1920.

——. *Rain Making and Other Weather Vagaries*. New York: AMS Press, 1926.

——. "Volcanic Dust and Other Factors in the Production of Climatic Changes, and Their Possible Relation to Ice Ages." *Journal of the Franklin Institute* 176 (1913): 131–172.

Intergovernmental Panel on Climate Change. "IPCC History." http://www.ipccfacts.org/history.html.

"Interim Report on Artificially Induced Precipitation." April 21, 1947. Records of the Defense Research and Development Board, National Archives.

Izrael, Yu., V. M. Zakharov, N. N. Petrov, A. G. Ryaboshapko, V. N. Ivanov, A. V. Savchenko, Yu. V. Andreev, Yu. A. Puzov, B. G. Danelyan, and V. P. Kulyapin. "Field Experiment on Studying Solar Radiation Passing Through Aerosol Layers." *Russian Meteorology and Hydrology* 34 (2009): 265–273. http://www.springerlink.com/content/l4n10470500l3048/

Jacoby, Henry D. "Climate Favela." In Joseph E. Aldy and Robert N. Steavins, eds., *Architectures for Agreement: Addressing Global Climate Change in the Post-Kyoto World*, 270–279. Cambridge: Cambridge University Press, 2007.

Jankovic, Vladimir. *Reading the Skies: A Cultural History of English Weather, 1650–1820*. Chicago: University of Chicago Press, 2001.

Jenkins, Gary. *The Wizard of Sun City: The Strange True Story of Charles Hatfield, the Rainmaker Who Drowned a City's Dreams*. New York: Thunder's Mouth Press, 2005.

Johnson, R.G. "Climate Control Requires a Dam at the Strait of Gibraltar." *EOS: Transactions of the American Geophysical Union* 78 (1997): 277, 280–281.

Jordan, David Starr. "The Art of Pluviculture." *Science* 62 (1925): 81–82.

Juda, Lawrence. "Negotiating a Treaty on Environmental Modification Warfare: The Convention on Environmental Warfare and Its Impact upon Arms Control Negotiations." *International Organization* 32 (1978): 975–991.

Kael, Pauline. "Reviews A–Z," s.v. "The Rainmaker." http://www.geocities.com/pauline kaelreviews.

Kant, Immanuel. *Critique of Pure Reason*. 1781. Translated by F. Max Müller. London: Macmillan, 1881.

———. *Grounding for the Metaphysics of Morals*. 1785. Translated by James W. Ellington. Indianapolis: Hackett, 1994.

Keith, David. "Engineering the Earth System." American Geophysical Union, fall meeting, 2005. Abstract No. U54A-02.

———. "Geoengineering the Climate: History and Prospect." *Annual Review of Energy and the Environment* 25 (2000): 245, 269–280.

———. "Why Capture CO_2 from the Atmosphere?" *Science* 325 (2009): 1654–1655.

Keith, David, and Hadi Dowlatabadi. "A Serious Look at Geoengineering." *EOS: Transactions of the American Geophysical Union* 73 (1992): 289, 292–293.

Kellogg, William W. "Review of Saturn High Water Experiment." Memorandum to Robert F. Fellows, NASA, February 19, 1962. Copy in Wexler Papers, box 18.

Kellogg, William W., and Stephen H. Schneider. "Climate Stabilization: For Better or Worse?" *Science* 186 (1974): 1163–1172.

Kennedy, John F. "Address to the United Nations General Assembly, Sept. 25, 1961." http://www.jfklibrary.org.

King, D. Whitney. "Can Adding Iron to the Oceans Reduce Global Warming? An Example of Geoengineering." In James Rodger Fleming and Henry A. Gemery, eds., *Technology and the Environment: Multidisciplinary Perspectives*, 112–135. Akron: Akron University Press, 1994.

Kintisch, Eli. "Giving Climate Change a Kick." *ScienceNOW*, November 9, 2007, 1.

———. "Tinkering with the Climate to Get Hearing at Harvard Meeting." *Science* 318 (2007): 551.

Kobler, John. "Stormy Sage of Weather Control." *Saturday Evening Post*, December 1, 1962, 68–70.

Kramer, Harris P., and Malcolm Rigby. "A Selective and Annotated Bibliography on Cloud Physics and 'Rain Making.'" *Meteorological Abstracts and Bibliography* 1 (1950): 191–193.

Krook, Max. "Interstellar Matter and the Solar Constant." In Harlow Shapley, ed., *Climatic Change: Evidence, Causes, and Effects*, 143–146. Cambridge, Mass.: Harvard University Press, 1953.

Lackner, Klaus S. "Capture of Carbon Dioxide from Ambient Air." *European Physical Journal, Special Topics* 176 (2009): 93–106.

——. "Submission from Professor Klaus S. Lackner, Columbia University." Memorandum 166. United Kingdom, House of Commons. Innovation, Universities, and Skills Committee (November 2008). http://www.publications.parliament.uk.

Lamb, Hubert H. "Climate-Engineering Schemes to Meet a Climatic Emergency." *Earth-Science Reviews* 7 (1971): 87–95.

Lambright, W. Henry. *Weather Modification: The Politics of an Emergent Technology*. Syracuse, N.Y.: Inter-University Case Program, 1970.

Lampitt, Richard, E. P. Achterberg, T. R. Anderson, J. A. Hughes, M. D. Iglesias-Rodriguez, B. A. Kelly-Gerreyn, M. Lucas, et al. "Ocean Fertilization: A Potential Means of Geoengineering?" *Philosophical Transactions of the Royal Society* A 366 (2008): 3919–3945.

Landsberg, Helmut E. "Memorandum for the Record—Briefing on Weather Control." November 5, 1951. Records of the Defense Research and Development Board, National Archives.

Langmuir, Irving. "The Growth of Particles in Smokes and Clouds and the Production of Snow from Supercooled Clouds." *Proceedings of the American Philosophical Society* 92 (1948): 167–185.

——. "Pathological Science." Original lecture note cards (December 18, 1953) and sound recording (March 8, 1954). Langmuir Papers. [Published as *Pathological Science*. Edited by Robert N. Hall. G.E. Report No. 68-C-035. Schenectady, N.Y.: General Electric, Research and Development Center, 1968. Reprinted as "Pathological Science." *Physics Today* 42 (1989): 36–48. http://www.cs.princeton.edu/~ken/Langmuir/langmuir.htm]

——. "The Production of Rain by a Chain Reaction in Cumulus Clouds at Temperatures Above Freezing." *Journal of Meteorology* 5 (1948): 175–192.

——. "Report on Evaluation of Past and Future Experiments on Widespread Control of Weather." 1955. Advisory Committee on Weather Control. Copy in Schaefer Papers.

——. "A Seven Day Periodicity in Weather in the U.S. During April, 1950." *Bulletin of the American Meteorological Society* 31 (1950): 386–387.

——. *Summary of Results Thus Far Obtained in Artificial Nucleation of Clouds*. Final Report: Project Cirrus. G.E. Report No. RL-140. Schenectady, N.Y., 1948.

Latham, John, Philip Rasch, Chih-Chieh Chen, Laura Kettles, Alan Gadian, Andrew Gettelman, Hugh Morrison, Keith Bower, and Tom Choularton. "Global Temperature Stabilization via Controlled Albedo Enhancement of Low-Level Maritime Clouds." *Philosophical Transactions of the Royal Society* A 366 (2008): 3969–3987.

Launder, Brian, and J. Michael T. Thompson, eds. "Geoscale Engineering to Avert Dangerous Climate Change." Special issue, *Philosophical Transactions of the Royal Society* A 366 (2008).

Le Maout, Charles. *Effets du canon et du son ses cloches sur l'atmosphère*. Saint-Brieuc, 1861.

——. *Encore le canon et la pluie*. Saint-Brieuc, 1870.

——. *Lettre à M. Tremblay sur les moyens proposes pour faire cesser la sécheresse des six premiers mois de l'année 1870*. Edited by Emile Le Maout. Cherbourg, 1891.

Lenin, V. I. *Materialism and Empirio-Criticism: Critical Comments on a Reactionary Philosophy*. 1908. http://www.marxists.org.

Leslie, Eliza. "The Rain King, or, A Glance at the Next Century." *Godey's Lady's Book* 25 (1842): 7–11.

Liebling, A. J. *The Honest Rainmaker: The Life and Times of Colonel John R. Stingo*. San Francisco: North Point Press, 1989. [Originally published as "Profiles: Yea Verily," I–III. *New Yorker*, September 13–27, 1952]

——. *Just Enough Liebling: Classic Work by the Legendary "New Yorker" Writer*. New York: North Point Press, 2004.

Lodge, Oliver. "Electrical Precipitation." In *Physics in Industry*, 3:15–34. London: Oxford University Press, 1925.

Lomax, Benjamin. *Bells and Bellringers*. London: Infield, 1879.

Lovell, Bernard, and Martin Ryle. "Interference to Radio Astronomy from Belts of Orbiting Dipoles (Needles)." *Quarterly Journal of the Royal Astronomical Society* 3 (1962): 100–108.

Lovelock, James. "A Geophysiologist's Thoughts on Geoengineering." *Philosophical Transactions of the Royal Society* A 366 (2008): 3883–3890.

——. *The Revenge of Gaia: Earth's Climate in Crisis and the Fate of Humanity*. New York: Penguin, 2006.

——. *The Vanishing Face of Gaia: A Final Warning*. New York: Basic, 2009.

Lovelock, James E., and Chris G. Rapley. "Ocean Pipes Could Help the Earth to Cure Itself." *Nature* 449 (2007): 403.

Luginbuhl, Christian B., Constance E. Walker, and Richard J. Wainscoat. "Lighting and Astronomy." *Physics Today* 62 (2009): 32–37.

MacCracken, Michael C. "Beyond Mitigation: Potential Options for Counterbalancing the Climatic and Environmental Consequences of the Rising Concentrations of Greenhouse Gases." December 2008. Typescript prepared for the World Bank.

——. "Geoengineering the Climate." Report UCRL-JC-108014. Lawrence Livermore National Laboratory, Livermore, Calif., 1991.

——. "On the Possible Use of Geoengineering to Moderate Specific Climate Change Impacts." *Environmental Research Letters* 4 (2009): 045107.

——. "Working Toward International Agreement on Climate Protection." Paper presented at the Mid-Maine Global Forum, Waterville, Maine, September 25, 2009.

MacDonald, Gordon J. F. "How to Wreck the Environment." In Nigel Calder, ed., *Unless Peace Comes*, 181–205. New York: Viking, 1968.

——. "Statement." In House Committee on Foreign Affairs, Subcommittee on International Organizations. *Prohibition of Weather Modification as a Weapon of War: Hearings on H.R. 28*. 94th Cong., 1st sess., 1975.

Mackenzie, Donald. *The Flooding of the Sahara*. London: Sampson Low, Marston, Searle, and Rivington, 1877.

Manabe, Syukuro, and F. Möller. "On the Radiative Equilibrium and Balance of the Atmosphere." *Monthly Weather Review*, December 1961, 503–532.

Mann, Martin. "War Against Hail." Manuscript included in Mann to John von Neumann, April 10, 1947. Von Neumann Papers, box 15.

"Many Legal Entanglements Forecast for Man in New Role as Rain-Maker." *Harvard Law School Record*. Ca. 1947. Clipping in Records of the Defense Research and Development Board, National Archives.

Marchetti, Cesare. "On Geoengineering and the CO_2 Problem." *Climatic Change* 1 (1977): 59–68.

Markin, Arkadii Borisovich. *Soviet Electric Power: Developments and Prospects*. Moscow: Foreign Languages Publishing, 1956.

Marshall, Eliot. "The Greening of the National Labs." *Science* 260 (1993): 25.

Martin, John H. "Glacial–Interglacial CO_2 Change: The Iron Hypothesis." *Paleoceanography* 5 (1990): 1–13.

Martin, John H., and Steve E. Fitzwater. "Iron Deficiency Limits Phytoplankton Growth in the Northeast Pacific Subarctic." *Nature* 331 (1988): 341–343.

Martin, John H., R. Michael Gordon, and Steve E. Fitzwater. "The Case for Iron." *Limnology and Oceanography* 36 (1991): 1793–1802.

McAdie, Alexander. "The Control of Fog." *Scientific Monthly*, July 1931, 28–36.

——. "Natural Rain-makers." *Popular Science Monthly*, September 1895, 642–648. [Reprinted in *Alexander McAdie: Scientist and Writer*, compiled by Mary R. B. McAdie, 76–82. Charlottesville, Va., 1949]

McFadden, Manus. "Is Rainmaking Riddle Solved?" *Popular Science Monthly*, May 1923, 29–30. Copy in Bancroft Papers.

[McGavin, Darren.] "Darren's Theatre Page." http://www.darrenmcgavin.net/darren's_theatre_page.htm.

McNeill, William H. *The Pursuit of Power: Technology, Armed Force, and Society Since A.D. 1000*. Chicago: University of Chicago Press, 1982.

Mendelsohn, Everett. "Science, Scientists, and the Military." In John Kriege and Dominique Pestre, eds., *Science in the Twentieth Century*, 175–202. Amsterdam: Harwood Academic, 1997.

Merchant, Carolyn. *The Death of Nature: Women, Ecology, and the Scientific Revolution*. 1980. Reprint, San Francisco: Harper & Row, 1989.

Mergen, Bernard. *Weather Matters: An American Cultural History Since 1900*. Lawrence: University Press of Kansas, 2008.

Meyer, William B. *Americans and Their Weather*. New York: Oxford University Press, 2000.

Milton, John. *Paradise Lost. A Poem in Twelve Books*. 2d rev. ed. London: Simmons, 1674. http://www.dartmouth.edu/~milton/reading_room/pl/book_1/index.shtml.

Mindling, George W. "The Raymete and the Future." March 29, 1939. http://www.history.noaa.gov.

"Miracle." *Time*, November 10, 1924. http://www.time.com.

Molina, Mario J., and Frank S. Rowland. "Stratospheric Sink for Chlorofluoromethanes: Chlorine Atom–Catalyzed Destruction of Ozone." *Nature* 249 (1974): 810–812.

Moore, Sir William. "Famine: Its Effects and Relief." *Transactions of the Epidemiological Society of London*, n.s., 11 (1891–1892): 27–46.
Moore, Willis L. "Fake Rainmaking: A Letter from the Chief of the Bureau." *Monthly Weather Review*, April 1905, 152–153.
Morton, Oliver. "Climate Change: Is This What It Takes to Save the World?" *Nature* 447 (2007): 132–136.
"Motorpool: The Rest of the Story." http://www.awra.us/gallery-jan05.htm.
Munk, W., N. Oreskes, and R. Muller. "Gordon James Fraser MacDonald, 1930–2002: A Biographical Memoir." *Biographical Memoirs of the National Academy of Sciences* 84 (2004): 3–26.
Nash, N. Richard. *The Rainmaker: A Romantic Comedy*. New York: Random House, 1955.
National Academy of Sciences. *Critical Issues in Weather Modification Research*. Washington, D.C.: National Academy Press, 2003.
——. *Energy and Climate: Studies in Geophysics*. Washington, D.C.: National Academy Press, 1977.
——. "Geoengineering Options to Respond to Climate Change: Steps to Establish a Research Agenda." A workshop to provide input to America's Climate Choices suite of activities, Washington, D.C, June 15–16, 2009. http://americasclimatechoices.org/GeoEng Agenda 6-11-09.pdf.
——. *Policy Implications of Greenhouse Warming: Mitigation, Adaptation, and the Science Base*. Washington, D.C.: National Academy Press, 1992.
——. *Weather and Climate Modification: Problems and Progress*. Washington, D.C.: National Academy Press, 1973.
Nebeker, Frederik. *Calculating the Weather: Meteorology in the 20th Century*. San Diego: Academic Press, 1995.
"No Quick or Easy Technological Fix for Climate Change, Researchers Say." UCLA press release, December 17, 2008. http://www.newsroom.ucla.edu.
Nordhaus, William D. "The Challenge of Global Warming: Economic Models and Environmental Policy." 2007. http://nordhaus.econ.yale.edu/dice_mss_091107_public.pdf.
——. "An Optimal Transition Path for Controlling Greenhouse Gases." *Science* 258 (1992): 1315–1319.
"Novel Method of Precipitating Rainfalls." *Scientific American*, November 27, 1880, 342.
Oberth, Hermann. *Die Rakete zu den Planetenräumen*. Munich: Oldenbourg, 1923.
——. *Wege zur Raumschiffahrt*. Munich and Berlin: Oldenbourg, 1929.
Oceanus, January 11, 2008. http://www.whoi.edu/oceanus/viewArticle.do?id=35866.
O'Connor, William Douglas. "The Brazen Android." *Atlantic Monthly*, April 1891, 433–455.
Ogden, R. J. "Fog Dispersal at Airfields." *Weather* 43 (1998): 20–23, 34–38.
Orville, Howard T. "Weather Made to Order?" *Collier's*, May 28, 1954, 25–29.
"OUP Blog," s.v. "hypermiling." Oxford University Press USA. http://blog.oup.com/2008/11/hypermiling/; see also http://www.squidoo.com/ecohacking.
Ovid. *Metamorphoses*. Translated by A. D. Melville. Oxford: Oxford University Press, 1986.
Owning the Weather. Directed by Robert Greene. New York: 4th Row Films, 2009.
Owning the Weather. Written by Josh Rosen. San Francisco: Spine Films, 2005. http://www.youtube.com/watch?v=GsTYbIuCoWI.

Pacala, Stephen, and Robert Socolow. "Stabilization Wedges: Solving the Climate Problem for the Next 50 Years with Current Technologies." *Science* 205 (2004): 968–972.

Patterson, Thomas W. "Hatfield the Rainmaker." *Journal of San Diego History* 16 (1970): 2–27. http://www.sandiegohistory.org/journal/70winter/hatfield.htm.

Penner, S. S., A. M. Schneider, and E. M. Kennedy. "Active Measures for Reducing the Global Climatic Impacts of Escalating CO_2 Concentrations." *Acta Astronautica* 11 (1984): 345–348.

Petterssen, Sverre. *Cloud and Weather Modification: A Group of Field Experiments*. Meteorological Monographs 2, no. 11. Boston: American Meteorological Society, 1957.

——. *Weathering the Storm: Sverre Petterssen, the D-day Forecast, and the Rise of Modern Meteorology*. Edited by James Rodger Fleming. Boston: American Meteorological Society, 2001.

Phillips, Norman A. "The General Circulation of the Atmosphere: A Numerical Experiment." *Quarterly Journal of the Royal Meteorological Society* 82 (1956): 123–164.

Plutarch. *Plutarch's Lives*. Vol. 3. Translated by John Langhorne and William Langhorne. London: Valpy, 1832.

Polybius. *Universal History*. In Michael Curtis, ed., *The Great Political Theories*. Vol. 1, *From Plato and Aristotle to Locke and Montesquieu*. New York: Avon Books, 1961.

Porky the Rain-Maker. Directed by Tex Avery. Looney Tunes. Los Angeles: Leon Schlesinger Productions/Warner Brothers, 1936.

Powers, Edward. "Rain-Making." *Science* 19 (1892): 52–53.

——. *War and the Weather, or, the Artificial Production of Rain*. Chicago, 1871; Delavan, Wis., 1890.

President's Science Advisory Committee. *Restoring the Quality of Our Environment. Report of the Environmental Pollution Panel*. Appendix Y, *Atmospheric Carbon Dioxide*, 111–133. Washington, D.C.: Executive Office of the President, 1965.

Price, Saul, and Jack C. Pales. "Mauna Loa Observatory: The First Five Years." *Monthly Weather Review*, October 1963, 665–680.

"Project Cirrus." *Bulletin of the American Meteorological Society* 31 (1950): 286–287.

"Project Cirrus: The Story of Cloud Seeding." *G.E. Review*, November 1952, 8–26.

"The Question of Global Warming: An Exchange." *New York Review of Books*, September 25, 2008. http://www.nybooks.com/articles/21811.

Quinn, Daniel. *Ishmael*. New York: Bantam, 1992.

"Rainmaker Fails." *Monthly Weather Review*, December 1924, 591.

"The Rain Makers." *Life*, April 5, 1923, 24. American Periodical Series online. Copy in Bancroft Papers.

"Rain Making Ineffective." *Science News-Letter*, December 11, 1948, 375.

Rasch, Philip J., Simone Tilmes, Richard P Turco, Alan Robock, Luke Oman, Chih-Chieh (Jack) Chen, Georgiy L. Stenchikov, and Rolando R. Garcia. "An Overview of Geoengineering of Climate Using Stratospheric Sulphate Aerosols." *Philosophical Transactions of the Royal Society* A 366 (2008): 4007–4037.

Red Green Show. "Rain Man," season 15, episode 297. http://www.redgreen.com, episode online at http://youtube.com.

Reichelderfer, Francis W. Letter to the editor. *Fortune*, March 1948, 38.

"Review of Savannah Hurricane (Oct. 10–16, 1947), Exposure of a Colossal Meteorological Hoax." Reichelderfer Subject Files, Records of the U.S. Weather Bureau, National Archives.

Riker, Carroll Livingston. *Conspectus of Power and Control of the Gulf Stream: With Elaborated Plan*. Brooklyn, N.Y., 1913.

Roberts, Elizabeth Madox. *Jingling in the Wind*. New York: Viking, 1928.

Robinson, Kim Stanley. *Red Mars*. New York: Bantam, 1993.

——. *Green Mars*. New York: Bantam, 1994.

——. *Blue Mars*. New York: Bantam, 1996.

Robock, Alan. "The Need for Organized Research on Geoengineering Using Stratospheric Aerosols." Paper presented at the meeting of America's Climate Choices, hosted by the National Academy of Sciences, Washington, D.C., June 2009.

——. "20 Reasons Why Geoengineering May Be a Bad Idea." *Bulletin of the Atomic Scientists* 64 (2008): 14–18, 59.

Robock, Alan, Luke Oman, and Georgiy L. Stenchikov. "Regional Climate Responses to Geoengineering with Tropical and Arctic SO_2 Injections." *Journal of Geophysical Research* 113 (2008): D16101.

Rodin, M. B., and D. C. Hess. "Weather Modification." Argonne National Laboratory, Argonne, Ill., 1991. Copy in Wexler Papers, box 12.

Rosenfeld, Albert. *The Quintessence of Irving Langmuir*. Vol. 12 of *The Collected Works of Irving Langmuir*. Oxford: Pergamon, 1966.

Rossi, Paolo. *The Birth of Modern Science*. Translated by Cynthia De Nardi Ipsen. London: Blackwell, 2001.

Royal Academy of Engineering. "Submission." Memorandum 148. United Kingdom, House of Commons. Innovation, Universities, and Skills Committee (September 2008). http://www.publications.parliament.uk.

Royal Society of London. *Geoengineering the Climate: Science, Governance, and Uncertainty*. RS Policy document 10/09. London: Royal Society, 2009.

Ruddiman, William F. *Plows, Plagues, and Petroleum: How Humans Took Control of Climate*. Princeton, N.J.: Princeton University Press, 2005.

Ruggles, Daniel. *Memorial . . . asking an appropriation, to be expended in developing his system of producing rainfall*. U.S. Senate, Misc. Doc. 39, 46th Cong., 2nd sess., 1880.

——. "Method of Precipitating Rainfalls." U.S. Patent 230067, July 13, 1880.

Rusin, Nikolai Petrovich, and Liya Abramovna Flit. *Man Versus Climate*. Moscow: Peace Publishers, 1960.

——. *Methods of Climate Control*. 1962. Translated from the Russian. TT 64–21333. Washington, D.C.: Department of Commerce, Office of Technical Services, Joint Publications Research Services, 1964.

Salter, Stephen, Graham Sortino, and John Latham. "Sea-going Hardware for the Cloud Albedo Method of Reversing Global Warming." *Philosophical Transactions of the Royal Society* A 366 (2008): 3989–4006.

Sanford, Fernando. "Rain-Making." *Popular Science Monthly*, August 1894, 478–491.

Santer, Benjamin D., Karl E. Taylor, Tom M. L. Wigley, Joyce E. Penner, Philip D. Jones, and Ulrich Cubasch. "Towards the Detection and Attribution of an Anthropogenic Effect on Climate." *Climate Dynamics* 12 (1995): 77–100.

Schaefer, Vincent J. "Preliminary Report, Hurricane Study, Project Cirrus, Oct. 15, 1947." Typescript in Schaefer Papers.

———. "The Production of Ice Crystals in a Cloud of Supercooled Water Droplets." *Science* 104 (1946): 459.

Schelling, Thomas C. "Climatic Change: Implications for Welfare and Policy." In National Research Council, *Changing Climate: Report of the Carbon Dioxide Assessment Committee*, 449–482. Washington, D.C.: National Academy Press, 1983.

———. "The Economic Diplomacy of Geoengineering." *Climatic Change* 33 (1996): 303–307.

[Schmidt, Gavin.] "Geo-engineering in Vogue . . ." June 28, 2006. http://www.realclimate.org/index.php/archives/2006/06/geo-engineering-in-vogue.

Schneider, Stephen H. "Earth Systems Engineering and Management." *Nature* 409 (2001): 417–421.

———. "Geoengineering: Could—or Should—We Do It?" *Climatic Change* 33 (1996): 291–302.

———. "Geoengineering: Could We or Should We Make It Work?" *Philosophical Transactions of the Royal Society* A 366 (2008): 3843–3862.

Schwartz, Peter, and Doug Randall. "An Abrupt Climate Change Scenario and Its Implications for United States National Security." 2003. http://www.environmentaldefense.org/documents/3566_AbruptClimateChange.pdf.

The Science of Superstorms: Playing God with the Weather. London: British Broadcasting Corporation/Discovery Channel, 2007.

Seneca. *Naturales Quaestiones*. Translated by Thomas H. Corcoran. Loeb Classical Library 457. Cambridge, Mass.: Harvard University Press, 1972.

Servos, John. "Wilder D. Bancroft (1867–1953)." *Biographical Memoirs of the National Academy of Sciences* 65 (1994): 1–39.

Shachtman, Noah. "NASA Funds Sci-Fi Technology." *Wired*, May 7, 2004. http://www.wired.com/news/technology/0,63362–0.html.

Shaler, Nathaniel. "How to Change the North American Climate." *Atlantic Monthly*, December 1877, 724–731.

Shapin, Steven. *The Scientific Revolution*. Chicago: University of Chicago Press, 1996.

Shapley, Deborah. "Weather Warfare: Pentagon Concedes 7-Year Vietnam Effort." *Science* 184 (1974): 1059–1061.

Shvets, M. Ye. "Control of Climate Through Stratospheric Dusting." Library of Congress, Aerospace Information Division, "Climate Control."

Simpson, Robert H. and Joanne Simpson. "Why Experiment on Tropical Hurricanes?" *Transactions of the New York Academy of Sciences* 28 (1966): 1045–1062.

Sky King. "The Rainbird." Broadcast January 9, 1956. *Sky King* Video 7. http://www.amosnandyvideo.com/skykingx.htm.

Smagorinsky, Joseph. "The Beginnings of Numerical Weather Prediction and General Circulation Modeling: Early Recollections." *Advances in Geophysics* 25 (1983): 3–37.

Smetacek, Victor, and S. W. A. Naqvi. "The Next Generation of Iron Fertilization Experiments in the Southern Ocean." *Philosophical Transactions of the Royal Society* A 366 (2008): 3947–3967.

Smith, Karl F. "Dr. Warren—Rain Maker." Memorandum, July 1, 1922. Bancroft Papers.

"Space Mirror." *Time*, April 5, 1954. http://www.time.com.

Spence, Clark C. *The Rainmakers: American "Pluviculture" to World War II*. Lincoln: University of Nebraska Press, 1980.

Stapler, Colonel Wendall T. "The New Face of the Pacific." *Air Force Weather*, August 6, 2006. http://www.afweather.af.mil/news/story.asp?id=123030943.

Steinberg, Theodore. *Slide Mountain, or the Folly of Owning Nature*. Berkeley: University of California Press, 1995.

Stix, Thomas H. "Removal of Chlorofluorocarbons from the Atmosphere." *Journal of Applied Physics* 66 (1989): 5622–5626.

Stone, Christopher D. "The Environment in Moral Thought." *Tennessee Law Review* 56 (1988): 1–13.

Stone, G. H. "Rain-Making by Concussion in the Rocky Mountains." *Science* 19 (1892): 52.

Suits, C. Guy. "Statement on Weather Control." Testimony to U.S. Senate, March 1951. Copy in Houghton Papers.

Suits, C. Guy, and Miles J. Martin. "Irving Langmuir." *Biographical Memoirs of the National Academy of Sciences* 45 (1974): 215–247.

Summers, Lawrence H. "The Memo." December 12, 1991. http://www.whirledbank.org/ourwords/summers.html.

"Sykes Sells Sunshine, Say Shrewd Sport Satraps." *Literary Digest*, September 27, 1930, 38.

Takano, Kenzo, and Keiji Higuchi. "A Numerical Experiment on the General Circulation in the Global Ocean." In Hidetoshi Kato, ed., *Challenges from the Future: Proceedings of the International Future Research Conference*, 3:361–363. Tokyo: Kodansha, 1970.

Talman, Charles Fitzhugh. "Can We Control the Weather?" *Outlook*, March 14, 1923, 493–495.

Tate, Ken, and Janice Tate. *Good Old Days Country Wisdom*. Big Sandy, Tex.: House of White Birches, 2001.

Teller, Edward. *Memoirs*. Cambridge, Mass.: Perseus, 2001.

Teller, Edward, Lowell Wood, and Roderick Hyde. "Global Warming and Ice Ages: I. Prospects for Physics-Based Modulation of Global Change." UCRL-JC-128715. Lawrence Livermore National Laboratory, Livermore, Calif., 1997.

"Terraforming Information Pages." http://www.users.globalnet.co.uk/~mfogg.

Thompson, G. A. "Plan for Converting the Sahara Desert into a Sea." *Scientific American*, August 10, 1912, 114.

Thompson, Philip D. "A History of Numerical Weather Prediction in the United States." *Bulletin of the American Meteorological Society* 64 (1983): 755–769.

Tilmes, Simone, Rolf Müller, and Ross Salawitch. "The Sensitivity of Polar Ozone Depletion to Proposed Geoengineering Schemes." *Science* 320 (2008): 1201–1204.

Train, Arthur, and Robert Williams Wood. *The Man Who Rocked the Earth*. Garden City, N.Y.: Doubleday, 1915.

Tuan, Yi-fu. *Landscapes of Fear*. New York: Pantheon, 1979.

Turco, Richard P., Owen B. Toon, Thomas Ackerman, James B. Pollack, and Carl Sagan (TTAPS). "Nuclear Winter: Global Consequences of Multiple Nuclear Explosions." *Science* 222 (1983): 1283–1293.

Tuthill, Barbara. "Hatfield the Rainmaker." *Western Folklore* 13 (1954): 107–112.

Twain, Mark. *The American Claimant*. New York: Webster, 1892.

Tyndall Centre for Climate Change Research. "The Potential of Geo-engineering Solutions to Climate Change." Memorandum 149. United Kingdom, House of Commons. Innovation, Universities, and Skills Committee (September 2008). http://www.publications.parliament.uk.

Tyson, Neil deGrasse. "The Five Points of Lagrange." *Natural History*, April 2002, 44–49. http://research.amnh.org/~tyson/18magazines_five.php.

"An Unfortunate Rain-maker." *Harper's Weekly*, August 5, 1893, 735.

U.S. Army Air Forces, Air Technical Service Command. "Memorandum Report on German Liquid Rocket Development." RG-341, Series 163, TSEAL-6-X-2. Records of Headquarters, National Archives.

U.S. Army Medical Department. "Regulations for the Medical Department." In *Military Laws and Rules and Regulations for the Army of the United States*, 227–228. Washington, D.C., 1814.

U.S. Congress. House of Representatives. *Production of Rain by Artillery-Firing*. Ex. Doc. 786. 43rd Cong., 1st sess., June 23, 1874.

——. Committee on Foreign Affairs. Subcommittee on International Organizations. *Prohibition of Weather Modification as a Weapon of War: Hearings on H.R. 28*. 94th Cong., 1st sess., 1975.

——. Committee on Foreign Affairs. Subcommittee on International Organizations and Movements. *Weather Modification as a Weapon of War: Hearings on H.R. 116 and 329*. 93rd Cong., 2d sess., 1974.

——. Committee on Science and Technology. *Geoengineering: Assessing the Implications of Large-Scale Climate Intervention*. 111th Cong., 2d sess., November 5, 2009. http://science.house.gov/publications/hearings_markups_details.aspx?NewsID=2668.

U.S. Congress. Senate. Committee on Foreign Relations. Subcommittee on Oceans and International Environment. *Prohibiting Military Weather Modification: Hearings on S.R. 281*. 92nd Cong., 2d sess., 1972.

U.S. Joint Numerical Weather Prediction Unit. "Facts Sheet, 1955." Rare Book Room, NOAA Central Library, Silver Spring, Md.

U.S. Library of Congress. Congressional Research Service. *Weather Modification: Programs, Problems, Policy, and Potential*. 95th Cong., 2nd sess. Washington, D.C.: Government Printing Office, 1978.

——. Legislative Reference Service. *Weather Modification and Control . . . for the Use of the Committee on Commerce, United States Senate*. Washington, D.C.: Government Printing Office, 1966.

U.S. National Aeronautics and Space Administration. *Workshop Report on Managing Solar Radiation*. NASA/CP-2007-214558. April 2007.

U.S. National Oceanic and Atmospheric Administration. "Hurricane Research Division History." http://www.aoml.noaa.gov/hrd/hrd_sub/beginning.html.

U.S. Navy. Bureau of Naval Weapons, Research, Development, Test and Evaluation Group, Meteorological Management Group. "Technical Area Plan for Weather Modification and Control." TAP No. FA-4. January 1, 1965.

United Nations. "Convention on the Prohibition of Military or Any Other Hostile Use of Environmental Modification Techniques." http://daccess-ods.un.org/TMP/6130753.html.

——. "Framework Convention on Climate Change" (UNFCCC). Rio de Janeiro, 1992. http://unfccc.int/resource/docs/publications/guideprocess-p.pdf.
——. "International Co-operation in the Peaceful Uses of Outer Space." Resolution adopted by the General Assembly 1721 (XVI), December 21, 1961. http://www.oosa.unvienna.org/oosa/ar/SpaceLaw/gares/html/gares_14_1472.html.
——. *Multilateral Treaties Deposited with the Secretary-General: Status as of 31 December 1982*. New York, 1983.
Van Bibber, Andrew. "Rain Not Produced by Cannonading." *Scientific American*, December 25, 1880, 405.
Veraart, A.W. *Meer zonneschijn in het nevelige Noorden, meer regen in de tropen*. Amsterdam: Seyffardt, 1931.
Verne, Jules. *De la terre à la lune, trajet direct en 97 heures*. Paris: Hetzel, 1865.
——. *From the Earth to the Moon Direct in Ninety-seven Hours and Twenty Minutes, and a Trip Round It*. Translated by Louis Mercier and Eleanor E. King. New York: Sampson Low, Marston, Low, and Searle, 1873.
——. *The Illustrated Jules Verne*. 1889. http://jv.gilead.org.il/rpaul/Sans%20dessus%20dessous.
——. *L'Invasion de la mer*. Paris: Hetzel, 1905
——. *Invasion of the Sea*. Translated by Edward Baxter. Edited by Arthur B. Evans. Middletown, Conn.: Wesleyan University Press, 2001.
——. *The Purchase of the North Pole: A Sequel to "From the Earth to the Moon."* London: Sampson Low, Marston, 1890.
Vince, Gaia. "One Last Chance to Save Mankind" [interview with James Lovelock]. *New Scientist*, January 23, 2009. http://www.newscientist.com.
Voigt, Wolfgang. *Atlantropa: Weltbauen am Mittelmeer: Ein Architektentraum der Moderne*. Hamburg: Dölling und Galitz, 1998.
von Neumann, John. "Can We Survive Technology?" *Fortune*, June 1955, 106–108.
Vonnegut, Bernard. "The Nucleation of Ice Formation by Silver Iodide." *Journal of Applied Physics* 18 (1947): 593–595.
Vonnegut, Kurt, Jr. *Cat's Cradle*. New York: Delacorte, 1963.
——. *Fates Worse Than Death*. New York: Putnam, 1991.
——. *Player Piano*. New York: Delacorte, 1952.
——. "Report on the Barnhouse Effect." *Collier's*, February 11, 1950. [Reprinted in *Welcome to the Monkey House: A Collection of Short Works*, 173–188. New York: Delacorte, 1968]
Walker, Gabrielle, and Sir David King. *The Hot Topic: What We Can Do About Global Warming*. London: Harcourt, 2008.
Ward, Robert DeCourcy. "Artificial Rain: A Review of the Subject to the Close of 1889." *American Meteorological Journal*, May 1891–April 1892, 484–493.
——. "How Far Can Man Control His Climate?" *Scientific Monthly*, January 1930, 5–18.
Warren, L. Francis. *Facts and Plans. Rainmaking—Fogs and Radiant Planes*. January 1925. Copy in Bancroft Papers.
Washington, Warren M. "Computer Modeling the Twentieth- and Twenty-first-Century Climate." *Proceedings of the American Philosophical Society* 150 (2006): 414–427.
Weart, Spencer. "Climate Modification Schemes." In *The Discovery of Global Warming*. http://www.aip.org/history/climate.

"The Weather Weapon: New Race with the Reds." *Newsweek*, January 13, 1958, 54–56.

Webb County Commissioners. Official Minutes, Court Meetings of April 14, 2003, and December 8, 2003. www.webbcountytx.gov.

Weinberg, Alvin. "Can Technology Replace Social Engineering?" *University of Chicago Magazine*, October 1966, 6–10. [Reprinted many times]

Westmoreland, William. *A Soldier Reports*. New York: Da Capo Press, 1976.

Wexler, Harry. "Deozonizer." Memorandum to weather bureau colleagues, November 24, 1961. Wexler Papers, box 13.

——. "Further Justification for the General Circulation Research Request for FY 63." February 9, 1962. Draft in Wexler Papers, box 18.

——. "Modifying Weather on a Large Scale." *Science*, n.s., 128 (1958): 1059–1063.

——. "On the Possibilities of Climate Control." 1962. Manuscript and notes in Wexler Papers, box 18.

——. "U.N. Symposium on Science and Technology for Less Developed Countries." May 21, 1962. Draft remarks in Wexler Papers, box 14.

White, Andrew Dickson. *A History of the Warfare of Science with Theology in Christendom*. 2 vols. New York: Appleton, 1896.

Whitnah, Donald R. *A History of the United States Weather Bureau*. Urbana: University of Illinois Press, 1961.

"Who Owns the Clouds?" *Stanford Law Review* 1 (1948): 43–63.

Williams, Geoffrey. *Flying Through Fire: FIDO—The Fogbuster of World War Two*. Stroud: Sutton, 1995.

Williams, John. *The Climate of Great Britain: Or Remarks on the Change It Has Undergone, Particularly Within the Last Fifty Years*. London: Baldwin, 1806.

Williams, Paul K. "The Bizarre & Unusual Will of Robert St. George Dyrenforth." Scenes from the Past. *InTowner*, April 2005, 12. http://www.washingtonhistory.com/ScenesPast/index.html.

Williamson, Jack. *Terraforming Earth*. New York: Tor, 2001.

Willoughby, H. E., D. P Jorgensen, R. A. Black, and S. L. Rosenthal. "Project STORMFURY: A Scientific Chronicle, 1962–1983." *Bulletin of the American Meteorological Society* 66 (1985): 505–514.

Wilson, Margaret Adelaide. "The Rain-Maker." *Scribner's Magazine*, April 1917, 503–509.

Wood, Lowell. "Stabilizing Changing Climate via Stratosphere-Based Albedo Modulation." PowerPoint presentation at NASA-Ames–CIW Workshop on Managing Solar Radiation, 2006.

Wylie, Francis E. *M.I.T. in Perspective: A Pictorial History*. Boston: Little, Brown, 1975.

Yalda, Sepideh. "Harry Wexler." In *Complete Dictionary of Scientific Biography*, 25:273–276. Detroit: Scribner/Thomson Gale, 2007.

Yudin, M. I. "The Possibilities for Influencing Large-Scale Atmospheric Processes." *Contemporary Problems in Climatology* (1966). Translated by Foreign Technology Division, Wright-Patterson Air Force Base, Ohio, 1967.

Zikeev, Nikolay T., and George A. Doumani, comps. *Weather Modification in the Soviet Union, 1946–1966: A Selected Annotated Bibliography*. Washington, D.C.: Library of Congress, Science and Technology Division, 1967.

Zubrin, Robert. *The Case for Mars: The Plan to Settle the Red Planet and Why We Must*. New York: Free Press, 1996.

Zworykin, Vladimir K. "Outline of Weather Proposal." RCA Laboratories, Princeton, N.J., October 1945. Copy in Wexler Papers, box 18. [Reproduced in *History of Meteorology* 4 (2008): 57–78]

INDEX

Numbers in italics refer to pages on which illustrations appear.